Concepts in Electron Correlation

NATO Science Series

A Series presenting the results of scientific meetings supported under the NATO Science Programme.

The Series is published by IOS Press, Amsterdam, and Kluwer Academic Publishers in conjunction with the NATO Scientific Affairs Division

Sub-Series

I. Life and Behavioural Sciences	IOS Press
II. Mathematics, Physics and Chemistry	Kluwer Academic Publishers
III. Computer and Systems Science	IOS Press
IV. Earth and Environmental Sciences	Kluwer Academic Publishers
V. Science and Technology Policy	IOS Press

The NATO Science Series continues the series of books published formerly as the NATO ASI Series.

The NATO Science Programme offers support for collaboration in civil science between scientists of countries of the Euro-Atlantic Partnership Council. The types of scientific meeting generally supported are "Advanced Study Institutes" and "Advanced Research Workshops", although other types of meeting are supported from time to time. The NATO Science Series collects together the results of these meetings. The meetings are co-organized bij scientists from NATO countries and scientists from NATO's Partner countries – countries of the CIS and Central and Eastern Europe.

Advanced Study Institutes are high-level tutorial courses offering in-depth study of latest advances in a field.
Advanced Research Workshops are expert meetings aimed at critical assessment of a field, and identification of directions for future action.

As a consequence of the restructuring of the NATO Science Programme in 1999, the NATO Science Series has been re-organised and there are currently Five Sub-series as noted above. Please consult the following web sites for information on previous volumes published in the Series, as well as details of earlier Sub-series.

http://www.nato.int/science
http://www.wkap.nl
http://www.iospress.nl
http://www.wtv-books.de/nato-pco.htm

Series II: Mathematics, Physics and Chemistry – Vol. 110

Concepts in Electron Correlation

edited by

Alex C. Hewson
Department of Mathematics,
Imperial College, London, United Kingdom

and

Veljko Zlatić
Institute of Physics,
Zagreb, Croatia

Kluwer Academic Publishers

Dordrecht / Boston / London

Published in cooperation with NATO Scientific Affairs Division

Proceedings of the NATO Advanced Research Workshop on
Concepts in Electron Correlation
Hvar, Croatia
September 29–October 3, 2002

A C.I.P. Catalogue record for this book is available from the Library of Congress.

ISBN 1-4020-1418-X (HB)
ISBN 1-4020-1419-8 (PB)

Published by Kluwer Academic Publishers,
P.O. Box 17, 3300 AA Dordrecht, The Netherlands.

Sold and distributed in North, Central and South America
by Kluwer Academic Publishers,
101 Philip Drive, Norwell, MA 02061, U.S.A.

In all other countries, sold and distributed
by Kluwer Academic Publishers,
P.O. Box 322, 3300 AH Dordrecht, The Netherlands.

Printed on acid-free paper

Printed in the Netherlands.

Contents

Preface

The NATO sponsored Advanced Research Workshop on "Concepts in Electron Correlation" took place on the Croatian island of Hvar during the period from the 29th of September to the 3rd of October, 2002. The topic of electron correlation is a fundamental one in the field of condensed matter, and one that is being very actively studied both experimentally and theoretically at the present time. The manifestations of electron correlation are diverse, and play an important role in systems ranging from high temperature superconductors, heavy fermions, manganite compounds with colossal magnetoresistance, transition metal compounds with metal-insulator transitions, to mesoscopic systems and quantum dots. The aim of the workshop was to provide an opportunity for a dialogue between experimentalists and theoreticians to assess the current state of understanding, and to set an agenda for future work. There was also a follow-up workshop on the same topic where the presentations included more background and introductory material for younger researchers in the field.

The papers presented in these proceedings clearly demonstrate the diversity of current research on electron correlation. They show that real progress is being made in characterising systems experimentally and in developing theoretical approaches for a quantitative comparison with experiment. The more one learns, however, the more there is to understand, and many of the contributions help to map out the territory which has yet to be explored. We hope that the articles in this volume will be a stimulus for such future work.

February, 2003 A. C. Hewson, V. Zlatic

A.C. Hewson and V. Zlatić (eds.), Concepts in Electron Correlation, 5.

SUPERCONDUCTIVITY BY KINETIC ENERGY SAVING?

Proceedings of the ARW NATO Workshop Hvar, Croatia, October 2002

D. VAN DER MAREL, H. J. A. MOLEGRAAF, C. PRESURA and I. SANTOSO
Materials Science Centre, University of Groningen, 9747 GA Groningen, The Netherlands

Abstract. A brief introduction is given in the generic microscopic framework of superconductivity. The consequences for the temperature dependence of the kinetic energy, and the correlation energy are discussed for two cases: The BCS scenario and the non-Fermi liquid scenario. A quantitative comparison is made between the BCS-prediction for d-wave pairing in a band with nearest neighbor and next-nearest neighbor hoppping and the experimental specific heat and the optical intraband spectral weight along the plane. We show that the BCS-prediction produces the wrong sign for the kink at T_c of the intraband spectral weight, even though the model calculation agrees well with the specific heat.

Key words: Optical Conductivity, Spectral Weight, Specific Heat, Pair Correlation, Kinetic Energy, Correlation Energy, Internal Energy

1. Model independent properties of the superconducting state

1.1. INTERNAL ENERGY

When we cool down a superconductor below the critical temperature, the material enters a qualitatively different state of matter, manifested by quantum coherence over macroscopic distances. Because the critical temperature represents a special point in the evolution of the internal energy versus temperature, the internal energy departs from the temperature dependence seen in the normal state when superconductivity occurs. Because for $T < T_c$ the superconducting state is the stable equilibrium state, the internal energy in equilibrium at $T = 0$ is an absolute minimum. Hence cooling down from above the phase transition one would expect a drop in the internal energy when at the critical temperature. This drop of internal energy stabilizes the superconducting phase among all alternative states of

A.C. Hewson and V. Zlatić (eds.), Concepts in Electron Correlation, 7–16.

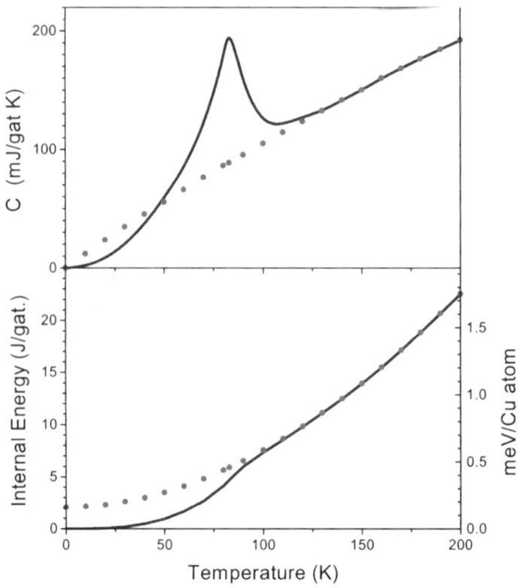

Figure 1. Experimental internal energy and the specific heat obtained by Loram *et al* (1). The extrapolations are obtained following the procedure of Ref. (3)

matter. An experimental example of this well known behavior is displayed in Fig. 1, where the internal energy was calculated from the electronic specific heat data (1) according to the relation $E_{int}(T) = \int_0^T c(T')dT'$. The broadened appearance of the phase transition suggests that the superconducting correlations disappear rather gradually when the temperature is increased above the phase transition, an effect which remains noticeable in the specific heat graph up to 125 K.

Understanding the mechanism of superconductivity means to understand what stabilizes the internal energy of the superconducting state. In BCS theory superconductivity arises as a result of a net attractive interaction between the quasi-particles of the normal state. Note, that implicitly this approach is firmly rooted in the paradigm of a Fermi-liquid type normal state. On the other hand, the school based on Anderson's original work (2) asserts, that a strong on-site *repulsive* interaction can also give rise to high T_c superconductivity. The latter models typically require that the material is not a Fermi liquid when superconductivity is muted, either by raising the temperature or by other means.

1.2. PAIR-CORRELATIONS

Without loss of generality, *i.e.* independent of the details of the mechanism which leads to superconductivity, it is possible to provide a microscopic definition of the superconducting state. For this purpose let us consider the correlation function defined as

$$G(r, R_1, R_2) = \left\langle \psi_\uparrow^\dagger(R_1 + r)\psi_\downarrow^\dagger(R_1)\psi_\downarrow(R_2)\psi_\uparrow(R_2 + r)\right\rangle \tag{1}$$

where $\psi_\sigma^\dagger(R_j)$ is a single electron creation operator. This function defines two types of correlation: (i) the electron pair-correlation as a function of the relative coordinate **r**, and (ii)v the correlation between two pairs located at center of mass coordinates $\mathbf{R}_1$ and $\mathbf{R}_2$. In the normal state there is no correlation of the phase of $G(r, R_1, R_2)$ over long distances $|R_1 - R_2|$ due to the finite mean free path of the electrons. As a result the integral over the center of mass coordinates of the correlation function

$$g(r) = \frac{1}{V^2}\int d^3R_1 \int d^3R_1 G(r, R_1, R_2) \tag{2}$$

averages to zero in the normal state. In contrast the superconducting state is characterized by long range phase coherence of the center of mass coordinates, implying (among other things) that the correlation function averaged over all center of mass coordinates, $g(r)$, is a finite number.

2. Internal energy and its decomposition using the BCS model

2.1. BCS: CORRELATION FUNCTION, AND CORRELATION ENERGY FOR D-WAVE PAIRING

In the weak coupling scenario of BCS theory the electrons have an effective attractive interaction, as a result of which they tend to form pairs. For the purpose of the present discussion we will assume that the interaction is of the form

$$H^i = V(r - r')\hat{n}(r)\hat{n}(r') \tag{3}$$

where $\hat{n}(r)$ is the electron density operator. The interaction energy in the superconducting state becomes lower than in the normal state, due to the fact that the effective attractive interaction favors a state with enhanced pair-correlations. The value of the interaction energy of the superconducting state, relative to the normal state is

$$\left\langle H^i\right\rangle_s - \left\langle H^i\right\rangle_n = \int d^3r g(r)V(r) = \sum_k g_k V_k \tag{4}$$

where g_k and V_k are the Fourier transforms of $g(r)$ and $V(r)$ respectively. Using the Bogoliubov transformation the correlation function can be expressed (4) in terms of the gap-function Δ_k and the single particle energies $E_k = \{(\epsilon_k - \mu)^2 + \Delta_k^2\}^{1/2}$.

$$g_k = \sum_q \frac{\Delta_{q+k}\Delta_q^*}{4E_{q+k}E_q} \tag{5}$$

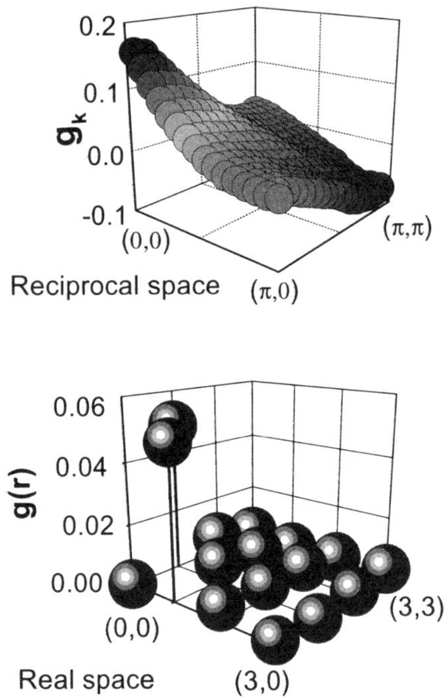

Figure 2. The k-space (top panel) and coordinate space (bottom panel) representation of the superconductivity induced change of pair-correlation function for d-wave symmetry (bottom panel). Parameters: $\Delta/W = 0.2$, $\omega_D/W = 0.2$. Doping level: x = 0.25

This corresponds to the conversion of a pair $(q, -q)$ to a pair with quantum numbers $(q+k, -q-k)$. In the expression for the correlation energy, Eq. 4, the transferred momentum k is carried by interaction kernel V_k.
Starting from a model expression for the single electron energy-momentum dispersion ϵ_k, and the gap-function Δ_k, it is a straightforward numerical exercise to calculate the summations in Eq. 5. Adopting the nearest neighbor tight-binding model with a d-wave gap, and adopting the ratio $\Delta(\pi,0)/W = 0.2$, where W is the bandwidth, the correlation function g_k can be easily calculated, and the result is shown in Fig. 2. We see from this graph that a negative value of $\langle H^i \rangle_s - \langle H^i \rangle_n$ requires either (i) $V_k < 0$ for k in the neighborhood of the origin, or (ii) $V_k > 0$ for k in the vicinity of (π,π). The corresponding representation in real space, $g(r)$, shown in Fig. 2, illustrates that the dominant correlation of the d-wave superconducting state is of pairs where the two electrons occupy a nearest neighboring site, while the on-site amplitude is zero.
Combining the information of Fig. 2 with Eq. 2, it is clear that the strongest saving of correlation energy is expected if the electrons interact with an interaction of the form $V(r_1, r_2) = V_0 \sum_a \delta(\mathbf{r}_1 - \mathbf{r}_2 + \mathbf{a})$ where the vector a

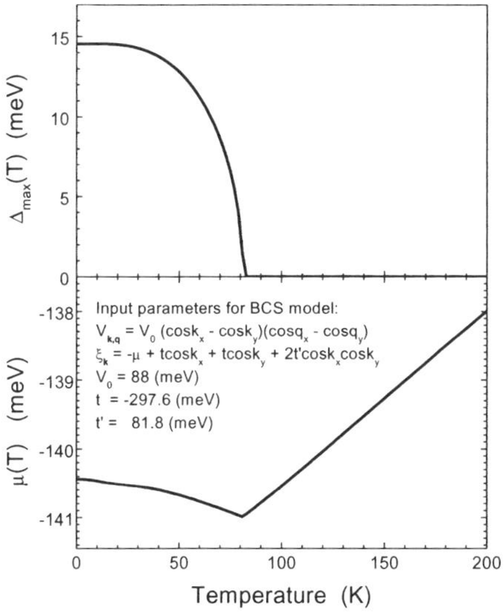

Figure 3. BCS prediction of the d-wave gap-function and the chemical potential.

runs over nearest neighbor sites, and V_0 is a negative (meaning attractive) interaction between electrons on nearest neighbor sites.

2.2. BCS: THE GAP-EQUATION, SPECIFIC HEAT AND INTERNAL ENERGY

To illustrate the predictions of BCS theory for the temperature dependence of the correlation energy and the kinetic energy, we start by solving the gap equation

$$\Delta_k = \sum_q \frac{V_{k-q}\Delta_q}{2E_q} \tanh\left(\frac{E_k}{2k_B T}\right) \tag{6}$$

which must solved together with the constraint that the average number of particles is temperature independent. This requires that the chemical potential must be adjusted to keep the thermal average of $\sum_k \langle \hat{n}_k \rangle = N$ at a constant value. In the numerical examples of this paper we have adopted $N = 0.85$ corresponding to a hole doping of 0.15. This solution of the gap equation is shown in Fig. 3. Notice, that the temperature dependence of the chemical potential follows closely the experimental observations reported for Y123 in Ref. (7).

To check that the band-parameters used here are reasonable, we display in Fig. 4 the corresponding prediction for the specific heat and the internal energy, using the same parameters as for Fig. 3. If we compare this to Fig. 1, we see that the band-parameters adopted here quantitatively reproduce the observed specific heat. Hence the present set of band-parameters (t,t') and

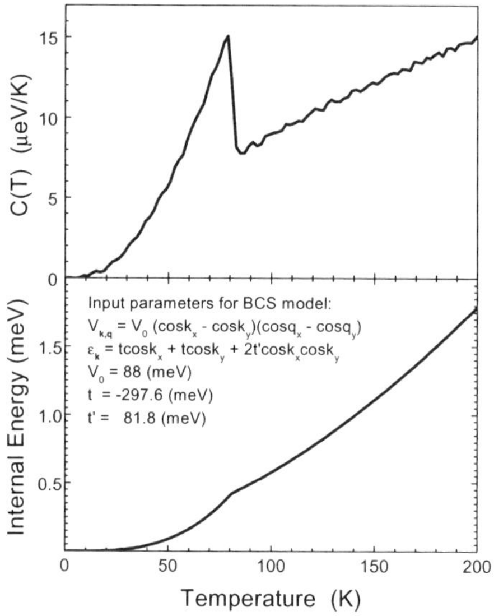

Figure 4. BCS prediction of the internal energy and the specific heat.

the coupling parameter V_0 represent the best phenomenological choice for quantitative testing of the BCS-model.

2.3. BCS: TEMPERATURE DEPENDENCE OF THE CORRELATION ENERGY AND THE KINETIC ENERGY

The BCS prediction for the temperature dependence of the average interaction energy follows from Eq. 4. We then use the BCS variational wave-function for the statistical average of Eq. 5, resulting in

$$\left\langle H^i \right\rangle_s - \left\langle H^i \right\rangle_n = -\sum_k \frac{|\Delta_k|^2}{2E_k} \tanh\left(\frac{E_k}{2k_B T}\right) \tag{7}$$

Simultaneously there is an increase of the 'kinetic' energy

$$\left\langle H^{kin} \right\rangle = \sum_k \epsilon_k \left\{1 - \frac{\epsilon_k - \mu}{E_k} \tanh\left(\frac{E_k}{2k_B T}\right)\right\} \tag{8}$$

In Fig. 5 this is displayed, using the same parameters as for Fig. 3.

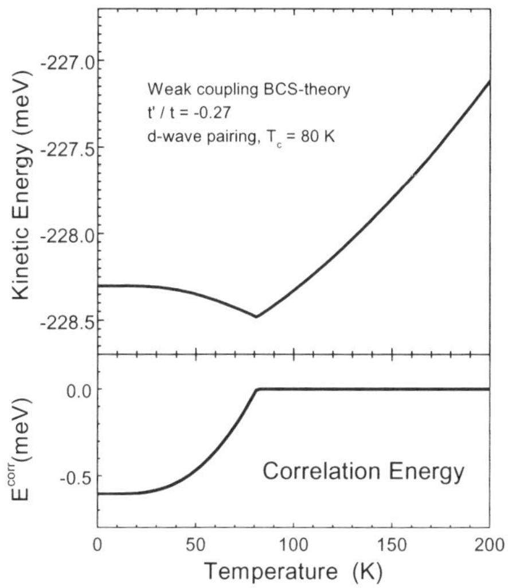

Figure 5. BCS prediction of the kinetic energy and the correlation energy.

3. Relationship between intra-band spectral weight and kinetic energy

A measure of the kinetic energy is provided by the following relation (8, 9, 10)

$$\int_{-\Omega}^{\Omega} d\omega \mathrm{Re}\sigma(\omega) d\omega = \pi \frac{e^2}{\hbar^2 V} \sum_{k,\sigma} \langle \hat{n}_{k,\sigma} \rangle \frac{\partial^2 \epsilon(\mathbf{k})}{\partial k^2} \tag{9}$$

where the high frequency limit indicates that the integral should include only the intra-valence band transitions, and the condensate peak at $\omega = 0$ if the material is a superconductor. The integral over negative and positive frequencies (note that $\sigma(\omega) = \sigma^*(-\omega)$) avoids ambiguity about the way the spectral weight in the condensate peak should be counted. If the band structure is described by a nearest neighbor tight-binding model, Eq. 9 leads to the simple relation

$$\rho_L \equiv \frac{\hbar^2}{a^2 \pi e^2} \int_{-\Omega}^{\Omega} \mathrm{Re}\sigma(\omega) d\omega = \langle -H_{kin} \rangle \tag{10}$$

Hence in the nearest neighbor tight-binding limit the *partial* f-sum provides the *kinetic energy* contribution, which depends both on the number of particles and the hopping parameter t (11, 12, 13, 14).
However, if the band-structure has both nearest neighbor hopping and next nearest neighbor hopping, Eq. 10 is not an exact relation, and instead Eq. 9 should be compared directly to the experiments. In Fig. 6 we compare the spectral weight, calculated directly using Eq. 9 to the result of Eq. 10, using the same parameters as for Fig. 3. Note that the kinetic energy has to be

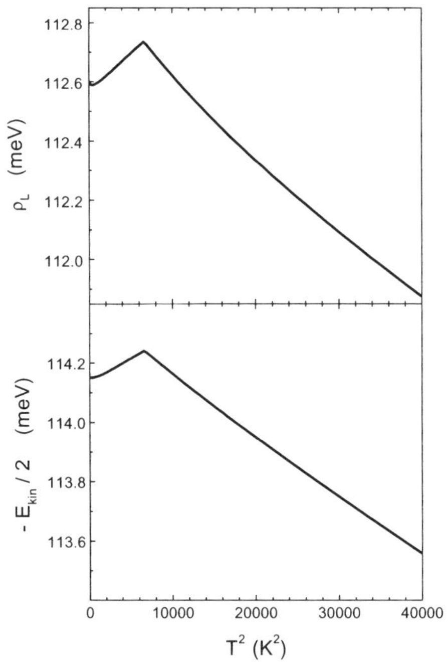

Figure 6. BCS prediction of the spectral weight function.

divided by a factor two, as we are interested in the projection along one of the two axes in the ab-plane, which can be compared directly to the experimental value of ρ_L. From Fig. 6 we can conclude, that the effect of including t' in the calculation is rather small, and it is still OK to identify ρ_L with the kinetic energy apart from a minus sign.

4. Experimental determination of the intra-band spectral weight

In two recent experimental papers measurements of ρ_L in Bi2212 have been reported (15, 16). The values of the kinetic energy change in the superconducting state were in quantitative agreement with each other, and both papers arrived at the same conclusion: Contrary to the BCS prediction, the kinetic energy of the superconducting state is *lower* than in the normal state (taking into account a correction for the temperature trends of the normal state). In Fig. 7 the data of Ref. (15) have been reproduced. Comparing this with the BCS prediction clearly demonstrates the large qualitative discrepancy between theory and experiment. Clearly the type of mechanism assumed in BCS theory is not at work here.

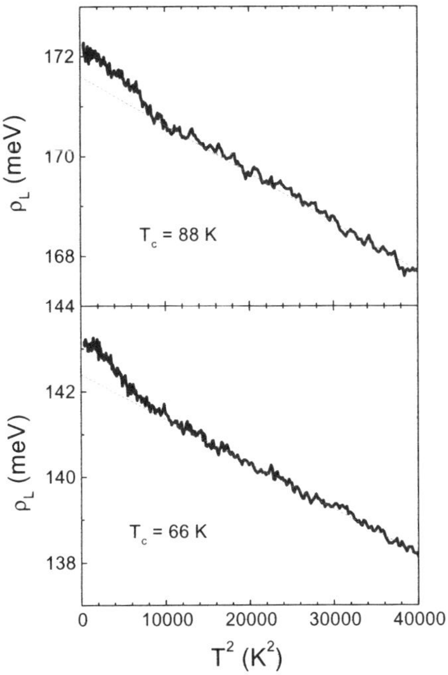

Figure 7. Experimental values of the ab-plane spectral weight function, taken from Ref. (15)

5. Implications of the experimental data

The trend seen in the experimental data has been predicted by Hirsch in 1992 (17, 18, 19). The model assumption made by Hirsch was, that the hopping probability of a single hole between two sites becomes larger if one of the two sites is already occupied by a hole. Although this model provides good qualitative agreement with the optical experiments, it has one serious deficiency: It also predicts s-wave symmetry for the order parameter, in sharp contrast to a large body of experimental data which show that the superconducting gap in the cuprates has d-wave symmetry.
In a recent set of calculations based on the Hubbard model, Jarrell et al (20) obtained a similar effect as seen in our experiments, both for underdoped and optimally doped samples. Crudely speaking the mechanism is believed due to the frustrated motion of single carriers in a background with short-range (RVB-type) spin-correlations, which is released once pairs are formed.

6. Conclusions

We have made a quantitative comparison between the BCS-prediction for d-wave pairing in a band with nearest neighbor and next-nearest neighbor hoppping and various experiments, in particular specific heat and measurements of the optical ab-plane sumrule. We have shown that the BCS-prediction produces the wrong sign for the kink at T_c of the ab-plane intraband spectral weight, while the model calculation is in good agreement with the experimental specific heat data.

Acknowledgements

DvdM gratefully acknowledges M. Norman, N. Bontemps, J. Hirsch, and M. Jarrell for stimulating discussions, and J. W. Loram for making his data files of the specific heat of Bi2212 available.

References

1. J. W. Loram, J. Luo, J. R. Cooper, W. Y. Liang and J. L. Tallon, J. Physics and Chemistry of Solids **62**, 59-64 (2001).
2. P.W. Anderson, Science **235**, 1196 (1987).
3. D. van der Marel, A.J. Leggett, J.W. Loram, J.R. Kirtley Physical Review B Rapid Communications **66**, 140501R (2002).
4. There is an additional term, due to the fact that the chemical potential of the superconducting state is slightly shifted relative to the normal state (5, 6): $0.25 \sum_p |v_{p+k} v_p|^2$. In the numerical examples in this paper this additional term has been taken into account. Since the effect is small, we will not discuss it further here.
5. D. van der Marel, Physica C **165** (1990) 35-43;
6. D. van der Marel, and G. Rietveld, Phys. Rev. Lett. **69** (1992) 2575-2577;
7. G. Rietveld, N. Y. Chen, D. van der Marel, Phys. Rev. Lett. **69** (1992) 2578-2581;
8. M. R. Norman, and C. Pepin, Phys. Rev. B **66**, 100506(R) (2002).
9. R. Kubo, J. Phys. Soc. Japan **12**, 570 (1957).
10. H. D. Drew, P. Coleman, Phys. Rev. Lett. **78**, 1572 (1997).
11. Pierre F. Maldague, Phys. Rev. **16**, 2437 (1977).
12. J. Carmelo and D. Baeriswyl, Phys. Rev. B **37**, 7541 (1988).
13. D. N. Basov, S. I. Woods, A. S. Katz, E. J. Singley, R. C. Dynes, M. Xu, D. G. Hinks, C. C. Homes, M. Strongin, Science **283**, 49 (1999).
14. S. Chakravarty, H. Kee, E. Abrahams, Phys. Rev. Lett. **82** , 2366 (1999).
15. H. J. A. Molegraaf, C. Presura, D. van der Marel, P. H. Kes , M. Li, Science **295**, 2239 (2002).
16. A.F. Santander-Syro, R.P.S.M. Lobo, N. Bontemps, Z. Konstantinovic, Z.Z. Li, H. Raffy, cond-mat/0111539 (2002).
17. J. E. Hirsch, Physica C **199**, 305 (1992).
18. J. E. Hirsch, Physica C **201** (1992) 347.
19. J. E. Hirsch, and F. Marsiglio, Physica C **331** (1999), 150-156.
20. Th. A. Maier, M. Jarrell, A. Macridin, C. Slezak, cond-mat/0211298 (2002).

JOSEPHSON EFFECT IN THE CUPRATES: MICROSCOPIC IMPLICATIONS

Proceedings of the ARW NATO Workshop Hvar, Croatia, October 2002

R. HLUBINA
Department of Solid State Physics, Comenius University, Mlynská dolina F2, SK-842 48 Bratislava, Slovakia

Abstract. In the tunnel limit, the current-phase relation of Josephson junctions can be expanded as $I(\phi) = I_1 \sin\phi + I_2 \sin 2\phi$. Standard BCS theory predicts that $I_1 R_N \sim \Delta/e$ and $I_2/I_1 \sim D$, where R_N is the resistance of the junction in the normal state, Δ is the superconducting gap, and $D \ll 1$ is the junction transparency. In the cuprates, the experimental value of $I_1 R_N$ (I_2/I_1) is much smaller (larger) than the BCS prediction. We argue that both peculiarities of the cuprates can be explained by postulating quantum fluctuations of the pairing symmetry.

Key words: High Temperature Superconductivity, Josephson Effect, Josephson Product, Current-Phase Relation, Quantum Fluctuations, D-Wave Pairing, S-Wave Pairing, RVB

1. Introduction

The proximity to the metal-insulator transition is well known to lead to an anomalous normal state of the high T_c superconductors. It is therefore interesting to ask whether, apart from the d-wave symmetry of pairing, also the superconducting state of the cuprates can be regarded as unconventional. Most studies attempt to answer this question by considering the properties of *quasiparticle excitations*. For instance, photoemission experiments seem to support the conventional alternative, since sharp spectral functions have been observed at low temperatures (1). On the other hand, the effect of strong correlations on the *condensate* has remained largely unexplored so far. This is surprising, since large quantum phase fluctuations are to be expected in the cuprates as a combined effect of the suppression of charge fluctuations and of the uncertainty principle. In particular, such fluctuations have been suggested to stabilize the RVB state and the presumably related pseudogap phase of the cuprates (2). In this paper we argue that by a

A.C. Hewson and V. Zlatić (eds.), Concepts in Electron Correlation, 17–25.

detailed analysis of the Josephson effect, new insights into the nature of quantum fluctuations in the cuprates can be obtained.

Josephson junctions involving the cuprate superconductors have been studied mainly because they enable phase-sensitive tests of pairing symmetry (3). Attention has been paid especially to two particular types of Josephson junctions: grain boundary (4) and intrinsic Josephson junctions (5). The best studied type of grain boundary Josephson junctions involves junctions built on c-axis oriented films, where the weak link forms at the boundary of grains which are rotated around the [001] axis with respect to each other. Idealized junctions of this type are characterized by a planar interface and two angles θ_1 and θ_2 between the interface normal and the crystallographic directions in the grains forming the junction. The properties of the junctions depend dominantly on the misorientation angle $\theta = \theta_2 - \theta_1$. It is well known (4) that the transparency D of grain boundary junctions decays exponentially with increasing misorientation angle θ, $D(\theta) \propto \exp(-\theta/\theta_0)$. Thus, for $\theta > \theta_0 \approx 5°$, grain boundary junctions are in the tunnel limit and (for θ not too close to 45°, see Section 3) their current-phase relation $I(\phi) = I_1 \sin\phi + I_2 \sin 2\phi + \ldots$ can be well approximated by $I(\phi) = I_1 \sin\phi$, neglecting the higher-order harmonics.

2. The Josephson product

A useful quantity characterizing the superconducting electrodes forming the Josephson junction is the product of the first harmonic I_1 with the junction resistance in the normal state, R_N. According to standard theory (for homogeneous featureless barriers), this so-called Josephson product is independent of the junction area and of the barrier transparency, thus giving an intrinsic information about the superconducting banks. It is well known that at interfaces between d-wave superconductors, anomalous bound Andreev levels may form (6). At temperatures larger than the energy of such Andreev levels, BCS-like theory for rough interfaces between d-wave superconductors (7) predicts

$$(I_1 R_N)_{\mathrm{BCS}} = (\pi/4)(\Delta/e)\cos 2\theta, \tag{1}$$

where Δ is the maximal superconducting gap. The measured Josephson product of cuprate grain boundary junctions (4) can be well described by $I_1 R_N = \alpha^2 (I_1 R_N)_{\mathrm{BCS}}$ with a θ-independent renormalization factor $\alpha^2 \sim 10^{-1}$. In addition to the $\theta > \theta_0$ data of (4), this functional form can be tested also for $\theta = 0$ (which case can be realized in break junctions) with the result that average Josephson products of such junctions (8) are fully consistent with the grain boundary data. Moreover, in (8) it has been shown that Δ is not depressed in the junction region, thus explicitly demonstrating the breakdown of the BCS prediction for $I_1 R_N$ in the cuprates.

There exists no generally accepted explanation of the small renormalization factor α^2. One of the reasons is that the microstructure of Josephson junctions is typically quite complicated. In fact, it is well known that small angle grain boundaries can be modelled by a sequence of edge dislocations, while at larger misorientation angles the dislocation cores start to overlap and no universal picture applies to the structure of the grain boundary. For large-angle grain boundaries, Halbritter has proposed (9) that the junction can be thought of as a nearly impenetrable barrier with randomly placed highly conductive channels across it. If due to strong Coulomb repulsion only the normal current (and no supercurrent) is supported by these channels, the small value of $I_1 R_N$ follows quite naturally.

In this paper we shall argue that the smallness of the Josephson product does not follow from the particular properties of the barrier, but is rather an intrinsic property of the cuprates. Such a point of view has been first advocated in (10). However, that paper did not consider alternative more conventional explanations. In order to support our point of view, let us begin by discussing the Josephson product for intrinsic Josephson junctions in the c-axis direction. Such junctions can be viewed as an analogue of ab-plane break junctions (since the misorientation angle vanishes for both), but are preferable because of simpler geometry of the interface. Moreover, zero energy surface bound states which may develop at ab-plane surfaces because of the d-wave symmetry of the pairing state (6) do not form in the c-axis direction, simplifying the analysis of intrinsic Josephson junctions.

Standard BCS theory applied to the case of tunneling between two-dimensional superconductors (11) predicts (for coherent c-axis tunneling) that the c-axis critical Josephson current density is $j_1 = (2e/\hbar)N(0)\langle z_k t_k^2 \rangle$, where $\langle \ldots \rangle$ denotes an average along the (two dimensional) Fermi line, z_k is the wavefunction renormalization, and t_k is the matrix element for c-axis tunneling between neighboring CuO_2 planes. $N(0)$ is the bare (unrenormalized) density of states, $N(0) = \oint dk(4\pi^2 \hbar v_k)^{-1}$, where the integration runs along the two dimensional Fermi line and v_k is the bare Fermi velocity at the Fermi surface point k. Note that although we have assumed that the self-energy is only frequency dependent, for this geometry z_k enters the expression for j_1 (and also the normal-state resistance, see below).

Let us also note that the use of ordinary perturbation theory in deriving an expression for j_c has been criticized recently (12). However, our formula yields the correct answer for $t_k < \Delta$. This can be shown either by an explicit solution of a 4×4 Bogoliubov problem for two coupled planes with phase differences 0 and π between the planes, or by considering solutions to the gap equation in an infinite layered system with a finite c-axis total momentum of the Cooper pairs (for a similar calculation, see e.g. (13)).

The conductance per square in the normal state is given by $G_N = (2e^2/\hbar)N(0)\langle z_k t_k^2 \Gamma_k^{-1}\rangle_N$, where Γ_k is the inverse lifetime of the quasiparticles and the index N in the Fermi line average means that the quantities are to be evaluated in the normal state. Therefore standard theory predicts for the Josephson product of intrinsic Josephson junctions $I_1 R_N = j_1 G_N^{-1} \approx e^{-1}\langle z_k t_k^2\rangle / \langle z_k t_k^2 \Gamma_k^{-1}\rangle_N$. In conventional superconductors R_N can be measured at low temperatures in a sufficiently large magnetic field. This is impossible for the cuprates and thus R_N is usually defined as the c-axis resistivity at T_c.

Unfortunately, due to the unknown temperature dependence of z_k, the theoretical $I_1 R_N$ can not be directly tested by experiment. In order to overcome this problem, in (14) instead of the usual Josephson product a related characteristic of intrinsic Josephson junctions has been studied, namely the product of the critical current I_1 and of the resistivity R_S in the resistive mode of the junction (at low temperatures). Since the conductance per square in the resistive mode of the superconductor is (14) $G_S \approx (8e^2/h)N(0)(z_k t_k^2)_{\rm node}/\Delta$, standard BCS-like theory predicts $I_1 R_S = j_1 G_S^{-1} \approx (\pi/2)(\Delta/e)\langle z_k t_k^2\rangle/(z_k t_k^2)_{\rm node}$. In (14), $I_1 R_S \sim \Delta/e$ has been found experimentally and good agreement with theory has been claimed, since momentum-independent t_k and z_k were assumed. However, according to band structure calculations (15), t_k is strongly suppressed in the nodal directions. If this modulation of the tunnel matrix element t_k is taken into account and the presumably only moderate k-space dependence of z_k is neglected, the experimental $I_1 R_S$ is seen to be drastically reduced with respect to the theoretical predictions.

Thus we have shown that although the barriers in grain boundaries and in intrinsic Josephson junctions are of very different nature, both types of junctions exhibit a suppressed Josephson product. Therefore we believe that this suppression is not due to specific barrier properties as suggested in (9), but rather due to some intrinsic property of the high-T_c superconductors.

3. The second harmonic of the current-phase relation

Since the second harmonic I_2 is not forced by symmetry to depend on the angles θ_i, its Josephson product can be estimated (for temperatures larger than the energy of anomalous Andreev levels) using the standard BCS theory as $I_2 R_N \sim D\Delta/e$. Comparison with Eq. (1) implies that for junctions with $\theta \approx 45°$ the d-wave symmetry of pairing leads to a suppression of I_1, and I_2 may become comparable to I_1. This has in fact been observed in two different types of $45°$ grain boundary Josephson junctions (16, 17).

However, the results of (16, 17) are quite mysterious, if we take into account the actual experimental setup. In fact, standard BCS theory with ideal featureless barriers implies that in order that $|I_2| \sim |I_1|$, the average misorientation angle θ would have to be given with a precision $\sim D$, where $D \sim 10^{-3}$. This is not realistic and therefore two alternative explanations have been proposed, in both of which the origin of the anomalously large $|I_2/I_1|$ ratio has been sought in the barrier properties.

(i) Faceted scenario has been considered as an alternative explanation for symmetric 45° junctions (i.e. junctions with nominal geometry $\theta_1 = 0°$ and $\theta_2 = 45°$), in which $|I_2| > |I_1|$ has been found (16). It takes into account the faceting of the grain boundary and also the twinned nature of the (orthorhombic) YBCO thin films. Due to both of these features, the junction can be viewed as a parallel set of 0 and π junctions (18). It has been shown (19) that in such a case spontaneous currents are generated along the interface, the ground state energy of the junction is minimized at a macroscopic phase difference $\pm\pi/2$, and consequently the current-phase relation is dominated by the second harmonic I_2.

In what follows we analyze quantitatively whether the faceted scenario can apply to the results of (16). Let us denote the current densities corresponding to the harmonics I_i (with $i = 1, 2$) as j_i and introduce the Josephson penetration depth of the junction, $\Lambda_J = (\Phi_0/4\pi\lambda\mu_0 j_2)^{1/2}$. Moreover, let the local critical current density in the 0 and π junctions be $\pm j_0$, their typical length a, and the bulk penetration depth be λ. Then, since $a \approx 0.01 - 0.1\mu$m, $\lambda \approx 0.15\mu$m, and $\Lambda_J \sim 3\mu$m (estimated making use of (4) $j_2 \sim 10^4$ A/cm^2), the inequalities $\pi\lambda \gg a$ and $\Lambda_J^2 \gg a\lambda$ are well satisfied. Following (19) it is easy to show that these inequalities guarantee that the spontaneously generated currents along the Josephson junction can be calculated within perturbation theory. If we denote the total junction length by L, then a straightforward calculation yields $j_2/j_0 \approx \mu_0 j_0 a\lambda^2/\Phi_0$ and $j_1/j_0 \approx (a/L)^{1/2}$. The equation for j_1 is a random walk-type formula, indicating that j_1 averages to zero in a sufficiently long junction. After some algebra the above equations are seen to imply $j_2/j_1 \approx \sqrt{L\lambda/(4\pi\Lambda_J^2)}$. Therefore, standard theory predicts that $I_2 > I_1$ can be realized only in sufficiently long junctions with $L > 4\pi\Lambda_J^2/\lambda$. This requires $L > 500\mu$m, whereas in (16) much shorter junctions with $L \sim 1\mu$m were studied.

(ii) Pinhole scenario has been proposed in (17) as an explanation for symmetric 45° junctions (i.e. junctions nominally characterized by $\theta_2 = -\theta_1 = 22.5°$). It views the barrier as basically impenetrable, the conduction being due to randomly placed highly conductive pinholes. This explains quite naturally the small value of the effective barrier transmission and, at the same time, the large value of $|I_2/I_1|$. Note that in order to explain the small value of the Josephson product, in addition to pinholes also Halbrit-

ter's conductive channels (9) have to be postulated, which are assumed to be highly conductive only in the normal and not in the superconducting channel. Because of this ad hoc nature of the pinhole scenario, and mainly because of the absence of higher harmonics in $I(\phi)$ at 4 K (17) whose presence it predicts, we believe that the pinhole picture should be discarded.

Thus we conclude that the large value of the second harmonic I_2 (compared with predictions of the standard BCS theory) is most probably not an extrinsic (barrier-related) effect, but rather an intrinsic property of the cuprates.

4. Microscopic implications

The two apparently unrelated experimental facts, namely the suppressed Josephson product I_1R_N and the enhanced ratio $|I_2/I_1|$, can be explained by a single assumption that in the cuprates some mechanism is operative which leads to a suppression of I_1, while leaving R_N and I_2 intact. In what follows we describe one such mechanism which we believe to be the most promising one. Namely, we suggest that at low temperatures the superconducting state of the cuprates supports fluctuations of pairing symmetry towards s-wave pairing (which pairing is expected to be locally stable within several microscopic models of the cuprates). Such fluctuations presumably do not affect R_N, while they do influence the Josephson current. In simplest terms, if we denote the phases of the superconducting grains forming the junction as ϕ_i, then the fluctuations renormalize the first and second harmonics by the factors $\langle e^{i\phi_1}\rangle\langle e^{i\phi_2}\rangle$ and $\langle e^{2i\phi_1}\rangle\langle e^{2i\phi_2}\rangle$, respectively, where $\langle\ldots\rangle$ denotes a ground-state expectation value. Thus experiment requires that the fluctuations have to be of such type that $|\langle e^{i\phi}\rangle| = \alpha \approx 0.3$ and $|\langle e^{2i\phi}\rangle| \approx 1$. Precisely this behavior is expected if the d-wave order parameter fluctuates towards s-wave pairing.

Now we proceed by introducing a minimal model of such fluctuations. Unlike in standard literature on this subject (see, e.g., (20) and references therein), we assign a phase field φ_i to each bond i of the square Cu lattice. In other words, φ_i lives on the sites of the dual lattice. Since large phase fluctuations are expected, the compactness of phase fluctuations is explicitly taken into account and the model reads

$$H = H_{\text{fluct}} + \sum_{\langle i,j\rangle} \left[-V\cos(2\varphi_i - 2\varphi_j) + W\cos(\varphi_i - \varphi_j)\right], \qquad (2)$$

where $H_{\text{fluct}} = -m^{-1}\sum_i \partial^2/\partial\varphi_i^2$. The second term in Eq. (2) (with $\langle i,j\rangle$ denoting a pair of nearest neighbor sites) describes for $0 < W < V$ a superconductor with a dominant d-wave pairing $V+W$ and subdominant s-wave pairing $V-W$. Since we concentrate on the $\mathbf{q} \approx (\pi,\pi)$ fluctuations

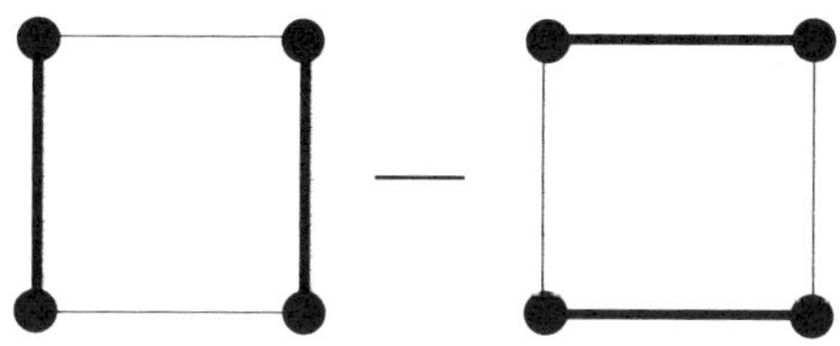

Figure 1. Ground state of the Heisenberg model on an elementary plaquette.

from d to s-wave pairing, as a first approximation we do not take into account the coupling of phase fluctuations to electromagnetism.

In order to gain insight into the microscopics of H_{fluct}, let us recall that the RVB fluctuations in an elementary plaquette (in a hole-free region) favor a minus sign between the two different valence bond configurations (see Fig. 1). If these configurations are thought of in terms of Bose condensates of valence bonds, this means that energy is gained for a relative phase of the x and y condensates $\arg(\sqrt{-1}) = \pm\pi/2$. RVB processes are thus seen to frustrate the phase ordering dictated by the second term in Eq. (2).

Since the other singlet state of the elementary plaquette (with a plus sign between the valence bond configurations) lies higher in energy, the effective Hamiltonian in the singlet sector is $H_{\text{RVB}} \propto \Delta_i^\dagger \Delta_{i+x+y}^\dagger \Delta_{i+x} \Delta_{i+y} + \text{H.C.}$, where $\Delta_i^\dagger$ creates a singlet on bond i. Neglecting the amplitude fluctuations of Δ_i, we can write $H_{\text{RVB}} = J \sum_i \cos(\varphi_i - \varphi_{i+x} + \varphi_{i+x+y} - \varphi_{i+y})$. In what follows we replace H_{fluct} by H_{RVB} in the quantum model Eq. (2), thereby obtaining a completely classical toy model.

The toy model permits a simple mean field analysis: We consider two types of solution, both of which live on two sublattices A and B, and we explicitly disregard four-sublattice solutions, since they correspond to translation-symmetry breaking states on the original Cu lattice. In the first type of solution $\varphi_A = \varphi$ and $\varphi_B = 0$ and there are two possibilities: either $\varphi = \pi$ (d-wave), or $\varphi < \pi$ (a complex mixture of d and s, to be called $d+is$). In the second type of solution we assume a disordered state of such type that on sublattice A, the phase equals 0 and π with probabilities $(1+\alpha)/2$ and $(1-\alpha)/2$, respectively, and on sublattice B the values 0 and π are interchanged. In this case the macroscopic symmetry is of the d-wave type, with renormalized averages $\langle e^{i\varphi_A}\rangle = -\langle e^{i\varphi_B}\rangle = \alpha$ and $\langle e^{2i\varphi_A}\rangle = \langle e^{2i\varphi_B}\rangle = 1$. Minimization of energy (with respect to φ or α) leads to the phase diagram shown in Fig. 2. The renormalization factor in the disordered d-wave state (which presumably corresponds to a homogeneous but strongly fluctuating d-wave phase in the model Eq. (2)) is $\alpha = (W/J)^{1/2}$. Thus the cuprates correspond to the region $W \approx J/10$ and $J < 4V$ in Fig. 2.

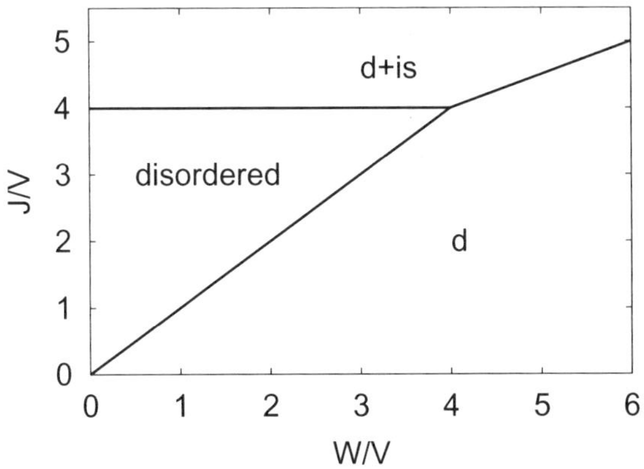

Figure 2. Mean field phase diagram of the toy model.

5. Conclusions

The anomalous effects observed in cuprate grain boundary and intrinsic Josephson junctions can be explained by a single assumption of strong quantum phase fluctuations at $\mathbf{q} = (\pi, \pi)$, which leads to $|\langle e^{i\phi}\rangle| = \alpha \approx 0.3$ and $|\langle e^{2i\phi}\rangle| \approx 1$. In addition, our picture implies that the Josephson product for junctions between the cuprates and low-T_c superconductors is renormalized by the factor α, in semiquantitative agreement with experiment (21). It also may be relevant for the experiment (22), where a large second harmonic has been found in a c-axis Josephson junction between YBCO and Nb.

Acknowledgements

I thank M. Grajcar for numerous stimulating discussions about the Josephson effect. I also thank T. V. Ramakrishnan for insightful remarks on the d-s phase fluctuations. This work was supported by the Slovak Scientific Grant Agency under Grant No. VEGA-1/9177/02 and by the Slovak Science and Technology Assistance Agency under Grant No. APVT-51-021602.

References

1. A. Damascelli, Z. X. Shen, and Z. Hussain, cond-mat/0208504.
2. P. W. Anderson, *The Theory of Superconductivity in the High-T_c Cuprates* (Princeton Univ. Press, Princeton, 1997).
3. M. Sigrist and T. M. Rice, J. Phys. Soc. Jpn. **61**, 4283 (1992).
4. H. Hilgenkamp and J. Mannhart, Rev. Mod. Phys. **74**, 485 (2002).
5. R. Kleiner and P. Müller, Phys. Rev. B **49**, 1327 (1994).

6. C. R. Hu, Phys. Rev. Lett. **72**, 1526 (1994).
7. M. Grajcar, R. Hlubina, E. Il'ichev and H.-G. Meyer, Physica C **368**, 267 (2002).
8. N. Miyakawa *et al.*, Phys. Rev. Lett. **83**, 1018 (1999).
9. J. Halbritter, Phys. Rev. B **46**, 14 861 (1992).
10. G. Deutscher, Nature **397**, 410 (1999).
11. L. N. Bulaevskii, Sov. Phys. JETP **37**, 1133 (1973).
12. S. Chakravarty, Eur. J. Phys. B **5**, 337 (1998).
13. R. Hlubina, Acta Physica Slovaca **45**, 611 (1995).
14. Yu. I. Latyshev *et al.*, Phys. Rev. Lett. **82**, 5345 (1999).
15. O. Andersen, O. Jepsen, A. Liechtenstein, I. Mazin, Phys. Rev. B **50**, 4145 (1994).
16. E. Il'ichev *et al.*, Phys. Rev. B **60**, 3096 (1999).
17. E. Il'ichev *et al.*, Phys. Rev. Lett. **86**, 5369 (2001).
18. J. Mannhart *et al.*, Phys. Rev. Lett. **77**, 2782 (1996).
19. A. J. Millis, Phys. Rev. B **49**, 15 408 (1994).
20. A. Paramekanti, M. Randeria, T. V. Ramakrishnan, S. S. Mandal, Phys. Rev. B **62**, 6786 (2000).
21. A. G. Sun *et al.*, Phys. Rev. B **54**, 6734 (1996).
22. P. V. Komissinski *et al.*, Europhys. Lett. **57**, 585 (2002).

RUTHENATES: UNCONVENTIONAL SUPERCONDUCTIVITY AND MAGNETIC PROPERTIES

Proceedings of the ARW NATO Workshop Hvar, Croatia, October 2002

MANFRED SIGRIST
Theoretische Physik, ETH-Hönggerberg, 8093 Zürich, Switzerland

Abstract. A review is given of the unconventional superconductivity of Sr_2RuO_4 and the magnetic properties of $Ca_{2-x}Sr_xRuO_4$. The electronic properties are determined by the three $4d$-t_{2g}-electrons of Ru. It is shown that superconductivity is due to the d_{xy}-electrons, while the magnetic behavior is more likely a feature of the other two orbitals, d_{yz} and d_{zx}.

Key words: Ruthenate, Unconventional Superconductivity, Magnetism, Spin-Orbit Systems, P-Wave Superconductivity, Spin-Orbit Coupling, Multi-Band Effects

1. Introduction

For the last two decades transition metal oxides have been among the leading topics in condensed matter physics. Best known are the cuprates which are high-temperature superconductors and the manganese oxides displaying the colossal magnetoresistence phenomena. Among these classes of materials also ruthenates play an important role as systems with intriguing properties. The interest in this material mainly arose from the discovery of unconventional superconductivity in Sr_2RuO_4 with p-wave pairing (1, 2). Unlike many other stoichiometric compounds Sr_2RuO_4 is not a Mott insulator, but is a very good metal with clear Fermi liquid properties. This Fermi liquid shows features of strong correlation and is quasi-two-dimensional due to the layered crystal structure. It was suggested that this system could be considered as the electronic analog of ^{3}He with the natural implication that the superconducting state originates from spin triplet p-wave pairing (3, 4).

The electronic properties are dominated by the nearly degenerate $4d$-t_{2g}-orbitals of the Ru^{4+}-ion, which introduces four electrons per Ru. These or-

A.C. Hewson and V. Zlatić (eds.), Concepts in Electron Correlation, 27–34.

bital disperse via π-hybridization with the intermediate oxygen $2p$-orbitals leading to three bands with two electron-like and one-hole-like Fermi surfaces. The form of the orbitals suggests to distinguish two electronic subsystems, the electron-like γ-band corresponding to the d_{xy}-orbital and the weakly hybridized quasi-one-dimensional α (hole-like) and β (electron-like) band originate from the degenerate d_{yz}/d_{zx}-orbitals. There is growing evidence that the superconductivity has its origin in the γ-band. On the other hand, magnetic properties are dominated by the α-β-bands which are responsible for the strongest feature in the spin fluctuations, an incommensurate spin correlation due to strong nesting properties - reminiscent of the basically one-dimensional nature of these bands. We would like to briefly review some aspects of both, the superconducting and the magnetic properties providing evidence for the very distinct character of the different orbitals.

2. Chiral p-wave superconductivity

There are two key experiments determining the symmetry of the pairing state. The measurement of the spin susceptibility via NMR-Knight shift addresses the spin part of pair wave function, and indicates that there is equal-spin-pairing with parallel spins in the RuO_2-planes (5). Furthermore, the zero-field muon spin relaxation shows the appearance of spontaneous intrinsic magnetism below the superconducting phase transition (6). This magnetism originates most likely from the net angular momentum of the Cooper pairs, determining the orbital part of the pair wave function. For a quasi-two-dimensional system the orbital angular momentum is naturally oriented perpendicular to the basal plane. The resulting pairing state is entirely consistent with the symmetry classification including spin-orbit coupling, and can be represented in the vector notation as

$$\mathbf{d}(\mathbf{k}) = \hat{\mathbf{z}}(k_x \pm i k_y), \tag{1}$$

the analog to the A-phase of ^{3}He (3, 7). Note that the orientation of the $\mathbf{d}$-vector is perpendicular to the equal-spin orientation. This state possesses chirality corresponding to the angular momentum, and is also called *chiral p-wave state*. This phase has two order parameter components which are degenerate. Competing states based on the same symmetry classification are non-chiral, with the form $\mathbf{d} = \hat{\mathbf{x}}k_x + \hat{\mathbf{y}}k_y$, $\hat{\mathbf{y}}k_x - \hat{\mathbf{x}}k_y$, $\hat{\mathbf{x}}k_x + \hat{\mathbf{y}}k_y$ and $\hat{\mathbf{y}}k_x + \hat{\mathbf{x}}k_y$, whose equal spin-orientation is parallel to the z-axis, inconsistent with the NMR data.

Interestingly, all these states possess the same condensation energy in a simple weak-coupling theory, which is entirely determined by the quasiparticle gap magnitude $|\mathbf{d}(\mathbf{k})|$ (7). Hence, we need a mechanism to lift this degeneracy. One class of mechanisms is based on feedback effects. Unconventional pairing implies that the attractive interaction is likely to be of purely electronic origin. Consequently, the properties of the pairing interaction would be modified, once the system undergoes the phase transition to the superconducting state. Some pairing states may benefit from this change others do not. Various of these feedback effects have been investigated and exclusively favor the state $\mathbf{d} = \hat{\mathbf{z}}(k_x \pm ik_y)$ (11). Feedback is a higher order correction and is here, unlike in ^{3}He, not so important. Actually spin-orbit coupling plays the deciding role directly affecting the transition temperature of the different states.

Multi-band effects: Sr_2RuO_4 represents a special opportunity to study the effects of spin-orbit coupling on a spin-triplet superconducting phase. The three $4d$-t_{2g}-bands are coupled by spin-orbit coupling, making it necessary to treat the complete multi-band problem:

$$H_{s.o.} = i\lambda \sum_i \sum_{\mu,\nu,\rho} \epsilon_{\mu\nu\rho} \sum_{s,s'} c^\dagger_{i\mu s} \sigma^\rho_{ss'} c_{i\nu s'} \tag{2}$$

with $c^\dagger_{i\mu s}$ creating an electron on site i in the orbital μ with spin s ($\epsilon_{\mu\nu\rho}$ is the completely antisymmetric tensor). It is possible to fit the band structure with a simple tight-binding model. Using a simple phenomenological model for the pairing interaction Ng and Sigrist concluded that the γ-band originating from $4d_{xy}$ would have to be dominant for superconductivity in order to stabilize the chiral p-wave phase (9, 10). This result was recently confirmed by Yanase and Ogata (17) using a microscopic pairing mechanism based on a perturbative approach of the extended three-band Hubbard model as introduced by Nomura and Yamada (19, 18). The latter group had found that within their theory indeed the γ-band would play the leading role for spin-triplet superconductivity.

The assumption that the pairing symmetry is related with the so-called orbital-dependent superconductivity (ODS) and pairing symmetry via spin-orbit coupling is very consistent with several experimental and theoretical facts. Spin-triplet superconductivity is believed to be supported by ferromagnetic spin fluctuations. The analysis of NMR-experiments suggests that the γ-band is closer to ferromagnetism than the α-β-bands which show strong incommensurate spin correlations due to Fermi surface nesting features (12, 13).

ODS plays also an important role for the thermodynamics, since multiple gaps introduce significant modifications in the temperature dependence (8). Power-law behavior of various quantities over a wide temperature range, which have been attributed to line nodes, can be well understood also within a multi-band picture as shown by Zhitomirsky and Rice (14), Nomura and Yamada (19) for the specific heat. Kusunose and Sigrist showed that the analysis of the behavior of the London penetration depth requires one furthermore to take non-local effects into account (15, 16). Nevertheless, the fitting of the thermodynamic quantities does not lead to the unique determination of gap shapes and it may remain even difficult to settle the question whether the gaps in some bands have nodes.

3-Kelvin phase: A surprising feature has been the onset of inhomogeneous superconductivity in some samples of Sr_2RuO_4 at higher temperatures than the bulk-T_c. In accordance with the approximate onset temperature this phase is called the "3 K-phase" (20). The analysis of the material revealed the presence of excess-Ru in form of micrometer-sized Ru-metal inclusion embedded into the pure Sr_2RuO_4-metal. Since Ru on its own becomes superconducting below 0.5K only, it has been suggested that the actual nucleation region of the 3 K phase is the interface between the Ru-inclusions and Sr_2RuO_4. It is possible that the transition temperature of the p-wave superconductivity is slightly enhanced at the interface due to crystal lattice deformations. Based on this assumption a Ginzburg-Landau theory of the nucleated phase has been developed (21). Without going into the calculational details we would like to review here some key results of this analysis and compare with experimental results as far as available.

1. The interface corresponds to a region of reduced symmetry, so that the two p-wave components with momentum direction parallel and perpendicular to the interface ($k_{\|}$ and $k_{\perp}$) have different onset temperatures. It is found that p-wave component parallel to the interface is more stable, because its coherence length along the normal direction of the interface is shorter, so that the order parameter is better confined to the beneficial region. Thus the 3 K phase is time reversal symmetry conserving and before the onset of the bulk phase a second phase transition has to occur breaking time reversal symmetry.

2. The superconducting interface state corresponds to a gap function with phase 0 and π depending on momentum direction. This is a condition to generate zero-energy Andreev bound states in the Ru-inclusions analogous to the zero-energy surface states in d-wave superconductors. These states are observable in quasiparticle tunneling experiments. Recent data by Mao et al in break junctions, where tunneling is likely to be dominated by inter-Ru-inclusion tunneling, show the presence of zero-bias voltage anomalies in the 3 K phase, consistent with the nucleation of the a single-component p-wave state at the interface (22).

3. The critical magnetic field for the 3 K phase is highest, if the magnetic field lies parallel to the interface, since in this case orbital depairing is reduced. Interestingly under this condition the temperature dependence of H_{c2} is different from the standard linear bulk behavior. For an infinitely planar interface there is a square-root behavior, $H_{c2} \propto \sqrt{T - T^*}$ with T^*

as the zero-field nucleation temperature. If the field is in addition parallel to the z-axis, it couples to the angular momentum of the Cooper pair ($k_{\|}, k_{\perp} \rightarrow k_{\|} + ik_{\perp}$). As a consequence H_{c2} should increase and display an up-turn at lower temperatures. The sublinear temperature dependence $H_{c2} \propto (T - T^*)^{\alpha}$ with $\alpha \approx 0.6 - 0.8$ have been observed as well as the expected up-turn for the z-axis oriented field (23).

The existence of the 3 K phase together with these experiments provides strong evidence for the scenario based on interface superconductivity and the realization of the chiral p-wave phase in the bulk.

3. Magnetic properties of $Ca_{2-x}Sr_xRuO_4$

Ca_2RuO_4: Replacing Sr by the isovalent Ca in Sr_2RuO_4 leads to drastic changes. At temperatures below 350 K, the compound is a Mott insulator and shows antiferromagnetic order below 110 K (24). The origin for this contrasting behavior between the Ca- and Sr-compound lies in the fact that the small Ca-ion induces a lattice deformation, including the rotation and tilting of the Ru-O-octahedra as well as their flattening. The latter is essential, since it gives a decisive change of the $4d$-t_{2g}-orbital levels arrangement. For Sr_2RuO_4 the elongation of the octahedra along the z-axis put the d_{xy}-orbital above the other two, while the flattening in Ca_2RuO_4 leads to the opposite situation. The lower d_{xy}-orbital is then occupied by two electrons filling the "γ-band" completely and the remaining two electrons are localized on the two other orbitals forming an $S = 1$ spin degree of freedom due to Hund's rule coupling, which order antiferromagnetically because of superexchange (24). Obviously, these magnetic properties originate from the electrons of the α-β-bands.

Intermediate doping: It is interesting to analyze the problem of the continuously doped system $Ca_{2-x}Sr_2RuO_4$. Nakatsuji and Maeno found experimentally that we can distinguish for $0 < x < 2$ three basic regions (24). Region I ($0 < x < 0.2$) corresponds to the antiferromagnetic Mott insulator discussed above. Region II ($0.2 < x < 0.5$) is a metallic phase with antiferromagnetic correlations at low temperature. Region III ($0.5 < x < 2$) shows an increasing tendency towards ferromagnetism as x approaches 0.5.

There are two basic approaches to the behavior in region III. Fang and Terakura (25), and Nomura and Yamada (26) consider the effect of Ca-doping into Sr_2RuO_4 as a change of the band structure in a way that, in particular, the γ-band is narrowed and its Fermi level is pushed closer to a Van Hove singularity. This gives rise to the increase of the spin susceptibility. All bands remain metallic. In contrast, Anisimov et al. investigated the possibility that a redistribution of charge happens transfering 3 electrons into the d_{yz}/d_{zx}-orbital which undergo Mott localization, while the single electron in the γ-band remains itinerant. This gives rise to a localized spin $S = 1/2$ degree of freedom. This would explain the fact that for x close to 0.5 the spin susceptibility indicates the presence of such a localized spin per site which is difficult to explain within the picture of entirely itinerant electrons (25, 26). On the other hand, it is not clear whether this orbital-selective Mott insulator would give rise to the experimentally observed degree of metallicity.

The scenario of the orbital-selective Mott insulator introduces with the three electrons in the d_{yz}/d_{zx}-orbitals beside the spin also an orbital degree of freedom (isospin 1/2). Thus the effective model of the localized degrees of freedom corresponds to a Kugel-Khomskii type of model for coupled spin 1/2 and isospin 1/2 (27). It has been shown that this effective model in region III would give rise to staggered orbital order, i.e. having a single hole in the d_{yz} (d_{zx})-orbital on the sublattice A (B) of the Ru square lattice. This in turn leads to a ferromagnetic spin interaction via the Goodenough-Kanamori rule.

Further doping leads into region II where the crystal changes symmetry from tetragonal to orthorhombic. This Jahn-Teller distortion introduces a bias on the localized orbital degrees of freedom of the d_{yz}/d_{zx}-orbitals, suppressing the staggered correlation in favor of an alignment of the orbitals. Eventually the spin-spin interaction turns antiferromagnetic with growing lattice deformation in the effective model. This behavior is in qualitative agreement with experiments. The mean field phase diagram of the effective spin-orbital model including the lattice deformation is shown in Fig.1. For small deformation the staggered orbital order is prevailing leading to a ferromagnetic phase at low enough temperature. There is, however, a first order phase transition to the competing antiferromagnetic order, which at sufficiently strong deformation truncates abruptly the staggered orbital (antiferro-orbital) order.

The role of Ca-Sr alloying is to introduce lattice deformations through octahedra rotation and tilt. In the alloy there is naturally disorder and the properties show a certain degree of spatial variation. This means that in the region $0.2 < x < 0.5$ the two phases separated by the first order transition, FM-AFO (ferromagnetic-antiferro-orbital) and AFM-FO (antiferromagnetic-ferro-orbital), coexist in domains with rather sharp boundaries. This aspect allows us to connect the metamagnetic transition and the anomalous behavior of the magnetoresistance, observed for $Ca_{1.8}Sr_{0.2}RuO_4$ (28). At temperature of about 2 K a smooth metamagnetic transition has been observed at about $B_{ex} = 2.7T$ and the longitudinal magnetoresistance passes a pronounced maximum at approximately the same field value. In a finite magnetic field the domain structure of the material is modified by increasing the FM-AFO domains. The metamagnetic transition is associ-

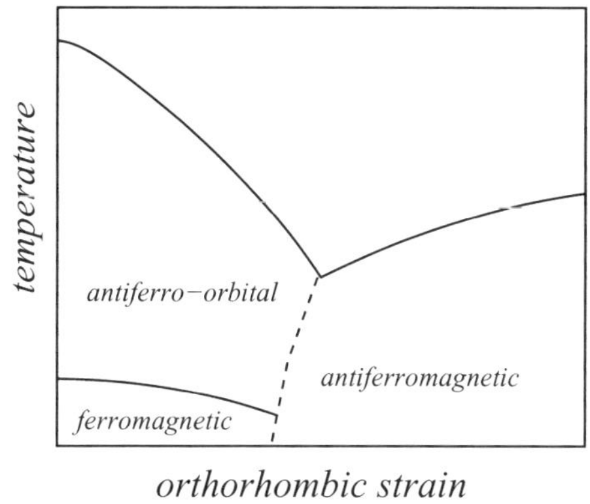

Figure 1. Schematic phase diagram for temperature versus orthorhombic strain for the Kugel-Khomskii type model with spin-orbital degrees of freedom: The solid (dashed) lines represent second (first) order phase transitions (29).

ated with the percolation of the domain boundaries. Domain boundaries influence also the transport properties of the remaining metallic band (γ) as strong scattering regions. In this way naturally the maximal impediment for conductance occurs when the domain boundaries percolate. In this way the metamagnetic transition and the maximal magnetoresistance are connected (29).

Although the scenario of the orbital-selective Mott insulator with localized spin and orbital degrees of freedom offers an explanation for a variety of phenomena in region II, it is difficult at the present stage to conclude on its validity. Further experimental tests are necessary, in particular, concerning the distinction of localized and itinerant portions of the t_{2g}-electrons.

4. The distinct roles of the different orbitals

The surprising contrasts of electronic behavior observed in the alloy $Ca_{2-x}Sr_xRuO_4$ are connected with the fact that the character of the $4d$-t_{2g}-orbitals is different. There is good evidence that the chiral p-wave superconductivity has its origin in the electrons of the γ-band, which therefore determines the low-temperature behavior of Sr_2RuO_4. On the other hand, the tendency towards antiferromagnetism in the alloy and Ca_2RuO_4 is clearly dominated by the α-β-bands which give rise to strong finite $\mathbf{Q}$ spin fluctuations.

Acknowledgements

I would like to thank to D.F. Agterberg, V. Anisimov, A. Furusaki, J. Goryo, H. Kusunose, M. Matsumoto, H. Monien, I.A. Nekrasov, D.E. Kondakov,

K.K. Ng, M. Troyer and, especially, T.M. Rice for the collaboration on the topics summarized here. I am also very grateful to Y. Maeno, S. Nakatsuji, H. Yaguchi and K. Wada for many helpful discussions. Furthermore I thank to M. Indergand for careful reading of the manuscript. These works had financial support from the Swiss Nationalfonds and a Grant-in-Aid by the Japanese Ministry of Education, Culture, Sports, Science and Technology.

References

1. Y. Maeno et al., Nature **372**, 532 (1994).
2. Y. Maeno, T.M. Rice and M. Sigrist, Physics Today **54**, 42 (2001).
3. T.M. Rice and M. Sigrist, J. Phys. Condens. Matter **7**, L643 (1995).
4. G. Baskaran, Physica **B 223-224**, 490 (1996).
5. K. Ishida et al., Nature **396**, 658 (1998).
6. G.M. Luke et al., Nature **394**, 558 (1998).
7. M. Sigrist et al., Physica **C 317-318**, 134 (1999).
8. D.F. Agterberg, T.M. Rice and M. Sigrist, Phys. Rev. Lett. **78**, 3374 (1997).
9. K.K. Ng and M. Sigrist, Europhys. Letts. **49**, 473 (2000).
10. M. Sigrist et al., J. Phys. Soc. Jpn **69** Suppl. B, 127 (2000).
11. J. Goryo and M. Sigrist, J. Phys. Condens. Matter **12**, L599 (2000); Europhys. Letts. **57**, 578 (2002).
12. T. Imai, A.W. Hunt, K.R. Thurber and F.C. Chou, Phys. Rev. Lett. **81**, 3006 (1998).
13. K. Ishida et al., Phys. Rev. B **64** 100501 (2001).
14. M.E. Zhitomirsky and T.M. Rice, Phys. Rev. Lett. **87**, 057001 (2001).
15. H. Kusunose and M. Sigrist, cond-mat/0205050.
16. I. Bonalde et al., Phys. Rev. Lett. **85**, 4775 (2000).
17. Y. Yanase and M. Ogata, preprint.
18. T. Nomura and K. Yamada, J. Phys. Soc. Jpn. **71**, 1993 (2002).
19. T. Nomura and K. Yamada, J. Phys. Soc. Jpn. **71**, 404 (2002).
20. Y. Maenoe et al., Phys. Rev. Lett. **81**, 3765 (1997).
21. M. Sigrist and H. Monien, J. Phys. Soc. Jpn. **70**, 2409 (2001).
22. Z.Q. Mao et al., Phys. Rev. Lett. **87**, 032003 (2001).
23. H. Yaguchi, M. Wada, T. Akima, Y. Maeno and T. Ishiguro, in preparation.
24. S. Nakatsuji and Y. Maeno, Phys. Rev. Lett. **84**, 2666 (2000); Phys. Rev. **B 62**, 6458 (2000).
25. Z. Fang and K. Terakura, Phys. Rev. **B 64**, 020509 (2001).
26. T. Nomura and K. Yamada, J. Phys. Soc. Jpn. **69**, 1856 (2000).
27. V.I. Anisimov et al., Eur. Phys. J. **B 25**, 191 (2002).
28. S. Nakatsuji, Ph.D. Thesis, Kyoto University (2000).
29. M. Troyer and M. Sigrist, in preparation.

REAL STRUCTURE OF PEROVSKITES LOOKED AT FROM THE BAND STRUCTURE POINT OF VIEW

Proceedings of the ARW NATO Workshop Hvar, Croatia, October 2002

K.–D. SCHOTTE and C.T. LIANG
Theoretische Physik, Freie Universität Berlin, Germany

Abstract. Although perovskites can be viewed as a dense packing of ionic spheres of three different sizes in a cubic structure, there are several deviations from this ideal. Most common is the tilting of the corner sharing octahedra like the SiO_6 octahedra in $MgSiO_3$. For transition metal complexes like CuF_6 in $KCuF_3$ Jahn-Teller distortions occur. With the symmetry lowered or the unit cell enlarged, gaps in the density of states grow or additional ones are opened. With local density band structure calculations (LDA) the tilting angle of the octahedra can be determined by minimizing the energy, in fair accord with the experimental values. With the same LDA Jahn-Teller distortions, however, cannot be calculated reliably. Since the *d*–density of states of the transition metal is placed above the p–density of O or F ions, a metal instead of an insulator is found. With the remedy of the LDA+U method one regains a proper description of the experimental facts. However, there are also cases for which the unmodified approach seems to be valid, for instance the perovskite $SrRuO_3$ with 4*d*–electrons is correctly calculated as a ferromagnetic metal. In studying the restricted perovskite class we want to illustrate the quality of band structure calculations. The shortcomings are not so much linked to an improper description of the transition metal ion open shell structure but to the failure in estimating the potential differences between the ions correctly. This is probably due to the wrong accounting of the unwanted self-interaction of the electrons in the d-shells. We think that considering the real crystallographic structure is necessary to sort out what is left over to be accounted for by correlated electron physics and not to be misguided by some technical problems in the otherwise quite accurate LDA–methods.

Key words: Perovskite, Crystal Structure, Band Structure Calculation, Jahn-Teller Effect

1. What are the structural pecularities of perovskites?

The mineral $CaTiO_3$ found in the Ural mountains, gave its name *perovskite* to a large group of compounds investigated intensively (1, 2). In solid state physics their magnetic, ferro–electric and superconductivity properties were the incentive. In mineralogy the high pressure conditions of the earth's

A.C. Hewson and V. Zlatić (eds.), Concepts in Electron Correlation, 35–43.

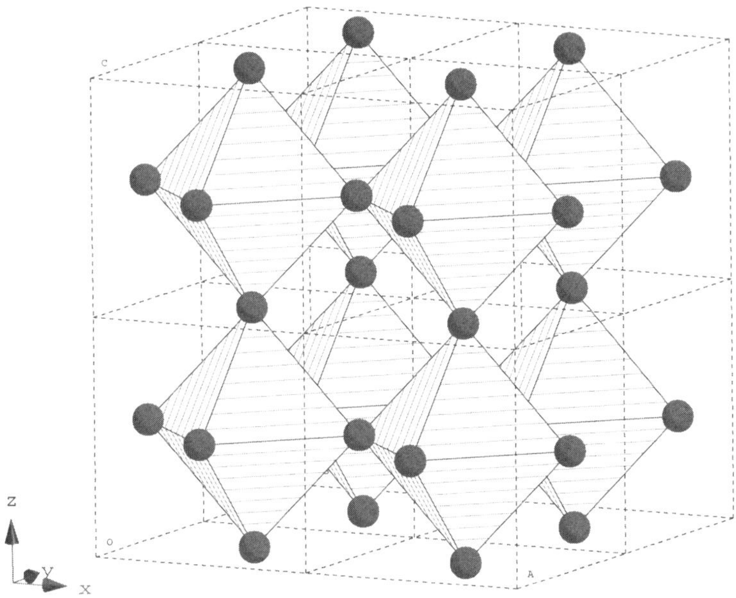

Figure 1. Structure of ReO_3 as an ideal network of ReO_6 octahedra

mantle favor a denser packing and make perovskites an abundant mineral. The silicates of the earth's crust transform there to the perovskite $MgSiO_3$ or $Mg_{1-x}Fe_xSiO_3$, see ref (3).

The basic unit of the perovskite ABX_3 consists of corner–linked octahedra of X anions (usually oxygen or fluorine) with B cations at their center and cations A sitting in the empty spaces between them. In a cubic unit cell the positive ions A are at the corners, the negative ions X in the middle of the faces and cations B in the center of the cube (see fig.1 above). The Goldschmidt ratio, defined as

$$t = \frac{d_{A-X}/\sqrt{2}}{d_{B-X}} \tag{1}$$

is 1 geometrically. Using the ionic radii for the two distances d one obtains usually smaller values, so that Mg^{2+} in $MgSiO_3$ is too small for the space available. The tolerance ratio $t < 0.9$ is an indicator for quite a number of perovskites according to the data collection of Wyckoff (1) deviating from a simple cubic structure with one formula unit per cell.

The compound ABX_3 considered as a corner sharing network of octahedral complexes BX_6 indicates that the octahedra taken as rigid units have the freedom to rotate or tilt. Since there are many possibilities for such displacements a large variety in the structure of perovskites is possible (4).

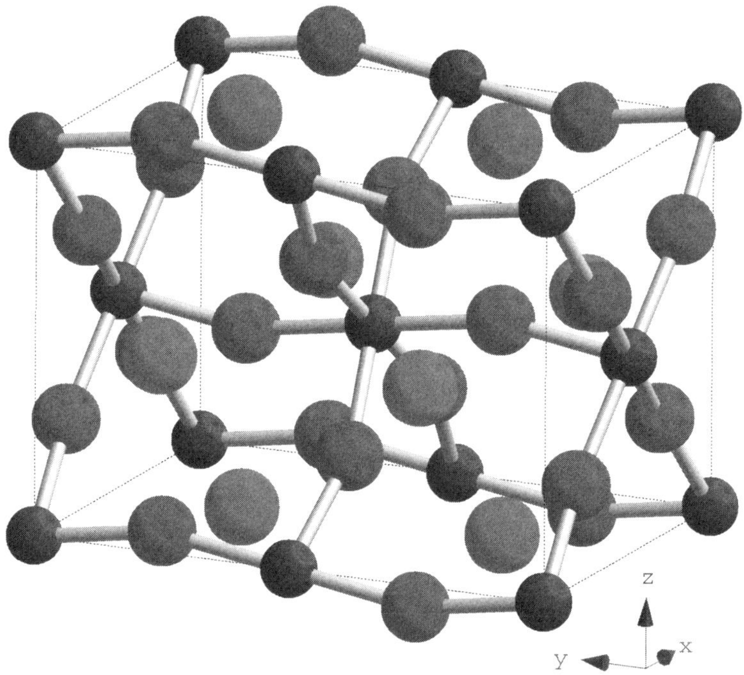

Figure 2. One unit cell of orthorhombic $SrRuO_3$. The rotations can be classified as $b^+a^-a^-$ after (4), that is rotations around the x, y, z directions of an octahedron having an opposite sense a^- or the same b^+ in a adjecent octahedra along the rotation axes. Ru (smallest) and O are connected by bonds, the largest spheres are Sr.

The assumption of rigidity is no longer valid, if the central atom B is Jahn–Teller active, like Cu^{2+} in $KCuF_3$ or Mn^{3+} in $GdMnO_3$. In the ferro–electric substances like $BaTiO_3$, where Ti is no longer in the center of O_6–octahedron, the distortions of the octadron are relatively small. The influence on the structure is also small compared to tilting of octahedra or the Jahn-Teller effect. In the following sections we want to adress the question whether this can be calculated in a band structure approach. The tilting will be discussed first.

2. The tilting of octahedra

Why octadra in perovskites tilt can be answered by looking at the compound ReO_3 which has a network of corner sharing ReO_6 complexes depicted in fig.1. However, there is no cation A between the octahedra as in the perovskite structure. For this substance and also for the few others listed under ScF_3 type in (1) tilting does not occur.

In a perovskite ABX_3 the positive A ions attract the negative X ion and only the repulse forces from the core of the A ion can stabilize the

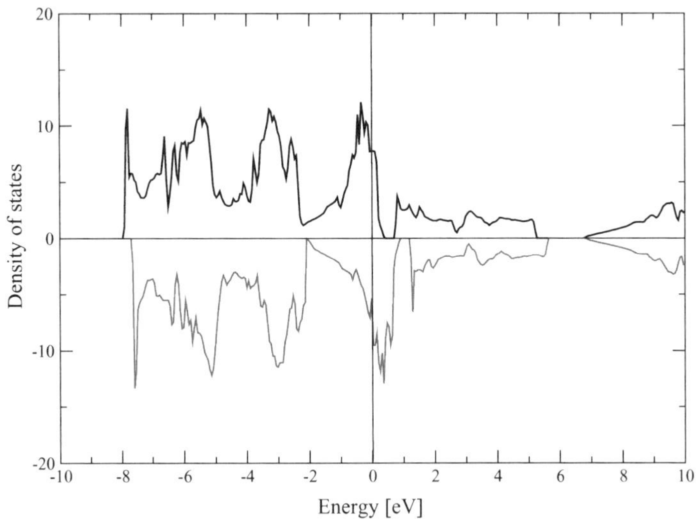

Figure 3. Density of states of orthorhombic $SrRuO_3$. Spin–up upper curve, spin–down lower one. F p and Ru d states are mixed, around and above $E_F = 0$ mostly d states.

structure. However, to fix the X–ion at the high symmetry position is not possible if the core is too small indicated by the tolerance ratio, see eq(1), $0.75 < t < 0.9$. Since further the BX_6 octahedron is a rigid unit only a collective movement of all the X–ions is possible, as classified by Glazer (4). A typical example can be seen in fig.2, where the structure of $SrRuO_3$ is displayed as determined by neutron scattering (5). The earth's mantle material $MgSiO_3$ (3) has the same structure like many others found first in the ferrite and "parent" substance $GdFeO_3$ (6). Four molecules ABX_3 are contained in a orthorhombic unit cell with the space group *Pnma* (7).

Provided one has a so called "full–potential" program that avoids spherical averaging of the ionic potentials, one can determine this structure with the present band structure codes using the local density approximation. However, to find a minimum of the total energy close to the tilted structure determined experimentally may be time consuming (see (3) and fig.4).

From band structure point of view a more interesting object than the insulator $MgSiO_3$ is a metal like $SrRuO_3$, especially since it is a ferromagnet. Due to the lower symmetry and larger unit cell compared to the cubic one, diffraction effects could lead to gaps in the density of states. If such changes are close to the Fermi energy, they influence the electronic transport and certainly the magneto–resistance.

We repeated the calculation of Singh (8) with the flpo2–program (9). As seen in fig.3 gaps appear, which for spin–up electrons are close to the

Fermi energy. The $4d$ states of Ru are split by crystal field effect into t_g and e_g. The e_g states are empty and the four d–electrons occupy only the t_g states. In an isolated ion three electrons would fill the up states completely and one of the down states, which would give spin 1. The band structure gives a lower value of 0.75. The small gap is due to a small hybridization of t_g and e_g, which would be absent for cubic $SrRuO_3$ without tilting. We suspect that correlation effects could pin this feature really to the Fermi level.

3. Jahn–Teller effect in the perovskites

Unlike the tilting of octahedra discussed, the band structure approach has difficulties to reproduce the Jahn–Teller distortions of the transition metal perowskites with Jahn–Teller active ions like Cu^{2+}, Mn^{3+} and Cr^{2+}. For these ions the Jahn–Teller effect is equivalent to the orbital ordering (10) of e_g states. These states are doubly degenerate in a perfect octahedral co-ordination of the transition metal ions. Lifting this degeneracy by lowering the symmetry is the essence of the Jahn–Teller effect (11).

Spin ordering appears similar to orbital ordering in that the two–fold degeneracy between up and down direction is lifted. This can be treated by modifying the local *density* approximation (LDA) into a local *spin* density approximation (LSDA). Otherwise ferromagnetism or anti–ferromagnetism would not occur in a band structure calculation (12). Similar modifications seem to be necessary to generate orbital ordering.

This has been done by adding to the standard LSDA an orbital dependent potential (14) which is called shortly LDA + U. That indeed an orbital ordering and a Jahn–Teller effect in fair agreement with the structural data occurs had been demonstrated by Liechtenstein (15).

We repeated his calculation for $KCuF_3$ using the "Wien2k"–code where a LDA + U version is included (19). In order to keep the computational effort low we have chosen the ferro–distortive $KCuF_3$(16) and imposed ferromagnetic spin ordering so that only two formula units were in a tetragonal cell. The type of magnetic ordering has little influence on the structure. $P4/mbm$ (7) the Wyckoff positions are $4h$: $(\frac{1}{2} - \delta,\ \frac{1}{2} + \delta,\ \frac{1}{2})$ for F in ab–plane around the Jahn–Teller active Cu at $2c$: $(0,\ \frac{1}{2},\ \frac{1}{2})$ as shown in fig.4. K has the position $2a$: (0, 0, 0) and the other F at $2d$: $(0,\ \frac{1}{2},\ 0)$. We have taken for $U = 0.59$, see eq(2), and $J = 0.07$ Ryd.[1] The lattice constants are $a = 5.8543$ Å and $c = 3.9303$ Å. The ordering of the orbitals can be seen in fig.5.

---

[1] To run the Wien2k–program the muffin–tin radii were set to $R_{Cu} = 2.0$, $R_K = 2.5$, $R_F = 1.35$ (in Bohr) and the number of k–points restricted to 210.

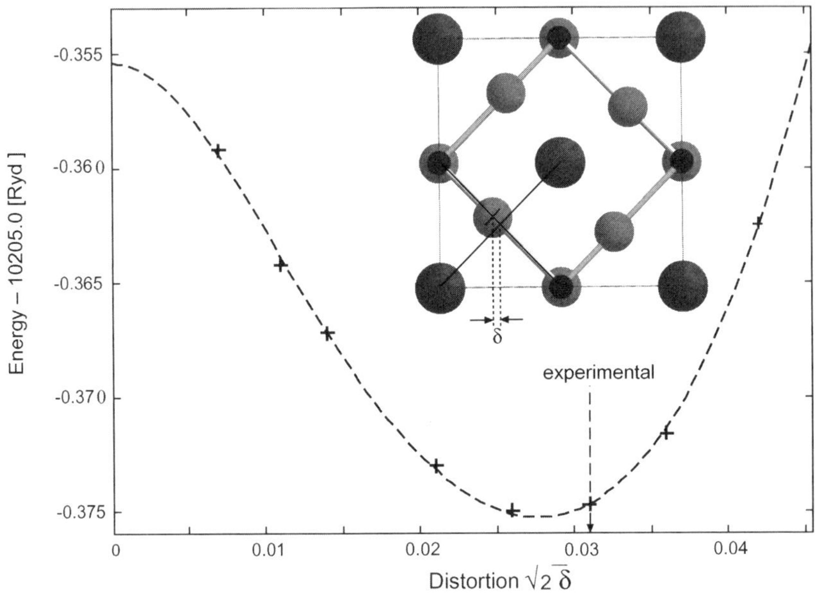

Figure 4. Total energy as a function of the Jahn–Teller distortion δ as indicated in the insert showing the tetragonal unit cell of $KCuF_3$ seen along the c–axis with $\delta_{exp} = 0.022$.

Also with a Hartree–Fock band structure calculation orbital ordering or cooperative Jahn–Teller leads to lowering of energy in the case of $KCuF_3$ as has been shown (13). $KCuF_3$ analyzed by LSDA is a nonmagnetic metal, whereas Hartree–Fock generates a gap of 1.3 Rydberg. In choosing the U in LDA + U approach one has the means to tune the gap to a more reasonable value and get a better agreement with the structural data.

A counterexample seems to be the ferromagnetic insulator K_2CuF_4 where a small gap appears and ferromagnetism is found using the unmodified LSDA (17). No minimum in energy for a distorted structure similar to fig.4 above could be determined. Possibly the precision of the programs in the milli Rydberg range is a limiting factor. With the Wien–code (19) we found a shallow minimum reduced by a factor of ten. This situation may be comparable to NiO where also a small gap is found with the LSDA (18). In order to get a larger and realistic gap using LDA + U (14) one has a remedy for the other transition metal oxydes where the unmodified approach fails to produce gaps.

4. Eliminating the self–interaction solves all problems?

Why does "Slater's dream machine"[2] fail so badly with the transition metal oxides or halides? "Correlation" is the standard answer. However, in Hartree–Fock without any additions and correlations, the orbital ordering and generating of gaps seem to be no problem (13). Following ref. (14) the LDA + U ansatz is constructed by adding a Hubbard–like term and subtracting its mean value, that is the energy is

$$E_{LDA+U} = E_{LDA} - \tfrac{1}{2}U\,N(N-1) + \tfrac{1}{2}U\sum_{i\neq j} n_i n_j\ , \tag{2}$$

where $N = \sum_i n_i$ is the total number of electrons in the d–shell and n_i the occupation of sublevel i. One assumes here that the standard LDA leads to a reasonable total energy but the energy of the level should be changed. By taking the derivative one obtains an orbital dependent potential

$$\frac{\partial E_{LDA+U}}{\partial n_i} = V_{LDA} + U\left(\tfrac{1}{2} - n_i\right)\ , \tag{3}$$

so that an occupied level is shifted downward in energy by $-\frac{1}{2}U$ with $n_i = 1$ but an unoccupied level is shifted upward by $+\frac{1}{2}U$ since $n_i = 0$. Summing the shifts of the occupied levels one arrives at $-\frac{1}{2}UN$, which could be interpreted as correcting the Coulomb self–interaction of N d–electrons.

Whether a self–interaction corrected (SIC) method (20) can also lead to orbital ordering in perovskites is not known. Eliminating the unwanted self–interaction one does not have the problem of a best choice for U in LDA + U, which is still under debate. However, the computational procedure for SIC is complicated as may be inferred from the method to solve Hartree's self–consistent equations where the self–interaction is also excluded (21).

Acknowledgements

We thank Jürgen Kübler, Helmut Eschrig and Peter Blaha for their help.

References

1. R.W.G. Wyckoff, *Crystal Structures*, Vol. 2, Inorganic Compounds RX_n, R_nMX_2, R_nMX_3, Interscience Publ. (1964).

---

[2] P.W. Anderson in his talk at the "International Conference of Theoretical Physics" in Paris 2002 referring the significant developments of solid state physics in the past.

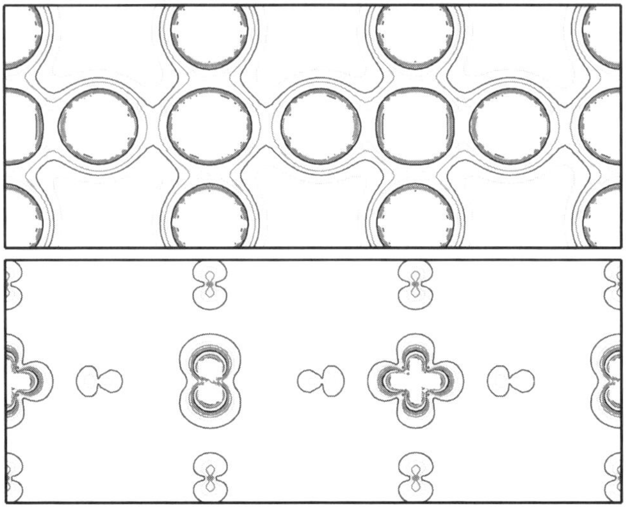

Figure 5. Density plots: above the sum and below the difference of spin up and down. The contour lines have values 0.2, 0.4 Å^{-3}.. The horizontal axis is in the *ab*–plane along the Jahn–Teller distorted Cu–F bonds, vertical is the *c*–axis. The e_g hole orbitals alternate as described in ref. (10). Above the Cu ions shrink where the hole orbitals are largest.

2. J.B. Goodenough & J.M. Longo, *Crystallograhpic and Magnetic Properties of Perovskites and Perovskite–Related Compounds*, Landolt–Börnstein *Neue Serie* III/4a, p.126, Springer Verlag (1970)
3. Lars Stixrude and R.E. Cohen, Nature **364**, 613 (1993).
4. A.M. Glazer, Acta Cryst. B **28**, 3384 (1972), see also Acta Cryst. A31,756 (1975).
5. C.W. Jones, P.T. Battle, P. Lightfoot and W.T.A. Harrison, Acta Cryst. C **45**, 365 (1989).
6. S. Geller, J. Chem. Phys. **24**, 1236 (1956).
7. *International Tables for X–Ray Crystallography*, Kynoch Press, Birmingham (1969).
8. D.J. Singh, *Electronic and Magnetic properties of the 4d itinerant ferromagnet $SrRuO_3$*, J. Appl. Phys. **79**, 4818 (1996).
9. K. Koepernik and H. Eschrig, *Full–Potential Nonorthogonal Local–Orbital minimum–basis band structure scheme*, Phys. Rev. B **59**, 1743 (1999).
10. K.I. Kugel and D.I. Khomskii, *The Jahn–Teller Effect and Magnetism: Transition metal compounds*, Sov. Phys. Usp. **25**, 231 (1982).
11. M.D. Sturge, *Jahn–Teller Effect in Solids*, Solid State Physics **20**, 91 (1967).
12. J. Kübler, *Theory of Itinerant Electron Magnetism*, Clarendon Press (2000).
13. M.D. Towler, R. Dovesi and V.R. Saunders, Phys. Rev. B **52**, 10 150 (1995).
14. V.I. Anisimov, I.V. Solovyev, M.A. Korotin, M.T.Czyżyk and Sawatzky, *Density–Functional Theory and NiO Photoemission Spectra*, Phys. Rev. **B48**, 16 929 (1993).
15. A.I. Liechtenstein, V.I. Anisimov, J. Zaanen, Phys. Rev. B **52**, R5467 (1995).
16. M.T. Hutchings, E.J. Samuelsen, G. Shirane & K. Hirakawa, Phys. Rev. B**188**, 919 (1969).
17. V. Eyert and K.-H. Höck, J. Phys.: Condens. Matter **5**, 2987 (1993).

18. K. Terakura, A.R. Williams, T. Oguchi and J. Kübler, *Transition–Metal Monoxides: Band or Mott Insulators*, Phys. Rev. Lett. **52**, 1830 (1984).
19. P. Blaha, K. Schwarz, G.K.H. Madsen, D. Kvasnicka and J. Luitz, *WIEN2k, An Augmented Plane Wave + Local Orbitals Program for Calculating Crystal Properties*, (Karlheinz Schwarz, Techn. Univ. Wien, Austria), ISBN 3-9501031-1-2 (2001), see also P. Novák, $WIENROOT/SRC/novak_lecture_on_ldau.ps (2001).
20. A. Svane, *Electronic Structure of La_2CuO_4 in the Self–Interaction–Corrected Density Functional Formalism*, Phys. Rev. Lett. **68**, 1900 (1992).
21. C. Edmiston and K. Ruedenberg, Rev. Modern Phys. **35**, 457 (1963).

TOWARDS MODELS OF MAGNETIC INTERACTIONS IN THE CUPRATES

Proceedings of the ARW NATO Workshop Hvar, Croatia, October 2002

H. ESCHRIG, K. KOEPERNIK and I. CHAPLYGIN
IFW Dresden, P.O.B. 27 0016, D-01171 Dresden, Germany

Abstract. The electronic structure of undoped cuprates is reanalyzed with the help of the LDA+U approach of density functional theory. It is found that in-plane oxygen orbitals forming π-bonds with copper are in energetic concurrence to those forming the Zhang-Rice singlet. The corresponding bands display a surprisingly large c-axis dispersion due to hybridization with unoccupied cation d-states in the adjacent layers. Hence, they must have relevance to c-axis magnetic coupling. In view of this situation, the Emery model is incomplete for the description of magnetism of undoped cuprates.

Key words: Electronic Structure, Undoped Cuprates, Density Functional Theory, LDA + U, Strong Correlations, Hubbard Model, Emery Model, Zhang-Rice Singlet, Magnetic Coupling, t-J Model

1. Density Functional Theory and Strong Correlation

The variational principle by Hohenberg and Kohn for the ground state energy E and spin density $\check{n}$ of electrons in a static external field $\check{u}$:

$$E[\check{u}, N] = \min_{\check{n}} \{H[\check{n}] + (\check{u} \mid \check{n}) \mid (\check{1} \mid \check{n}) = N\}, \tag{1}$$

where we use the short-hand notation

$$(\check{u} \mid \check{n}) = \sum_{ss'} \int d^3r u_{ss'} n_{s's} = \int d^3r (un - \boldsymbol{B} \cdot \boldsymbol{m}), \quad (\check{1} \mid \check{n}) = \sum_{s} \int d^3r n_{ss}, \tag{2}$$

holds true in any case of arbitrarily strong correlation. It is based on many-particle quantum theory by rigorous mathematics. Of course, the density functional $H[\check{n}]$ is unknown.

Since the external field $\check{u}$ must be static, it does not cope with non-adiabatic electron-lattice interaction.

A.C. Hewson and V. Zlatić (eds.), Concepts in Electron Correlation, 45–53.

The electronic quasiparticle excitations (if they exist) are obtained from the coherent part (pole term) of the single particle Green's function:

$$G_{ss'}(\boldsymbol{r},\boldsymbol{r}';\omega)=\sum_k\frac{\chi_s(\boldsymbol{r})\eta^*_{s'}(\boldsymbol{r}')}{\omega-\epsilon_k}+G^{\text{incoh}}_{ss'}(\boldsymbol{r},\boldsymbol{r}';\omega), \tag{3}$$

the pole ϵ_k of which is obtained from Dyson's equation for $\omega\to\epsilon$:

$$\sum_{s'}\int d^3r'\Big[\delta(\boldsymbol{r}-\boldsymbol{r}')\Big(-\frac{\boldsymbol{\nabla}^2}{2}+u(\boldsymbol{r})+u_{\text{H}}(\boldsymbol{r})\Big)+\Sigma_{ss'}(\boldsymbol{r},\boldsymbol{r}';\epsilon_k)\Big]\chi_{s'}(\boldsymbol{r}')=\chi_s(\boldsymbol{r})\epsilon_k \tag{4}$$

Here, the Hartree potential u_{H} and the self-energy Σ are functionals of the ground state density $\check{n}$ (we consider the case of temperature $T\to 0$).

Recall that the assumption of a pole in the Green's function is a working approximation of the theory. In principle from the full $\Sigma_{ss'}(\boldsymbol{r},\boldsymbol{r}';\omega)$ the total energy might also be obtained. We have only model theories for both $H[\check{n}]$ and $\Sigma[\check{n}]$.

2. Model Functionals

In the standard modeling of $H[\check{n}]$, the Kohn-Sham (KS) ansatz

$$n_{ss'}(\boldsymbol{r})=\sum_i\phi_i(\boldsymbol{r}s)\,n_i\,\phi^*_i(\boldsymbol{r}s'), \tag{5}$$

$$0\le n_i\le 1,\quad \langle\phi_i|\phi_j\rangle=\delta_{ij},\quad (1\,|\,\check{n})=\sum_i n_i=N \tag{6}$$

is used to split the density functional into an orbital variation expression $K[\check{n}]$ and a (possibly generalized by gradient terms) local density expression $L[\check{n}]$:

$$\begin{aligned}H[\check{n}]&=K[\check{n}]+L[\check{n}]\\ K[\check{n}]&=\min_{\{\phi_i,n_i\}}\Big\{k[\phi_i,n_i]\Big|\sum_i\phi_in_i\phi^*_i=\check{n}\Big\}\\ L[\check{n}]&=\int d^3rn(\boldsymbol{r})l(n_{ss'}(\boldsymbol{r}),\boldsymbol{\nabla}n,\ldots).\end{aligned} \tag{7}$$

The well known LSDA model is obtained with

$$k[\phi_i,n_i]=t_{\text{s}}+e_{\text{H}}=\sum_i n_i\langle\phi_i|\hat{t}|\phi_i\rangle+\frac{1}{2}\sum_{ij}n_in_j\langle\phi_i\phi_j|w|\phi_i\phi_j\rangle \tag{8}$$

and $l(\check{n})$ taken to be the exchange and correlation energy of the homogeneous electron liquid.

$K[\check{n}]$ works well for weakly interacting particles in an arbitrary external potential. $L[\check{n}]$ works well for weak external potential (nearly homogeneous system) with arbitrarily strong interaction provided the total energy of the homogeneous system as a function of the density is known.

The generalized KS equation

$$\frac{1}{n_k}\frac{\delta k}{\delta \phi_k^*} + \Big(\frac{\delta L}{\delta \check{n}} + \check{u}\Big)\phi_k = \phi_k \epsilon_k \tag{9}$$

is obtained by variation of ϕ_k^*, variation of the orbital occupation number n_k yields the aufbau principle.

2.1. THE ORBITAL FUNCTIONAL AND L(S)DA+U

In the L(S)DA+U model, the k-functional is extended by an e_U-functional which is expressed through occupation numbers $\tilde{n}_{\mu\sigma}$ of correlated local orbitals $|\boldsymbol{R}_i\mu\sigma)$ (centered in the unit cell at $\boldsymbol{R}_i$):

$$k = t_{\rm s} + e_{\rm H} + e_U, \quad e_U = e_U(\tilde{n}_{\mu\sigma}), \qquad \tilde{n}_{\mu\sigma} = \tilde{n}_{\mu\sigma}[\phi_k, n_k] \tag{10}$$

The corresponding U-potential is an operator containing a projector:

$$\frac{1}{n_k}\frac{\delta}{\delta\phi_k^*}e_U = \sum_{\boldsymbol{R}\mu\sigma} \underbrace{\frac{\partial e_U}{\partial \tilde{n}_{\mu\sigma}}}_{U\text{-potential}} \underbrace{\frac{1}{n_k}\frac{\delta \tilde{n}_{\mu\sigma}}{\delta \phi_k^*}}_{\text{projector}} = \hat{u}_U \phi_k. \tag{11}$$

2.2. THE ORBITAL OCCUPATION MATRIX

The L(S)DA+U models depend on the basis set of the solver of the KS equation. We will use for the latter a nonorthogonal local-orbital basis (1), since a local orbital representation is mandatory for considering strong correlations. For an LMTO solver see (2), for an LAPW solver (3).

Consider KS orbitals $|k\rangle = |\phi_k\rangle$ and orbital occupation numbers n_k as previously; they need not be eigenstates of spin. Let $\{|l)\} \ni |i)$ be a possibly non-orthogonal basis for KS orbitals: $|k\rangle = \sum_l |l)c_{lk}$, $S_{ll'} = (l|l')$. The occupation matrix $\tilde{n} = \tilde{n}[\phi_k, n_k]$ of correlated orbitals $|i)$ at site $\boldsymbol{R}_i$ in an orthogonal form is introduced as (see Fig. 1)

$$\tilde{n}_{mm'\sigma} = \sum_k \sum_{ll'} (S^{-1})_{(\boldsymbol{R}_i m\sigma),l}(l|k\rangle n_k \langle k|l')(S^{-1})_{l',(\boldsymbol{R}_i m'\sigma)} = \sum_k c_{ik}\, n_k\, c_{i'k}^*, \tag{12}$$

$$\frac{1}{n_k}\frac{\delta\tilde{n}_{mm'\sigma}}{\delta\langle k|} = \sum_{ll'} |l')(S^{-1})_{l'i'}(S^{-1})_{il}(l|k\rangle, \quad \Big(l_1\Big|\frac{1}{n_k}\frac{\delta\tilde{n}_{mm'\sigma}}{\delta\langle k|}\Big) = \delta_{l_1 i'}c_{ik} \tag{13}$$

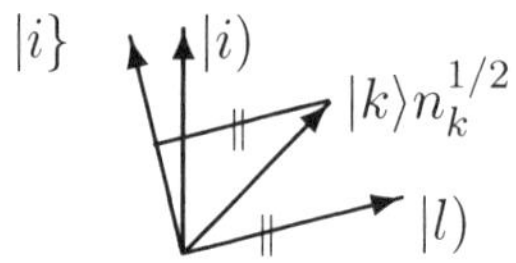

$$|i\} = \textstyle\sum_l |l)(S^{-1})_{li}, \ (l|i\} = \delta_{li}$$

Figure 1. Orthogonal projection on the contragredient orbital.

The orbital occupation matrix may be diagonalized with respect to m, m' at each lattice site $\boldsymbol{R}_i$ and for each spin value σ independently:

$$\tilde{n}_{mm'\sigma} = \tilde{U}^{(\sigma)}_{m\mu_\sigma}\, \tilde{n}_{\mu\sigma}\, \tilde{U}^{(\sigma)*}_{m'\mu_\sigma}. \tag{14}$$

Averages are:

$$\tilde{n}_\sigma = \frac{1}{2l+1}\sum_\mu \tilde{n}_{\mu\sigma}, \quad \tilde{n} = \frac{1}{2}(\tilde{n}_\uparrow + \tilde{n}_\downarrow) \tag{15}$$

2.3. THE ORBITAL POLARIZATION LSDA+U FUNCTIONAL

This functional was introduced under the name 'around the mean field' (AMF) in Ref. (4). It is zero if the orbitals of an atomic shell are equally occupied, hence it depends on orbital polarization. It is given by

$$l(n_{ss'}(\boldsymbol{r}), \ldots) = l_{\mathrm{LSDA}},$$

$$\begin{aligned}
e_U^{\mathrm{AMF}} &= \frac{1}{2}\sum_{\boldsymbol{R}\sigma\mu\mu'} \Big\{(\mu_\sigma\mu'_{-\sigma}|\tilde{w}|\mu_\sigma\mu'_{-\sigma})(\tilde{n}_{\mu\sigma} - \tilde{n}_\sigma)(\tilde{n}_{\mu'-\sigma} - \tilde{n}_{-\sigma}) \ + \\
&\quad +[(\mu_\sigma\mu'_\sigma|\tilde{w}|\mu_\sigma\mu'_\sigma) - (\mu_\sigma\mu'_\sigma|\tilde{w}|\mu'_\sigma\mu_\sigma)](\tilde{n}_{\mu\sigma} - \tilde{n}_\sigma)(\tilde{n}_{\mu'\sigma} - \tilde{n}_\sigma)\Big\} \\
&= \frac{1}{2}\sum_{\boldsymbol{R}\sigma\mu\mu'} \Big\{(\mu_\sigma\mu'_{-\sigma}|\tilde{w}|\mu_\sigma\mu'_{-\sigma})\, \tilde{n}_{\mu\sigma}\tilde{n}_{\mu'-\sigma} + \\
&\qquad +[(\mu_\sigma\mu'_\sigma|\tilde{w}|\mu_\sigma\mu'_\sigma) - (\mu_\sigma\mu'_\sigma|\tilde{w}|\mu'_\sigma\mu_\sigma)]\, \tilde{n}_{\mu\sigma}\tilde{n}_{\mu\sigma}\Big\} \\
&\qquad -\frac{1}{2}\sum_{\boldsymbol{R}\sigma}\Big\{U(N - \tilde{n}_\sigma) - J\Big(N_\sigma - \tilde{n}_\sigma\Big)\Big\}N_\sigma, \\
N_\sigma &= \sum_\mu \tilde{n}_{\mu\sigma} = (2l+1)\tilde{n}_\sigma.
\end{aligned} \tag{16}$$

The corresponding U-potential is

$$\begin{aligned}
\frac{\partial e_U^{\mathrm{AMF}}}{\partial \tilde{n}_{\mu\sigma}} &= \sum_{\mu'}\Big\{(\mu_\sigma\mu'_{-\sigma}|\tilde{w}|\mu_\sigma\mu'_{-\sigma})(\tilde{n}_{\mu'-\sigma} - \tilde{n}_{-\sigma}) + \\
&\quad +[(\mu_\sigma\mu'_\sigma|\tilde{w}|\mu_\sigma\mu'_\sigma) - (\mu_\sigma\mu'_\sigma|\tilde{w}|\mu'_\sigma\mu_\sigma)](\tilde{n}_{\mu'\sigma} - \tilde{n}_\sigma)\Big\}
\end{aligned} \tag{17}$$

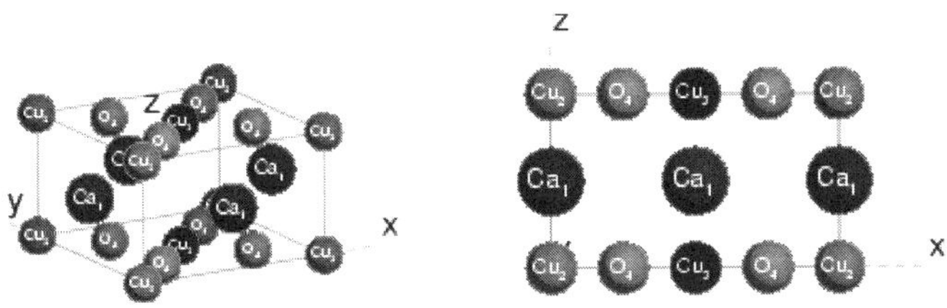

Figure 2. Unit cell of antiferromagnetic $CaCuO_2$

2.4. THE 'ATOMIC LIMIT' LSDA+U FUNCTIONAL

The AMF-functional yields nothing for a completely spin-polarized half-filled shell which is not so bad since the LSDA gives nearly the right spin polarization energy of Gd, although there is a problem with the right magnetic ground state. For this reason and aiming at a better description of the photoelectron spectra, Czyżyk and Sawatzky introduced in (4) another functional which they labeled 'atomic limit' (AL). At least regarding its relation to photoemission it should rather be considered a model for the quasiparticle self-energy Σ instead of being related to e_U. Nevertheless it was given in an e_U-form as (again $l(n_{ss'}(\boldsymbol{r}),\ldots) = l_{\mathrm{LSDA}}$)

$$
\begin{aligned}
e_U^{\mathrm{AL}} &= \frac{1}{2}\sum_{\boldsymbol{R}\sigma\mu\mu'}\Big\{(\mu_\sigma\mu'_{-\sigma}|\tilde{w}|\mu_\sigma\mu'_{-\sigma})\,\tilde{n}_{\mu\sigma}\tilde{n}_{\mu'-\sigma} + \\
&\qquad +[(\mu_\sigma\mu'_\sigma|\tilde{w}|\mu_\sigma\mu'_\sigma) - (\mu_\sigma\mu'_\sigma|\tilde{w}|\mu'_\sigma\mu_\sigma)]\,\tilde{n}_{\mu\sigma}\tilde{n}_{\mu\sigma}\Big\} \\
&\quad -\frac{1}{2}\sum_{\boldsymbol{R}}\Big\{UN(N-1) - J\sum_\sigma N_\sigma\big(N_\sigma - 1\big)\Big\} = \qquad (18) \\
&= e_U^{\mathrm{AMF}} + \frac{1}{2}\sum_{\boldsymbol{R}\sigma}(U-J)(1-\tilde{n}_\sigma)N_\sigma.
\end{aligned}
$$

$$
\frac{\partial e_U^{\mathrm{AL}}}{\partial \tilde{n}_{\mu\sigma}} = \frac{\partial e_U^{\mathrm{AMF}}}{\partial \tilde{n}_{\mu\sigma}} - (U-J)\Big(\tilde{n}_\sigma - \frac{1}{2}\Big) \qquad (19)
$$

3. Application to $CaCuO_2$ and to $Sr_2CuO_2Cl_2$

The structure of the 'infinite layer' cuprate $CaCuO_2$ is shown in Fig. 2.

There are two relevant molecular orbitals (MO) which are candidates for the highest occupied molecular orbital (HOMO) in the electronic structure of cuprates, that is, those MOs which are relevant for the valence band edge, see Fig. 3.

The LSDA+U KS bands of $CaCuO_2$ with the AMF and AL functionals are shown in Fig. 4. On Fig. 5 the orbital weight to the AMF bands for

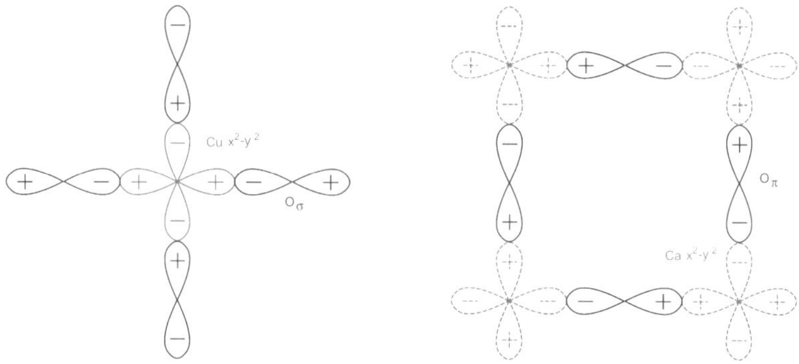

Figure 3. Left: Antibonding ($\boldsymbol{k} = (\pi, \pi, 0)$) $dp\sigma$-orbital commonly assumed as the HOMO that forms the Zhang-Rice singlet together with the nominal Cu-d hole (5, 6). Right: O-O antibonding ($\boldsymbol{k} = 0$) in-plane $p\pi$-orbital, lifted up by crystal field and weakly hybridized with Ca-d orbitals in adjacent layers: the true HOMO of the LSDA+U model (7, 8).

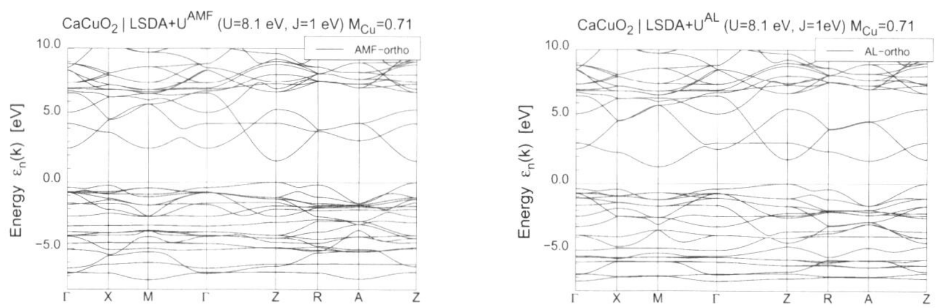

Figure 4. LSDA+U band structure of $CaCuO_2$. Left: AMF, right: AL.

several orbitals is shown, that is, the square of the coefficient in the expansion of the KS orbital into local basis orbitals is given by the linewidth. It is clearly seen that both the O $2p_\sigma$ and O $2p_\pi$ orbitals hybridize equally strongly with the Cu $3d_{x^2-y^2}$ orbital at the valence band edge. Thereby, the considerable dispersion in z-direction of the O $2p_\pi$ dominated bands comes from hybridization with the unoccupied Ca $3d_{x^2-y^2}$ orbitals.

$Sr_2CuO_2Cl_2$ is probably the most two-dimensional of all cuprates. The antiferromagnetic unit cell is shown in Fig. 6 and the same orbital weights as in the last section are shown on Fig. 7. The situation is quite similar except that there is practically no dispersion in z-direction.

4. Implications for Magnetic Interactions

Gross magnetic couplings may be obtained from total energy differences of LSDA+U results for ferromagnetic and antiferromagnetic order; in more detail they may be obtained from calculated energies of spin spiral states (9).

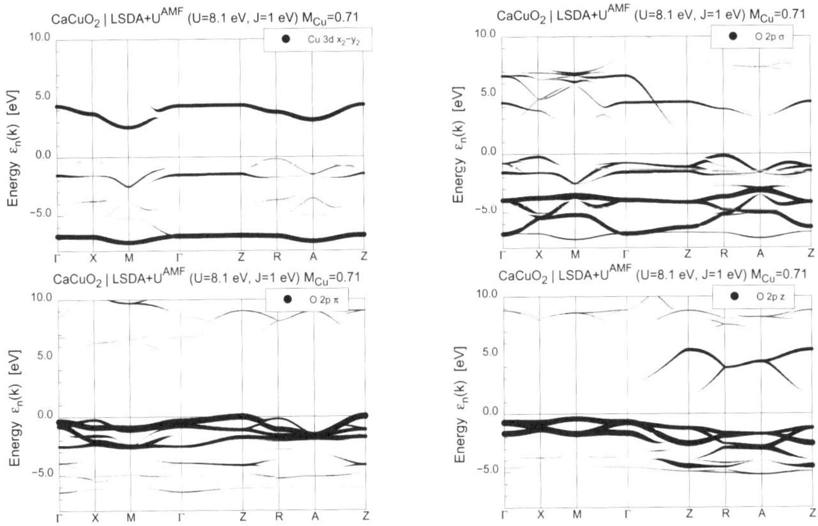

Figure 5. Upper panels: orbital weight of the Cu $3d_{x^2-y^2}$ orbital and the O $2p_\sigma$ orbitals, resp. Lower panels: the same for the O $2p_\pi$ and the O $2p_z$ orbitals.

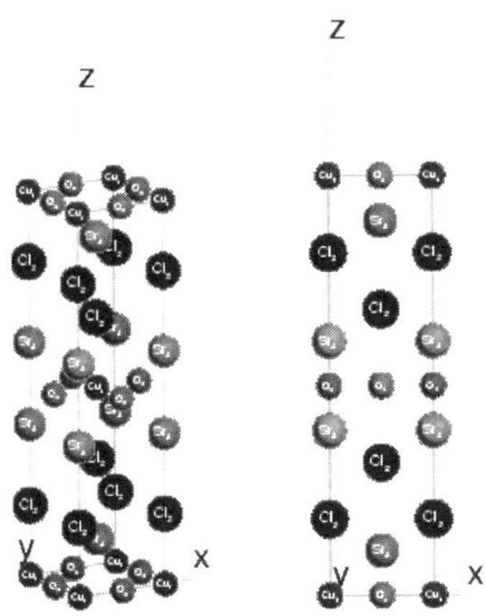

Figure 6. Unit cell of antiferromagnetic $Sr_2CuO_2Cl_2$

For a more detailed understanding of the physics, a tight-binding model of the kind of Emery's model should be extracted from the LSDA+U results, which then may be down-mapped to a kind of a $t-J$ model. For the ferromagnetic order such a tight-binding model was derived in Ref. (8).

The conclusion from the present orbital analysis is that in a large part of the Brillouin zone there is a strong hybridization of O-$2p_\sigma$ with O-$2p_\pi$ orbitals in bands hybridized with the Cu-$3d_{x^2-y^2}$ orbital. Hence, the O-$2p_\pi$ orbitals must be included in the Emery model in order to correctly describe the t-terms which determine the magnetic coupling.

In $CaCuO_2$ there is a sizable dispersion of the in-plane O-$2p_\pi$ bands

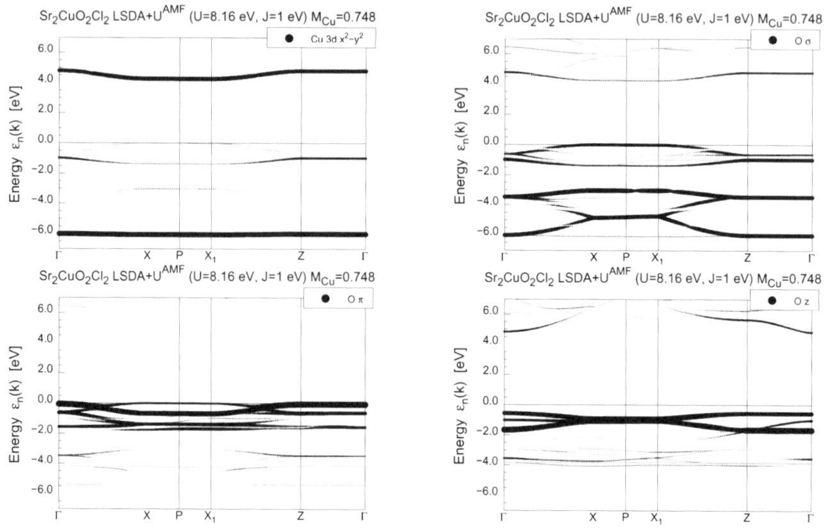

Figure 7. Same as Fig. 5 for $Sr_2CuO_2Cl_2$.

in z-direction mediated by some hybridization with Ca-$3d_{x^2-y^2}$ orbitals. This is in accordance with the experimental finding that $CaCuO_2$ has the highest Neel temperature, $T_N = 540$ K, of all layered cuprates indicating 3D magnetism (10, 11).

By contrast, in $Sr_2CuO_2Cl_2$ there is no dispersion of the corresponding bands in z-direction due to the Cl_2 buffer layers. There is only dipole-dipole coupling of the planes, compatible with the experimental findings, $T_N = 256$ K (12).

Acknowledgements

We acknowledge helpful discussions with R. Hayn, S.-L. Drechsler and H. Rosner as well as financial support by the DFG (SFB463) and by the GIF (project I-614-13.14/1999).

References

1. K. Koepernik and H. Eschrig, Phys. Rev. **B59**, 1743 (1999).
2. A. I. Liechtenstein, V. I. Anisimov and J. Zaanen, Phys. Rev. **B52**, R5467 (1995).
3. A. B. Shick, A. I. Liechtenstein and W. E. Pickett, Phys. Rev. **B60**, 10763 (1999).
4. M. T. Czyżyk and G. A. Sawatzky, Phys. Rev. **B49**, 14211 (1994).
5. V. J. Emery, Phys. Rev. Lett. **58**, 2794 (1987).
6. F. C. Zhang and T. M. Rice, Phys. Rev. **B37**, 3759 (1988).
7. J. J. M. Pothuizen *et al.*, Phys. Rev. Lett. **78**, 717 (1997).
8. R. Hayn *et al.*, Phys. Rev. **B60**, 645 (1999).

9. A. N. Yaresko *et al.*, Phys. Rev. **B65**, 115111 (2002).
10. D. Vaknin *et al.*, Phys. Rev. **B39**, 9122 (1989).
11. R. Pozzi *et al.*, Phys. Rev. **B56**, 759 (1997).
12. M. Greven *et al.*, Phys. Rev. Lett. **72**, 1096 (1994).

C-AXIS INTRA-LAYER COUPLINGS IN THE CUO_2 PLANES OF HIGH-T_C CUPRATES

Proceedings of the ARW NATO Workshop Hvar, Croatia, October 2002

J. RÖHLER
Universität zu Köln
Zülpicher Str. 77, D-50937 Köln, Germany

Abstract. We discuss the static deformations of the doped CuO_2 lattices in $YBa_2Cu_3O_x$. Intrinsic lattice effects driven by the strong correlations of the electron system are separated from extrinsic chemical and bandstructure effects. c-axis displacements of the planar copper atoms seem to be a generic property of the metallic CuO_2 lattices.

Key words: Superconductivity, Cuprates, $YBa_2Cu_3O_x$ CuO_2 Lattice, C-Axis, Buckling, Strongly Correlated Electrons, Doping Mechanism, X-Ray Diffraction, EXAFS

1. Introduction

Unconventional metals exhibit a plethora of unusual lattice effects not observed in "simple" band metals. In the physics of strongly correlated electron systems these puzzling "lattice instabilities" are usually ignored or relegated to the high energy scale of the crystal chemistry. This lecture focusses on the *static* distortions of the nearly square planar CuO_2 lattices in the under- and overdoped regimes of $YBa_2Cu_3O_x$, the superconducting cuprate with the best known lattice structure. We shall discuss the doping dependence of the basal lattice parameters, the CuO_2 dimples, and the orthorhombicity. The structural data suggest that doping into a Mott insulator leads to weak but significant deformations of the metallic CuO_2 lattice. Our focus is on the structural aspects of the normal state; it is not intended to suggest the inclusion of electron-lattice interactions into a specific model of high-T_c superconductivity.

A.C. Hewson and V. Zlatić (eds.), Concepts in Electron Correlation, 55–63.

2. Planar and perpendicular orbitals

The pioneering work of Müller and Bednorz (1) has been guided by the idea of high-T_c superconductivity (HTSC) in transition-metal chalcogenides with strong Jahn-Teller (JT) effects. In fact, the strong JT effect is confirmed to play a crucial role for the electronic structure of the superconducting cuprates selecting from the e_g doublet the antibonding $3d_{x^2-y^2}$ as the *only* relevant d orbital in the plane of the Cu–O polyhedron. It generates the enormous stability of the square-planar CuO_2 structure. An electronically active role of the perpendicular $3d_{z^2}$ orbital is however not confirmed.

But superconductivity is not a purely 2-dimensional effect. T_c depends strongly on interactions of the planes with the doping reservoir, and a mechanism linking the plane with its perpendicular surroundings is expected to be involved in the formation of the electronic ground state. Recent calculations of the one-electron bandstructure of $YBa_2Cu_3O_7$ (2) show that the generic low-energy layer-related features have to be described by the planar orbitals Cu$3d_{x^2-y^2}$, O$2p_{x,y}$ directing in the plane, *and* the isotropic Cu4s orbital. The chemical trend of T_{cmax} is found to be controlled by the energy of a "perpendicular orbital" , a hybrid between Cu4s, Cu$3d_{z^2}$, the apical O$2p_z$, and farther d_{z^2} or p_z orbitals. Interestingly materials with the perpendicular orbital more localized onto the CuO_2 layers, *i.e.* with pure Cu4s character, exhibit the highest T_{cmax}. It seems that the strong hybridization between Cu$3d_{x^2-y^2}$ and Cu4s allows for atomic displacements in the rigid covalent CuO_2 lattice of the parent Mott insulator such that doped holes are optimally accomodated in the background of the quantum liquid of spin singlets.

Notably the dependence of T_{cmax} from the structure and the chemistry perpendicular to the planes enters a single-band many electron Hamiltonian simply by the ratio of the *nnn* and *nn* hopping integrals, t'/t. It is worthwile to stress that the (rare) acknowledgement of the chemistry outside the planes as an integral part of the HTSC problem does not justify dynamically independent treatments of the Cu3d and O2p electrons in any kind of multiband many electron Hamiltonians.

3. Dimpled CuO_2 planes

The CuO_2 planes of the superconducting cuprates are not perfectly flat but dimpled. Many diffraction work, *e.g.* Ref. (3, 4, 6, 8), and some EXAFS studies (9, 10) yield evidence for dimples in the relevant CuO_2 planes of the multi-layer cuprates, *i.e.* the planes adjacent to the doping reservoirs, *cf.* table I. "Buckled" planes are also observed in single-layer materials, most

TABLE I. CuO_2 dimplings in typical HTSC, see text.

Compound	$P_\perp$	N_L	Dimpling (Å)	T_c	Ref.
$YBa_2Cu_3O_x$	$3^+ - 2^+$	2	0.28	92	(3, 9)
$YBa_2Cu_4O_8$	$3^+ - 2^+$	2	0.24	80	(8)
$Pb_2Sr_2YCu_3O_8$	$3^+ - 2^+$	2	0.23	70	(6)
$HgBa_2(Ca,Y)Cu_2O_x$	$2^+/3^+ - 2^+$	2	0.10	85-110	(4)
$HgBa_2CaCu_2O_x$	$2^+ - 2^+$	2	0.02	120	(4)
$HgBa_2Ca_2Cu_3O_x$	$2^+ - 2^+$	3	0.05	132	(4)
$HgBa_2Ca_3Cu_4O_x$	$2^+ - 2^+$	4	0.05	127	(4)

clearly in the doped La_2CuO_4 systems, but so far not in the single-layer Hg- and Tl-cuprates (4). As we shall see, the dimplings in the latter materials are expected to be *small*, and since their CuO_2 planes are crystallographic inflection planes the usual crystallographic refinements tend to average them to zero.

We distinguish "chemical" dimples driven by external electrostatic polarization along c, mixed valence of the dopants, lattice mismatches and other bandstructure effects, from "correlation-driven" dimples generic of the strongly correlated electron system. We show that the former define a *large* length-scale of ~ 0.3 Å, and the latter a *small* length-scale of ~ 0.03 Å, corresponding to buckling angles of $\sim 10°$, and $\sim 1°$, respectively. The contribution of the chemical dimples in many of the doped metallic phases can be traced back to the lattice of their antiferromagnetic parent phase, and thus be differentiated from the correlation-driven dimples. However in concentrated alloys, doped by varying ratios of one or two combinations of heterovalent cations, the correlation-driven dimples may be masked by the varying external electrostatic polarizations (12). Table I compares the CuO_2 dimplings in some typical multi-layer cuprates with each other: as a function of the static charge contrast polarizing the CuO_2 layer perpendicularly, $P_\perp$, the number of CuO_2 layers, N_L, and T_c. The *large* length-scale of ~ 0.3 Å shows clearly up in all compounds with $P_\perp = 2^+ - 3^+$, but is absent in compounds with $P_\perp = 2^+ - 2^+$. Materials alloyed with heterovalent cations, *e.g.* Ca^{2+}/ Y^{3+}, exhibit intermediate values. Notably CuO_2 layers sandwiched between isovalent layers seem to exhibit higher T_{cmax} than those between heterovalent layers, *cf.* Ref. (7).

4. Perpendicular and planar CuO_2 deformations in $YBa_2Cu_3O_x$

Figure 1b sketches a metallic CuO_2 layer of $YBa_2Cu_3O_x$ as a stack of copper and oxygen layers. The horizontal dashed lines in Figure 1a indicate a "chemical" offset of $\simeq 0.23$ Å, typical for Ba^{+2}/Y^{+3} polarization. The spacing between the Cu2 and O2,3 layers depends clearly on doping. While the Cu2 layer starts to move towards the Ba–O layer by about 0.05 Å between the onset of the metallic phase ($x \simeq 6.38$) and x_{opt}, the O2,3 layers seem to be almost unaffected. We show in figure 3 that the scatter in the Y–O2, O3 data points does not behave arbitrarily, and that the relatively weak perpendicular displacements of O2 and O3 are correlated with the planar deformations.

4.1. LATTICE PARAMETERS AND ORTHORHOMBIC DEFORMATIONS

The relatively strong orthorhombicity of its unit cell puts $YBa_2Cu_3O_x$ somewhat outside the lattice systematics of the most prominent high-T_c materials, which have mostly tetragonal unit cells. Weak orthorhombic strain might be however operative also on the nominally tetragonal lattices, if a correlation-driven anisotropy of the 2-D Fermi surface is a generic feature of HTSC, as theoretically suggested by renormalization group methods (11).

Figure 2a displays the doping dependence of the lattice parameters $a, b, c/3$ in the metallic phases of $YBa_2Cu_3O_x$. $a \neq b$ in the under- and overdoped regimes, but the inequality results clearly from different types of orthorhombic strain ellipsis. The quadrupolar α-ortho stresses a while straining b, and the monopolar β-ortho stresses both axis. "Plateaus" of a

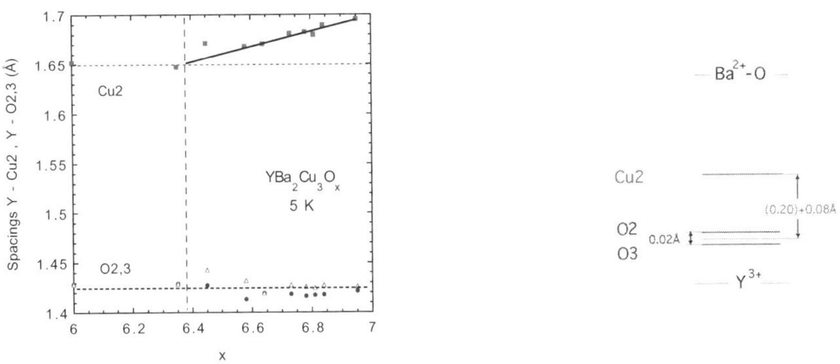

a. Y–Cu2, Y–O2,3 spacings　　　b. CuO_2 intralayer spacings

Figure 1. Layer spacings in $YBa_2Cu_3O_x$. a.: Y–Cu2, and Y–O2,3 interlayer spacings as a function of oxygen concentration, x, at 5 K. After Cava *et al.* (3). Open triangles: O2. Open circles: O3 . b.: CuO_2 plane sketched as a stack of copper and oxygen layers sandwiched between polarizing adjacent layers (arbitrary scale).

and b between $x \simeq 6.8$ and 6.9 connect the two regimes suggesting a thermodynamical instable regime. If the orthorhombic strain originated exclusively from Cu1–O4 hybridization in the 1-dimensional charge reservoir, we expect the b-, and the a-axis to exhibit similar orthorhombic deformations in both, the under- and overdoped regimes. The abrupt change from the quadrupolar β-ortho to the monopolar β-ortho deformation at x_{opt} points however to an additional and different mechanism to be operative.

Figure 2b displays the doping dependence of the basal area, $B(x) = a \cdot b$. Evidently accomodation of oxygen atoms in the reservoir layer does not increase $B(x)$ as might be expected from simple stereochemical grounds, but reduces the basal area. Hence the doping dependence of $B(x)$ is most likely controlled by the ground state of the quantum liquids in the metallic planes, and not by the anisotropic oxygen diffusion processes in the non-stoichiometric chain layer. To discuss $B(x)$ in more detail we find it useful to associate $\kappa_p \propto -\partial B/\partial n_h$ with a 2-dimensional "electronic compressibility", eliminating the almost doping independent bandstructure effects. Here $n_h \propto x$ denotes the hole concentration.

$\kappa_p \simeq 0$ in the insulating phase. The plane of the spin lattice in the lightly doped Mott insulator seems to be incompressible as indicated by the arrow for $x = 6.0 - 6.38$. The magnetic exchange energy J of order 1500 K determines the electron-electron interactions in the Cu2 planes.

$\kappa_p > 0$ in the metallic phases. While weakly decreasing in the underdoped α-ortho regime, $B(x)$ starts to collapse in the overdoped β-ortho regime. The plateau between $x = 6.8$ and 6.9 points to phase segregation, possibly into incommensurate stripe phases. In the metallic phase the doped

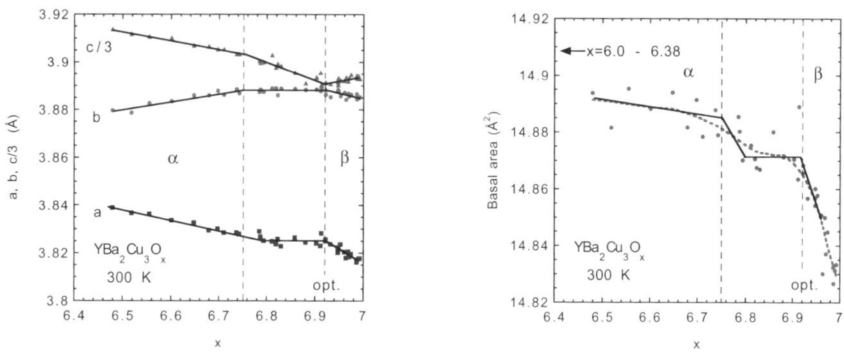

a. Lattice parameters b. Basal area

Figure 2. a.: Lattice parameters *vs.* oxygen concentration in near-equilibrium samples of metallic $YBa_2Cu_3O_x$ synthezised by direct oxydation of the elements (DO), after Conder *et al.* (5). Drawn out lines are guides to the eye. α and β label different orthorhombic strain ellipsis, see text. b.: Basal area, $a \cdot b$, from the data in a. Dashed line: smoothed average. The step indicates the onset of a possible two-phase regime.

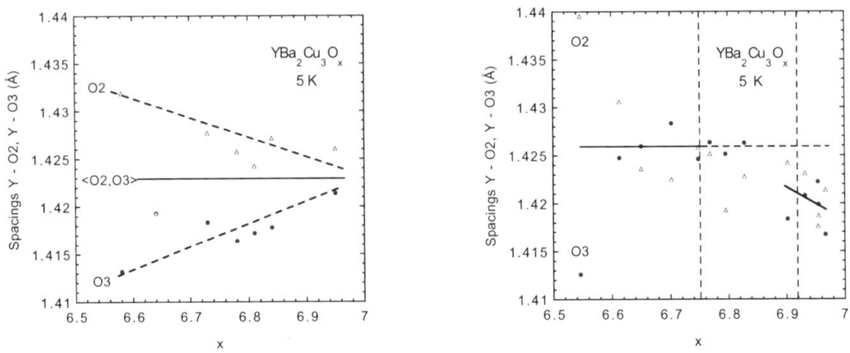

a. Without c-axis minimum (3) b. With c-axis minimum (13, 14)

Figure 3. Spacings Y–O2, Y–O3 *vs.* oxygen content in $YBa_2Cu_3O_x$ at 5 K from neutron diffraction in different samples. Triangles: O2 (a-axis). Dots: O3 (b-axis). All lines are guides to the eye. a.: In quenched samples from the carbonate route after Cava *et al.* (3). b.: In near equilibrium samples from the BaO route after Kaldis (13), Hewat *et al.* (14).

holes hop with a nn matrix element $t \propto d^{-n}$, $n \geq 2$, and thus shorter nn distances, d, increase strongly the kinetic energy of the holes. But the physics of the doped Mott insulator is that of competition between the exchange energy J and the kinetic energy per hole $n_h t$. The doped holes in the underdoped regime appear only as vacancies in the background of a spin singlet liquid. The lattice of such a strongly correlated $t - J$ type electron system is expected to be much harder than that of a nearly noninteracting electron liquid.

Hence the α- and β-ortho deformations may be identified as characteristic lattice responses to fundamentally different types of electron liquids in the metallic phase: a weakly compressible $t - J$ like in the under- and optimum doped, and a strongly compressible Fermi liquid-like in the overdoped regimes.

4.2. PERPENDICULAR O2,O3 DISPLACEMENTS

Figure 3 displays the doping dependence of the interlayer spacings Y–O2, Y–O3 from samples without (3) and with (14) a c-axis minimum. At the onset of the metallic phase the degenerate tetragonal positions of the planar oxygens are split into O2 along a, and O3 along b. Both are also displaced along c: O2 by ~ -0.05 Å below, and O3 by $\sim +0.05$ Å above the tetragonal reference value, *cf.* figure 4b. The anisotropic displacement of the planar oxygens along c is usually attributed to the anisotropic Coulomb repulsion between the 1-dimensional charge reservoir and the 2-dimensional planes.

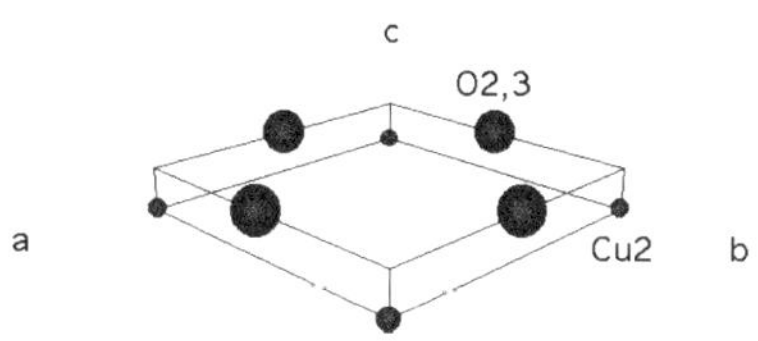

a. Tetragonal insulator

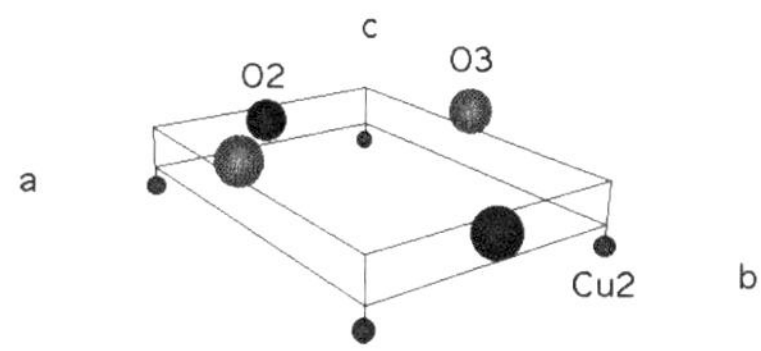

b. α-ortho underdoped metal

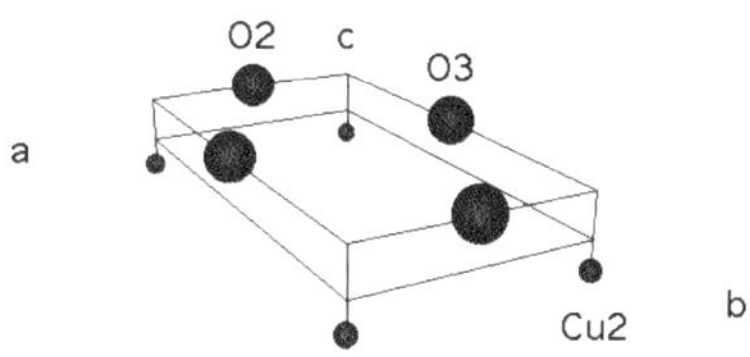

c. Optimum doped metal

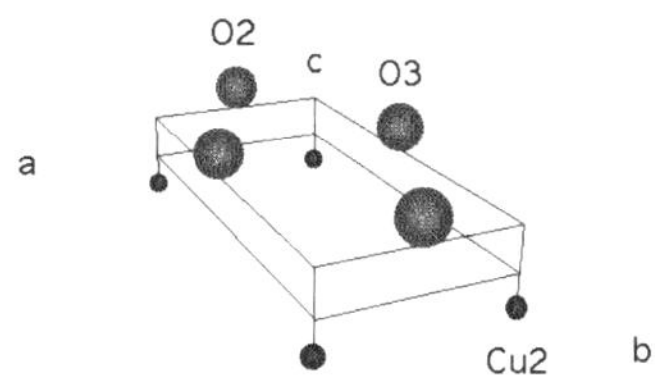

d. β-ortho overdoped metal

Figure 4. Sketches of the CuO_2 lattices in $YBa_2Cu_3O_x$. The frame indicates the dimpling due to the surrounding chemistry and bandstructure effects.

This is yet another example for Coulomb repulsion and hybridization are not able to correctly describe the doping-induced displacements in the CuO_2 layer: increasingly perfected chains in the reservoir layer relax the c-axis anisotropy of the planar O2, O3 instead of increasing them. Thus the average spacing Y–<O2,O3> matches the degenerate tetragonal reference value throughout the underdoped α-ortho regime (thick drawn out line in figures 3a,b). Meaningful crystallographic studies of the overdoped β-ortho regime require samples with a c-axis minimum around x_{opt} (13). Figure 3b displays the O2, O3 c-axis displacements in such samples (13, 14). Herein the transition into the overdoped regime seems to be connected with weak *unidirectional* displacements of both oxygens O2, O3 towards the Y-layer, *cf.* figure 4d. Studies of the local atomic structure by Y-EXAFS (9, 13) confirm this unidirectional c-axis shift of the planar oxygens in the overdoped regime also in standard samples without c-axis minimum (5).

We summarize the doping-induced displacements of Cu2 and O2,O3 in the schematic figures 4a-d:

i. Metallic hole concentrations increase the two-dimensional density of Cu2 by dimpling the planes. There is some similarity with the mechanism collapsing an umbrella (15). As Cu2 moves out of the plane the basal area shrinks.

ii. The underdoped regime is governed by the quadrupolar α-ortho strain. O2 and O3 shift oppositely along c such that the basal Cu2 area may adjust to a relative maximum.

iii. Close to optimum doping the quadrupolar α-ortho strain vanishes. O2 and O3 may achieve nearly degenerate c-axis positions.

iv. In the overdoped regime the deformations have changed from the quadrupolar α- to the monopolar β-ortho type stressing the a-, b-axes in the same direction therewith collapsing the basal Cu2 area. Both oxygens O2,3 shift perpendicularily in the *same* direction along c.

5. Concluding Remarks

We have undertaken an attempt to identify among the many doping-driven lattice effects in $YBa_2Cu_3O_x$ those which are most likely connected with the generic low energy electronic structure. Fermi surface driven lattice effects are well known from the classical metals. For instance about 3/4 of the metallic elements tend to maximize the gain of kinetic energy by crystallization in most closely packed structures hcp, fcc, bcc. In electron compounds and Hume-Rothery alloys the Fermi energy is well known to avoid maxima in the density of states and to drive structural transformations changing appropriately the symmetry of the Brillouin zone. These mechanisms are expected to be operative also in low dimensional and strongly correlated electron systems. The competition between exchange and kinetic energies in strongly correlated electron systems however gives rise to new and more intricate lattice effects which, once disentagled from bandstructure effects, may yield new insights into the many electron ground state.

Acknowledgements

I thank E. Kaldis for a stimulating and fruitful cooperation. The ESRF supported this work partially through project HE731 at BM29.

References

1. J. G. Bednorz and K. A. Müller, Z. Phys. B **64**, 189 (1986).
2. E. Pavarini, I. Dasgupta, T. Saha-Dasgupta, O. Jepsen and O. K. Andersen, Phys. Rev. Lett. **87**, 047003 (2001).
3. R. Cava, , A. Hewat, E. Hewat, B. Battlog, M. Marezio, I. K. K.M. Rabe, W. Peck Jr. and L. Rupp Jr., Physica C **165**, 419 (1990).
4. H. Schwer and J. Karpinski, In: A. Narlikar (ed.): Studies of High Temperature Superconductors, Vol. 24. p. 49 (1997).
5. K. Conder, D. Zech, C. Krüger, E. Kaldis, H. Keller, A. Hewat and E. Jilek: In: E. Sigmund and K. A. Müller (eds.): *Phase Separation in Cuprate Superconductors*. p. 210 (1994).
6. R. Cava, M. Marezio, J. Krajewski, W. Peck Jr., A. Santoro and F. Beech: Physica C **157**, 272 (1989).
7. S. Kambe, and O. Ishii, Physica C **341-348**, 555 (2000).
8. P. Fischer, J. Karpinski, E. Kaldis and S. R. E. Jilek, Solid State Comm. **69**, 531 (1989).

9. J. Röhler, , P. Loeffen, K. Conder and E. Kaldis, Physica C **282-297**, 182–195 (1997).
10. J. Röhler, and R. Crüsemann, In: E. Kaldis. K. A. Müller, D. Mihailovic, G. Ruani (ed.), *Anharmonic Properties of high-T_c Cuprates*. pp. 86–94 (1995).
11. C. J Halboth and W. Metzner, Phys. Rev. B **61**, 7364 (2000).
12. O. Chmaissem, J. D. Jorgensen, S. Short, A. Knizhnik, Y. Eckstein and H. Shaked, Nature **397**, 45 (1999).
13. E. Kaldis, In: K. A. Gschneidner Jr., L. Eyring and M. B. Maple (eds.), *Handbook on the Physics and Chemistry of Rare Earths*, Vol. 31 (2001).
14. A. W. Hewat, E. Kaldis, S. Rusiecki and E. J. P. Fischer, unpublished. Data sheets are included in Ref. 13 (1991).
15. Röhler, J.: 2000, Physica B **284-288**, 1041.

SPECTRAL FUNCTIONS AND PSEUDOGAP IN A MODEL OF STRONGLY CORRELATED ELECTRONS

Proceedings of the ARW NATO Workshop Hvar, Croatia, October 2002

P. PRELOVŠEK and A. RAMŠAK
Faculty of Mathematics and Physics, University of Ljubljana, 1111 Ljubljana, Slovenia
J. Stefan Institute, University of Ljubljana, 1111 Ljubljana, Slovenia

Abstract. The theoretical investigation of spectral functions within the single-band t-J model, as relevant for superconducting cuprates, is presented. For spectral functions the method of equations of motion is used, where for the self energy the decoupling of spin and single-particle fluctuations is performed. Longer-range spin fluctuations induce a pseudogap showing up at low doping in the effective truncation of the Fermi surface and in reduced electron and quasiparticle density of states at the Fermi level.

Key words: Cuprates, Superconductivity, t-J Model, Strong Correlations, Pseudogap, Spectral Functions, Fermi Surface, ARPES, Spin Fluctuations

1. Introduction

One of the central questions in the theory of strongly correlated electrons is the nature of the ground state and of low energy excitations. Experiments in many novel materials with correlated electrons reveal even in the 'normal' metallic state striking deviations from the usual Fermi-liquid universality as given by the phenomenological Landau theory involving quasiparticles (QP)(1). The attention in the last decade has been increasingly devoted to the underdoped cuprates, where experiments reveal characteristic 'pseudogap' temperatures $T > T_c$, which show up crossovers where particular properties change qualitatively. There seems to be an indication for two crossover scales T^* and T_{sg}. T^* scale (2) shows up clearly as the maximum of the spin susceptibility $\chi_0(T = T^*)$, the kink in the in-plane resistivity $\rho(T)$, in the anomalous Hall constant $R_H(T < T^*)$ and in the specific heat $\gamma = C_V/T$ (1). It seems plausible that the T^* crossover is

A.C. Hewson and V. Zlatić (eds.), Concepts in Electron Correlation, 65–73.

related to the onset of short-range antiferromagnetic (AFM) correlations for $T < T^*$, since this is clearly the case for an undoped AFM.

The single-particle spectral function $A(\mathbf{k},\omega)$ and its properties are of crucial importance. In recent years there has been an impressive progress in the angle-resolved photoemission spectroscopy (ARPES) experiments, in particular for cuprate materials, which in principle yield a direct information on $A(\mathbf{k},\omega)$. In most investigated $Bi_2Sr_2CaCu_2O_{2+\delta}$ (BSCCO) (3) ARPES shows quite well defined large Fermi surface in the overdoped and optimally doped samples at $T > T_c$. On the other hand, in the underdoped materials the quasiparticles (QP) dispersing through the Fermi surface (FS) are resolved by ARPES in BSCCO only in parts of the large FS, in particular along the nodal $(0,0)$-(π,π) direction (4, 3), indicating that the rest of the large FS is truncated (5). At the same time near the $(\pi,0)$ momentum ARPES reveals a hump at ~ 100 meV (4), which is consistent with large pseudogap scale T^*.

The prototype single-band model relevant for cuprates which takes explicitly into account strong correlations is the t-J model,

$$H = -\sum_{i,j,s} t_{ij} \tilde{c}^\dagger_{js} \tilde{c}_{is} + J \sum_{\langle ij \rangle} (\mathbf{S}_i \cdot \mathbf{S}_j - \frac{1}{4} n_i n_j), \tag{1}$$

where fermionic operators are projected ones, $\tilde{c}^\dagger_{is} = (1 - n_{i,-s}) c^\dagger_{is}$. We consider besides $t_{ij} = t$ for the n.n. hopping also $t_{ij} = t' < 0$ for the n.n.n. hopping on a square lattice.

There have been so far numerous analytical and numerical studies of the t-J model and related Hubbard model on a square lattice. The importance of AFM spin correlations for the emergence of the (large) pseudogap was found in some of these numerical studies (6, 7, 8) and in phenomenological model calculations (9, 10).

First, we describe some evidence for the pseudogap within the t-J model obtained via finite-size studies using the finite-temperature Lanczos method (FTLM) (8). A most straightforward evidence for a pseudogap appears in the uniform static spin-suceptibility $\chi_0(T)$. The maximum T^* being related to the AFM exchange $T^* \sim J$ in an undoped AFM gradually shifts to lower T with doping and finally disappears at 'critical' $c_h = c_h^* \sim 0.15$. Obtained results are qualitatively as well as quantitatively consistent with experiments in cuprates (1). Another quantity relevant for comparison with the anaytical approach is the single-particle density of states (DOS) $\mathcal{N}(\omega)$, which shows at smallest dopings a pronounced pseudogap at $\omega \sim 0$ which closes with increasing $T \sim J$. This again indicates the relation of this pseudogap with the AFM short-range correlations which dissolve for $T > J$. On the other hand, the pseudogap closes also on increasing doping being barely visible even at $c_h \sim 0.12$.

2. Spectral functions: Equation-of-motion approach

In our anaytical approach we analyse the electron Green's function for projected fermionic operators,

$$G(\mathbf{k},\omega) = \langle\langle \tilde{c}_{\mathbf{k}s}; \tilde{c}^{\dagger}_{\mathbf{k}s} \rangle\rangle_{\omega} = -i \int_0^{\infty} \mathrm{e}^{i(\omega+\mu)t} \langle \{ \tilde{c}_{\mathbf{k}s}(t), \tilde{c}^{\dagger}_{\mathbf{k}s} \}_{+} \rangle dt. \tag{2}$$

In an equations-of-motion (EQM) method one uses relations for general correlation functions $\omega\langle\langle A;B\rangle\rangle_{\omega} = \langle\{A,B\}_{+}\rangle + \langle\langle [A,H];B\rangle\rangle_{\omega}$, applying them to the propagator $G(\mathbf{k},\omega)$ (11),

$$G(\mathbf{k},\omega) = \frac{\alpha}{\omega + \mu - \zeta_{\mathbf{k}} - \Sigma(\mathbf{k},\omega)}. \tag{3}$$

We notice that the renormalization $\alpha = (1+c_h)/2 < 1$ is already the consequence of the projected basis, while $\zeta_{\mathbf{k}}$ represents the new 'free' propagation term

$$\zeta_{\mathbf{k}} = \frac{1}{\alpha}\langle\{[\tilde{c}_{\mathbf{k}s}, H], \tilde{c}^{\dagger}_{\mathbf{k}s}\}_{+}\rangle - \bar{\zeta} = -4\eta_1 t\gamma_{\mathbf{k}} - 4\eta_2 t'\gamma'_{\mathbf{k}}, \tag{4}$$

where $\eta_j = \alpha + \langle \mathbf{S}_0 \cdot \mathbf{S}_j \rangle / \alpha$ and $\gamma_{\mathbf{k}} = (\cos k_x + \cos k_y)/2$, $\gamma'_{\mathbf{k}} = \cos k_x \cos k_y$.

The central quantity for further consideration is the self energy

$$\Sigma(\mathbf{k},\omega) = \langle\langle C_{\mathbf{k}s}; C^{+}_{\mathbf{k}s} \rangle\rangle^{irr}_{\omega} / \alpha, \qquad iC_{\mathbf{k}s} = [\tilde{c}_{\mathbf{k}s}, H] - \zeta_{\mathbf{k}}\tilde{c}_{\mathbf{k}s}, \tag{5}$$

where only the 'irreducible' part of the correlation function should be taken into account in the evaluation of Σ. We express EQM in variables appropriate for a paramagnetic metallic state

$$\begin{aligned} [\tilde{c}_{\mathbf{k}s}, H] &= [(1 - \tfrac{c_e}{2})\epsilon^0_{\mathbf{k}} - Jc_e]\tilde{c}_{\mathbf{k}s} + \\ &+ \tfrac{1}{\sqrt{N}} \textstyle\sum_{\mathbf{q}} m_{\mathbf{kq}} [s S^z_{\mathbf{q}} \tilde{c}_{\mathbf{k-q},s} + S^{\mp}_{\mathbf{q}} \tilde{c}_{\mathbf{k-q},-s} - \tfrac{1}{2}\tilde{n}_{\mathbf{q}} \tilde{c}_{\mathbf{k-q},s}], \end{aligned} \tag{6}$$

where $\epsilon^0_{\mathbf{k}}$ is the bare band energy, $\tilde{n}_i = n_i - c_e$ and $m_{\mathbf{kq}}$ is the effective spin-fermion coupling,

$$m_{\mathbf{kq}} = 2J\gamma_{\mathbf{q}} + \epsilon^0_{\mathbf{k-q}}. \tag{7}$$

One important achievement of EQM method is that it naturally leads to an effective coupling between fermionic and spin degrees of freedom, which are essential for the proper description of low-energy physics in cuprates. Such a coupling is e.g. assumed as input in phenomenological models (9, 10) as the spin-fermion model. The essential difference in our case is that $m_{\mathbf{kq}}$ is strongly dependent on $\mathbf{k}$ just in the vicinity of most relevant 'hot' spots.

We assume that spin fluctuations remain dominant at the AFM wavevector $\mathbf{Q}$ with the characteristic inverse AFM correlation length $\kappa = 1/\xi_{AFM}$.

It is sensible to divide spin fluctuations into two regimes with respect to $\tilde{\mathbf{q}} = \mathbf{q} - \mathbf{Q}$: a) For $\tilde{q} > \kappa$ spin fluctuations are paramagnons, they are propagating like magnons and are transverse to the local AFM ordering. b) For $\tilde{q} < \kappa$ spin fluctuations are critically overdamped and the deviations from the long range order are essential.

For $\tilde{q} > \kappa$ the decoupling of spin and fermion degrees of freedom reproduces for an undoped AFM the evaluation of $\Sigma(\mathbf{k},\omega)$ within the self-consistent Born approximation (12) which we generalise within the linearized magnon theory into a paramagnon contribution at finite doping (at $T = 0$), (13)

$$\Sigma_{\mathrm{pm}}(\mathbf{k},\omega) = \frac{1}{N} \sum_{q,\tilde{q}>\kappa} [M^2_{\mathbf{kq}} G^-(\mathbf{k}-\mathbf{q},\omega+\omega_{\mathbf{q}}) + M^2_{\mathbf{k+q,q}} G^+(\mathbf{k}+\mathbf{q},\omega-\omega_{\mathbf{q}})], \tag{8}$$

where $G^{\pm}(\mathbf{k},\omega)$ refer to the electron ($\omega > 0$) and hole-part ($\omega < 0$) of the propagator, respectively. We are dealing with a strong coupling theory due to $t > \omega_{\mathbf{q}}$ and a selfconsistent calculation of Σ_{pm} is required.

At $c_h > 0$ besides the paramagnon excitations also the coupling to longitudinal spin fluctuations becomes crucial. The latter restore the spin rotation symmetry in a paramagnet and EQM (6) introduce such a spin-symmetric coupling. Within a simplest approximation that the dynamics of fermions and spins is independent, we get (13)

$$\Sigma_{\mathrm{lf}}(\mathbf{k},\omega) = \frac{1}{\alpha} \sum_{\mathbf{q}} \tilde{m}^2_{\mathbf{kq}} \int\int \frac{d\omega_1 d\omega_2}{\pi} g(\omega_1,\omega_2) \frac{\tilde{A}(\mathbf{k}-\mathbf{q},\omega_1)\chi''(\mathbf{q},\omega_2)}{\omega-\omega_1-\omega_2}, \tag{9}$$

where $g = (1/2)[\mathrm{th}(\beta\omega_1/2) + \mathrm{cth}(\beta\omega_2/2)]$ and χ is the dynamical spin susceptibility. Quite analogous treatment has been employed previously in the Hubbard model (14) and more recently within the spin-fermion model (9, 10).

If we want to use the analogy with the phenomenological spin-fermion model the effective coupling parameter $\tilde{m}_{\mathbf{kq}}$ should satisfy $\tilde{m}_{\mathbf{k},\mathbf{q}} = \tilde{m}_{\mathbf{k}-\mathbf{q},-\mathbf{q}}$ therefore we use instead the symmetrized coupling $\tilde{m}_{\mathbf{kq}} = 2J\gamma_{\mathbf{q}} + (\epsilon^0_{\mathbf{k}-\mathbf{q}} + \epsilon^0_{\mathbf{k}})/2$. In contrast to previous related studies of spin-fermion coupling (9, 10), $\tilde{m}_{\mathbf{kq}}$ is strongly dependent on $\mathbf{k}$, but also quite modest along the AFM zone boundary ('hot' spots), here determined solely by J and t'. For $\tilde{A}$ we first insert the unrenormalized A^0, i.e. without Σ_{lf}.

In the present theory spin response $\chi(\mathbf{q},\omega)$ is taken as input. The system is close to the AFM instability, so we assume the overdamped form

$$\chi''(\mathbf{q},\omega) \propto \frac{\omega}{(\tilde{q}^2+\kappa^2)(\omega^2+\omega_\kappa^2)}. \tag{10}$$

Nevertheless, the appearance of the pseudogap and the form of the FS are not strongly sensitive to the particular form of $\chi''(\mathbf{q},\omega)$, provided that characteristic κ and ω_κ are fixed.

3. Pseudogap

We first establish some characteristic features of the pseudogap and the development of the FS following a simplified analysis (13). $\Sigma_{\rm pm}$ induces a large incoherent component in the spectral functions and renormalizes the effective QP band, leading to $\epsilon^{\rm ef}_{\mathbf{k}}$ and reduced QP weight $Z^{\rm ef}_{\mathbf{k}}$. Nevertheless, the pseudogap can appear only via $\Sigma_{\rm lf}$. We restrict ourselves here to $T=0$ and to the regime of intermediate (not too small) doping, where $\epsilon^{\rm ef}_{\mathbf{k}}$ defines a large FS. The simplest case is the quasi-static and single-mode approximation (QSA) which is meaningful if $\omega_\kappa \ll t, \kappa \ll 1$, where we get

$$G^{QSA}(\mathbf{k},\omega) = \frac{\alpha Z^{\rm ef}_{\mathbf{k}}(\omega-\epsilon^{\rm ef}_{\mathbf{k}-\mathbf{Q}})}{(\omega-\epsilon^{\rm ef}_{\mathbf{k}-\mathbf{Q}})(\omega-\epsilon^{\rm ef}_{\mathbf{k}})-\Delta^2_{\mathbf{k}}}. \tag{11}$$

The spectral functions show in this approximation two branches $E^{\pm}$, separated by the gap which opens along the AFM zone boundary $\mathbf{k}=\mathbf{k}_{AFM}$ and the relevant (pseudo)gap scale is

$$\Delta^{PG}_{\mathbf{k}} = |\Delta_{\mathbf{k}_{AFM}}| = \frac{Z^{\rm ef}_{\mathbf{k}}}{2}\sqrt{r_s}|2J-4t'\cos^2 k_x|. \tag{12}$$

$\Delta^{PG}_{\mathbf{k}}$ does not depend on t, but rather on smaller J and in particular t'. For $t'<0$ the gap is largest at $(\pi,0)$, consistent with experiments.

Within the simplified effective band approach, it is not difficult to evaluate numerically $\Sigma_{\rm lf}$ beyond the QSA, by taking explicitly into account $\chi''(\mathbf{q},\omega)$, Eq. (10), with $\kappa>0$ and $\omega_\kappa \sim 2J\kappa$. In Fig. 1 we present results for $A(\mathbf{k},\omega=0)$ at $T=0$ for a broad range of $\kappa=0.01-0.6$. Curves in fact display the effective FS determined by the condition $G^{-1}(\mathbf{k}_F,0)=0$. At the same time, intensities $A(\mathbf{k},\omega=0)$ correspond to the renormalization factor Z_F. At very small $\kappa=0.01$ we see the hole-pocket FS which follows from the QSA. Already small $\kappa\sim 0.05$ destroys the 'shadow' side of the pocket. On the other hand, in the gap emerge now QP solutions with very weak $Z_F \ll 1$ which reconnect the FS into a large one. We are dealing nevertheless with effectively truncated FS with well developed arcs. The effect of larger κ is essentially to increase Z_F in the gapped region, in particular near $(\pi,0)$.

Note that within present theory the low-energy excitations corresponds to a Fermi liquid, although a very strange one, where QP exist (at low T) everywhere along the FS. It is quite remarkable to notice that in spite

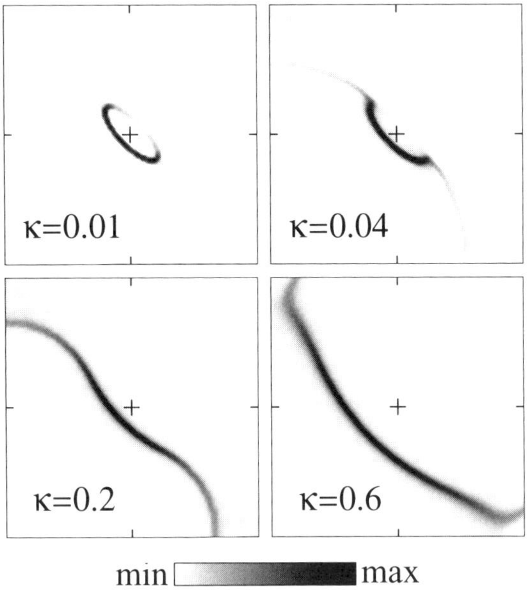

Figure 1. Contour plot of spectral functions $A(\mathbf{k},\omega=0)$ at $T=0$ for various κ in one quarter of the Brillouin zone.

of $Z_F \ll 1$ the QP velocity v_F is not diminished within the pseudogap. In fact it is even enhanced, as seen in in Fig. 2 where the contour plot of $A(\mathbf{k},\omega)$ is shown. Again, it is well evident in Fig. 2 that QP are well defined at the FS, while they become fuzzy at $\omega \neq 0$ merging with the solutions $E^{\pm}_{\mathbf{k}}$, respectively, away from the FS. The effect can be only explained with crucial $\mathbf{k}$ dependence of $\Sigma'(\mathbf{k},\omega)$ (13). Assuming that $\mathbf{k}$ enters only via $\epsilon(\mathbf{k})$ we can express v_F renormalization as

$$\frac{v_F}{v^{\mathrm{ef}}_{\mathbf{k}}} = (1+\frac{\partial\Sigma'}{\partial\epsilon})\frac{Z_F}{Z^{\mathrm{ef}}}. \qquad \frac{\partial\Sigma'}{\partial\epsilon}\Big|_{\omega=0} \sim \frac{\Delta^2}{w(\omega_\kappa+w)}, \tag{13}$$

While $Z_F/Z^{ef} \ll 1$, $\partial\Sigma'/\partial\epsilon$ compensates or even leads to an enhancement of v_F. In the case $\omega_\kappa w \ll \Delta^2$ we get $v_F/v^{\mathrm{ef}}_{\mathbf{k}} \sim \omega_\kappa/w \sim 2J/v_{\mathbf{k}}$. Final v_F is therefore not strongly renormalized, since $2J$ and $v^{\mathrm{ef}}_{\mathbf{k}}$ are of similar order. Furthermore, $\tilde{v}_F$ is enhanced near $(\pi,0)$ point. The situation is thus very different from 'local' theories where $\Sigma(\mathbf{k},\omega) \sim \Sigma(\omega)$ and the QP renormalization is governed only by Z_F.

From $A(\mathbf{k},\omega)$ we can calculate the DOS $\mathcal{N}(\omega)$. The contribution will come mostly from FS arcs near the zone diagonal while the gapped regions near $(\pi,0)$ will contribute less due to $Z(\mathbf{k}_F) \ll 1$. As expected the DOS

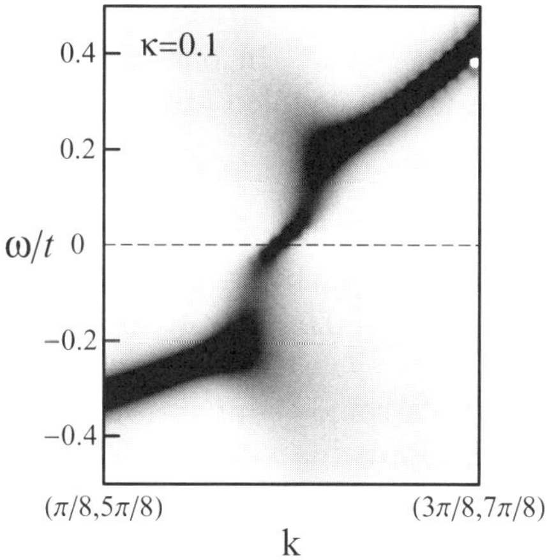

Figure 2. Contour plot of spectral functions $A(\mathbf{k}, \omega = 0)$ across the FS in the pseudogap regime.

decreases with decreasing κ, i.e. approaching an undoped AFM, consistent with experiments in cuprates (15, 16). We can as well define the QP DOS $\mathcal{N}_{QP} \propto \oint dS_F/v(\mathbf{k})$. It is quite important to notice a decreasing κ leads also to a decrease of $\mathcal{N}_{QP}$. This is consistent with the observation of the pseudogap also in the specific heat in cuprates (17), since $\mathcal{N}_{QP} \propto \gamma$. We note here that such a behavior is not at all evident in the vicinity of a metal-insulator transition. Namely, normally in a Fermi liquid one would drive the metal-insulator transition by $Z_{\mathrm{av}} \to 0$ and within an assumption of a local character $\Sigma(\omega)$ this would lead to $v_F \to 0$ and consequently to $\mathcal{N}_{QP} \to \infty$.

It is also important to understand the role of finite $T > 0$. The most pronounced effect is on QP in the pseudogap regime. The main conclusion is, that weak QP peak with $Z_F \ll 1$ at $T = 0$, as seen e.g. in Fig. 2, is not just broadened but entirely disappears (becomes incoherent) already by very small $T > T_s \ll J$ (13).

4. Conclusions

We have presented our results for spectral functions and pseudogap within the t-J model, which is the prototype model for strongly correlated electrons. The physics of the t-J model at lower doping is determined by the interplay of the magnetic exchange and the itinerant kinetic energy of fermions. At intermediate doping the system is frustrated, the effect showing up in large entropy, pronounced spin fluctuations, non-Fermi liquid effects etc. Evidently, this is one path towards the metal-insulator transition, but definitely not the only one possible. In our case, fermionic and spin degrees of freedom coexist but are coupled and both active and relevant for low-energy properties.

Within the present theory (13) the origin of the pseudogap feature is in the coupling to longitudinal spin fluctuations near the AFM wavevector $\mathbf{Q}$. It is important to note that apart from extremely small κ we are still dealing with large FS. Still, at $\kappa \ll \kappa^* \sim 0.5$ parts of the FS in the nodal direction remain well pronounced while the QP weight within the pseudogap (zone corners) region of the FS are strongly suppressed. QP within the pseudogap have small weight $Z_F \ll 1$ but due to nonlocal character of $\Sigma(\mathbf{k}, \omega)$ not diminished (or even enhanced) v_F. This gives an explanation for a well known theoretical challenge that approaching the magnetic insulator both electron and QP DOS decrease and vanish.

We presented results for $T = 0$, however the extension to $T > 0$ is straightforward. Discussing only the effect on the pseudogap, we notice that it is mainly affected by κ. So we can argue that the pseudogap should be observable for $\kappa(c_h, T) < \kappa^* \sim 0.5$. This effectively determines the crossover temperature $T^*(c_h)$ approximately as $T^* \sim T_0^*(1 - c_h/c_h^*)$ where $T_0^* \sim 0.6J$ and $c_h^* \sim 0.15$.

References

1. for a review see, e.g., M. Imada, A. Fujimori, and Y. Tokura, Rev. Mod. Phys. **70**, 1039 (1998).
2. B. Batlogg *et al.*, Physica C **235 - 240**, 130 (1994).
3. J.C. Campuzano and M. Randeria *et al.*, in *Proc. of the NATO ARW on Open Problems in Strongly Correlated Electron Systems*, Eds. J. Bonča, P. Prelovšek, A. Ramšak, and S. Sarkar (Kluwer, Dordrecht, 2001), p. 3.
4. D.S. Marshall *et al.*, Phys. Rev. Lett. **76**, 4841 (1996); H. Ding *et al.*, Nature **382**, 51 (1996).
5. M.R. Norman *et al.*, Nature **392**, 157 (1998).
6. R. Preuss, W. Hanke, C. Gröber, and H.G. Evertz, Phys. Rev. Lett. **79**, 1122 (1997).
7. J. Jaklič and P. Prelovšek, Phys. Rev. B **55**, R7307 (1997); P. Prelovšek, J. Jaklič, and K. Bedell, Phys. Rev. B **60**, 40 (1999).
8. for a review see J. Jaklič and P. Prelovšek, Adv. Phys. **49**, 1 (2000).
9. A.V. Chubukov and D.K. Morr, Phys. Rep. **288**, 355 (1997).
10. J. Schmalian, D. Pines, and B. Stojković, Phys. Rev. Lett. **80**, 3839 (1998); Phys. Rev. B **60**, 667 (1999).
11. P. Prelovšek, Z. Phys. B **103**, 363 (1997).
12. C.L. Kane, P.A. Lee, and N. Read, Phys. Rev. B **39**, 6880 (1989).
13. P. Prelovšek and A. Ramšak, Phys. Rev. B **63**, 180506 (2001); Phys. Rev. B **65**, 174529 (2002).
14. A. Kampf and J.R. Schrieffer, Phys. Rev. B **41**, 6399 (1990).
15. A. Ino *et al.*, Phys. Rev. Lett. **81**, 2124 (1998).

16. Ch. Renner, B. Revaz, K. Kadowaki, I. Maggio-Aprile, and O. Fischer, Phys. Rev. Lett. **80**, 3606 (1998).
17. J.W. Loram, K.A. Mirza, J.R. Cooper, and W.Y. Liang, Phys. Rev. Lett. **71**, 1740 (1993); J.W. Loram, J.L. Luo, J.R. Cooper, W.Y. Liang, and J.L. Tallon, Physica C **341 - 8**, 831 (2000).

MOTT INSULATOR TO SUPERCONDUCTOR VIA PRESSURE RVB THEORY & PREDICTION OF NEW SYSTEMS

Proceedings of the ARW NATO Workshop Hvar, Croatia, October 2002

G. BASKARAN
Institute of Mathematical Sciences
Madras 600 113, India

Abstract. This paper summarizes my new work, on the nature of Mott transition in orbitally non degenerate spin-$\frac{1}{2}$ system, motivated by observation of Mott insulator superconductor transition in real systems. The structure of superexchange perturbation theory for a repulsive Hubbard model at half filling is analyzed and it suggests, under some conditions, a resonating valence bond (RVB) mechanism of Mott insulator to superconductor transition; charge 'deconfinement' is accompanied by electron pair delocalization aided by a preexisting spin singlet correlations in the insulator. An RVB mean field theory at half filling illustrating our mechanism of Mott insulator-Superconductor transition is sketched. We identify some family of compounds as potential candidates for Mott insulator to superconductor transition under pressure: CuO, $BaBiO_3$, thin films of La_2CuO_4 or $CaCuO_2$ (infinite layer compound) under pressure or an effective ab-plane pressure/epitaxial compressive stress; synthesizing new compounds such as $La_2CuS_2O_2$, La_2CuS_4 or $CaCuS_2$ or compounds containing CuS_2 or $CuSe_2$ planes to mimic large ab-plane chemical pressure.

Key words: High Temperature Superconductivity, Mott Insulators, Pressure Induced Superconductivity, Resonating Valence Bond (RVB) Theory, Strongly Correlated Electron Systems, Organic Superconductors, Slave Boson Theory

1. Introduction

Following Bednorz-Muller's discovery (1) of high temperature superconductivity in doped La_2CuO_4 in 1986, Anderson's resonating valence bond (RVB) theory (2), brought out deep connections between the Mott insulating quantum state and high T_c superconductors. In RVB theory the pre-existing singlet correlations among the electron spins in a spin-$\frac{1}{2}$ Mott insulator readily become the superconducting correlations on doping. The RVB mean field theory (3), gauge theory (4) and the developments that

A.C. Hewson and V. Zlatić (eds.), Concepts in Electron Correlation, 75–84.

followed (5) over nearly 15 years have given a picture that is in qualitative agreements with a variety of experimental results.

Inspired by cuprates, RVB theory has focussed on metallization of Mott insulating state by means of hole or electron doping. However, it is also known that *undoped* commensurate systems such as the family of quasi-1 dimensional Bechgaard salts (6, 7), quasi-2 dimensional ET salts (7), and 3 dimensional fullerites (8, 9) do exhibit unconventional and high T_c superconductivity. For these narrow tight binding systems, *a half filled repulsive Hubbard model* provides a correct starting point and various theories (10, 11) show possibility of superconductivity near the Mott transition point. Recent theoretical work by Capone and collaborators (12) bring this out in a general context. Experimentally one does see, what may be termed as Mott insulating spin Peierls or antiferromagnetically ordered states that undergo a transition to a superconducting state as a function of pressure or chemical pressure. A striking example (9) is $K_3(NH_3)C_{60}$, a spin-$\frac{1}{2}$ antiferromagnet that becomes a high T_c superconductor with a T_c ($\approx$ 32 K) under pressure.

This has prompted us to reanalyze the issue of Mott insulator metal transition at commensurate fillings from the point of view of RVB theory. We focus on orbitally non degenerate spin-$\frac{1}{2}$ Mott insulators. We argue that strong local singlet correlations induced by higher order superexchange processes in the vicinity of Mott transition play a key role in the Mott insulator superconductor transition and RVB theory is the most natural way to understand the transition and the superconducting state itself.

Our paper is organized as follows. In section 2, starting from the large U repulsive Hubbard model at half filling we look at the higher order terms of the superexchange perturbation theory. The divergence of the perturbation theory suggests correlated singlet pair delocalization in the conducting state. In section 3, the above argument is substantiated by an RVB mean field theory directly applied to the repulsive Hubbard model at half filling, which shows a Mott insulator superconductor phase transition. In section 4 we suggest that some known Mott insulating systems should become high Tc superconductors under pressure. We also propose some new systems; they are predicted to be potential high temperature superconductors, either on their own or under pressure.

2. Mott Insulator Superconductor transition without Doping - a new RVB Route to high T_c Superconductivity

The low energy physics of an odd electron (per unit cell) system such as La_2CuO_4 is well described by a single band Hubbard model:

$$H = -\sum_{ij} t_{ij} c_{i\sigma}^{\dagger} c_{j\sigma} + h.c. + U \sum_{i} n_{i\uparrow} n_{i\downarrow} \tag{1}$$

Here t_{ij} are the hopping matrix elements and U is the on site repulsion. Our major focus will be on the half filled band situation. First we show that above a critical U, there is an insulating state with a true charge gap and then show how superconducting correlations develop as we decrease U, leading to an eventual Mott insulator to superconductor transition.

We work in a slave boson representation which helps us to separate charge and spin degrees of freedom (5, 4) in a manifest way: $c_{i\sigma}^{\dagger} \equiv s_{i\sigma}^{\dagger} d_i + \sigma s_{i\bar{\sigma}} e_i^{\dagger}$. Here the chargeons $d_i^{\dagger}$'s and $e_i^{\dagger}$'s are hard core bosons that create doubly occupied sites and empty sites respectively. The fermionic spinon operators $s_{i\sigma}^{\dagger}$'s create singly occupied sites with a spin projection σ. The local constraint, $d_i^{\dagger} d_i + e_i^{\dagger} e_i + \sum_{\sigma} s_{i\sigma}^{\dagger} s_{\sigma} = 1$, removes redundant Hilbert space.

In this representation Hubbard Hamiltonian takes a suggestive form:

$$\begin{aligned} H &= -\sum_{ij} t_{ij} (d_i^{\dagger} d_j \sum_{\sigma} s_{i\sigma} s_{j\sigma}^{\dagger} + e_i e_j^{\dagger} \sum_{\sigma} s_{i\sigma}^{\dagger} s_{j\sigma}) + h.c. + U \sum_{i} (d_i^{\dagger} d_i + e_i^{\dagger} e_i) \\ &- \sum_{ij} t_{ij} (d_i^{\dagger} e_j^{\dagger} \sum_{\sigma} \sigma s_{i\sigma} s_{j\bar{\sigma}} + e_i d_j \sum_{\sigma} \sigma s_{i\sigma}^{\dagger} s_{j\bar{\sigma}}^{\dagger}) + h.c. \end{aligned} \tag{2}$$

In the limit $|t_{ij}| >> U$ one derives an effective spin Hamiltonian for the low energy subspace by a degenerate perturbation theory - hopping parameter expansion, or equivalently a superexchange perturbation theory. The lowest energy eigenstates of the interaction Hamiltonian have one electron per site and a 2^N fold spin degeneracy. For large U/t we can treat the hopping, kinetic energy term (second line of equation 2) as a perturbation and get an effective Hamiltonian that lifts the 2^N fold spin degeneracy. The effective spin Hamiltonian is generated through virtual mixing to excited states having two, three etc. number of doubly occupied sites. The explicit form of the effective Hamiltonian up to fourth order in t_{ij} in the single occupancy subspace is shown to be (13):

$$H_{\text{eff}} = -\sum_{ij}' \frac{4t_{ij}t_{ji}}{U} b_{ij}^{\dagger} b_{ij} - \sum_{ijkl}' \frac{16 t_{ij} t_{jk} t_{kl} t_{li}}{U^3} b_{ij}^{\dagger} b_{jk} b_{kl}^{\dagger} b_{li} + O\left(\frac{t^6}{U^5}\right) \tag{3}$$

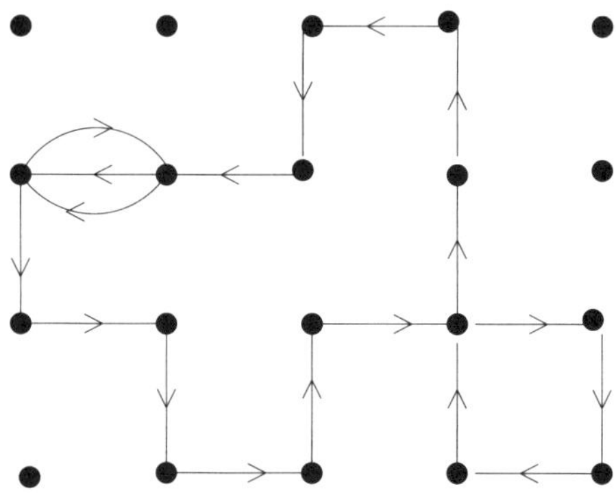

Figure 1. A higher order superexchange process that signals co-operative delocalization of charge $\pm 2d$ singlets.

where $b_{ij}^{\dagger} = \frac{1}{\sqrt{2}}(s_{i\uparrow}^{\dagger}s_{j\downarrow}^{\dagger} - s_{i\downarrow}^{\dagger}s_{j\uparrow}^{\dagger})$ is a spin singlet electron pair creation operator at the bond ij. In the singly occupied subspace the first term becomes the familiar superexchange term: $b_{ij}^{\dagger}b_{ij} \equiv -(\mathbf{S}_i \cdot \mathbf{S}_j - \frac{1}{4}n_i n_j)$. Similarly all the higher order terms also can be written purely in terms of the spin operators. The prime in the summation avoids double counting.

I have written the effective Hamiltonian in terms of spinon operators, rather than the customary Pauli spin operators, as it has some advantage in proving a theorem and helps in seeing the physics of Mott insulator to superconductor transition more transparently. Further a dynamically generated local U(1) gauge symmetry of H_{eff}, in the low energy subspace of the Mott insulator is also manifest (4): $b_{ij}^{\dagger} \rightarrow e^{i\theta_i}b_{ij}^{\dagger}e^{i\theta_j}$.

As far as I know there is no rigorous proof for the convergence of the above perturbation theory; such a proof should be simple as charge localization is a rather robust phenomenon in a Mott insulator. I present a rigorous proof for the convergence by using some of Takahashi's results (13) for half filled band repulsive Hubbard model.

Theorem: For the repulsive half filled band Hubbard model the hopping parameter expansion converges for $U/t >> 1$.

Proof: From Takahashi's result it is clear that (13) the nth order terms of the spin Hamiltonian corresponds to closed graphs drawn on the lattice, connected by n non-zero matrix elements t_{ij}; and n has to be even. A loop may intersect and also may self overlap (figure 1). As the effective Hamiltonian is also an operator cumulant expansion, there are no diagrams with disconnected graphs.

It can be shown that the most general term in the expansion has the form, $b_{i_1i_2}^{\dagger}b_{i_1i_2}b_{i_2i_3}^{\dagger}b_{i_3i_4}\ldots b_{i_{n-1}i_n}^{\dagger}b_{i_ni_1}$, with a coefficient $\sim \frac{t_{i_1i_2}t_{i_2i_3}\ldots t_{i_{n-1}i_n}t_{i_ni_1}}{U^{n-1}}$. They are closed loops on the lattice connected by the hopping parameters t_{ij}'s. It is also easy to see that there are prefactors of individual terms which are bounded by 2^n. Since the hopping matrix elements are short ranged the number of loops of size n is bounded by $(\alpha d)^n$, where α is a number of the order of unity and d is the spatial dimension.

The matrix elements of the operator $b^{\dagger}_{i_1 i_2} b_{i_1 i_2} b^{\dagger}_{i_2 i_3} b_{i_3 i_4} \ldots b^{\dagger}_{i_{n-1} i_n} b_{i_n i_1}$, in our single occupancy subspace is bounded by 1. Thus the maximum value of the sum of n-th order terms is bounded by $\frac{t(\max)^n}{U^{n-1}} 2^n (\alpha d)^n$. This term clearly becomes exponentially small for $\frac{U}{t} >> 1$. QED.

Mott transition is signalled by the divergence of the above expansion. Higher order terms that become important as we approach the Mott transition point correspond to very specific physical processes, namely *delocalization of bond singlet pairs over large closed loops*, as given by an n-th order term, $b^{\dagger}_{i_1 i_2} b_{i_1 i_2} b^{\dagger}_{i_2 i_3} b_{i_3 i_4} \ldots b^{\dagger}_{i_{n-1} i_n} b_{i_n i_1}$ (figure 1). Such higher order electron singlet pair delocalization processes establish a coherent superconducting state, if there are no other competing instabilities such as metallic antiferromagnetism.

The spin singlet correlations existing in the Mott insulator are the analogue of the preexisting neutral singlets of RVB theory for a doped Mott insulator. Across a Mott transition induced by pressure in real systems, the Mott-Hubbard gap does not close continuously - there is a first order transition to a state with a finite density of free carriers. Free carriers are produced in the ground state on top of the existing enhanced spin singlet correlations by *a kind of cooperative singlet pair delocalization process*, as described in the last paragraph. In other words, the charge delocalization is governed and guided by the background spin singlet correlations. This actually means a transition to a superconducting state as we will demonstrate within a mean field theory in the next section.

From weak coupling point of view, by avoiding Fermi surface nesting we can discourage umklapp, SDW and CDW instabilities. Nesting can be avoided by appropriate choice of the band hopping parameters t_{ij}. From various works in RVB theory (14) it is clear that even an ordered antiferromagnet has sufficient local singlet correlations and is indistinguishable from a spin liquid state as far as short range singlet correlations are concerned. Thus the ordered antiferromagnetic state of a Mott insulator has sufficient local spin singlet correlations to be able to induce electron pair delocalization and hence superconductivity in a Mott transition.

3. RVB mean field theory for Mott insulator to Superconductor Transition under Pressure: a direct analysis of Hubbard Model

As we mentioned before, early works on Mott transition focussed more on the charge degree of freedoms. In the present section we summarize the results of our recent work, where we give equal importance to spin and charge degree of freedom. We find that this leads to the possibility of Mott insulator to superconductor transition (without doping) as a function of pressure, in a natural way as superconductivity has been argued to arise in a Mott insulator on doping. From our work, which analysis the Hubbard model directly, we also reproduce known RVB mean field theory results (3, 5) for Heisenberg and t-J models as interesting and non-trivial limiting cases.

Superexchange processes in the Mott insulator is kinetic in origin; second term of equation 2, a part of the kinetic energy term of the Hubbard model, generates superexchange process. This enables us to perform a Hartree-Fock type of factorization of the kinetic energy term of the Hubbard model (equation 2),

$$e_i d_j \sum_\sigma \sigma s^\dagger_{i\sigma} s^\dagger_{j\bar{\sigma}} \rightarrow \Delta^*_{ij} e_i d_j + \eta_{ij} \sum_\sigma \sigma s^\dagger_{i\sigma} s^\dagger_{j\bar{\sigma}} \quad \text{etc.} \tag{4}$$

involving *singlet spinon pair amplitude*, $\Delta^*_{ij} \equiv \langle \sum_\sigma \sigma s^\dagger_{i\sigma} s^\dagger_{j\bar{\sigma}} \rangle, \chi_{ij} \equiv \langle \sum_\sigma s^\dagger_{i\sigma} s_{j\sigma} \rangle$, and *charge amplitudes* $\eta_{ij} \equiv \langle e_i d_j \rangle, e_{ij} \equiv \langle e^\dagger_i e_j \rangle$ and $d_{ij} \equiv \langle d^\dagger_i d_j \rangle$.

The resulting slave boson mean field Hamiltonian has a charge part:

$$\tilde{H}_{\text{ch}} = U \sum_i (d^\dagger_i d_i + e^\dagger_i e_i) - \sum_{ij} t_{ij} \chi_{ij} (e^\dagger_i e_j - d^\dagger_i d_j) - \sum_{ij} t_{ij} \Delta_{ij} (d^\dagger_i e^\dagger_j + e_i d_j) + h.c.$$

and a spin part

$$\tilde{H}_{\text{sp}} = -\sum_{ij\sigma} t_{ij} (d_{ij} + e_{ij}) s^\dagger_{i\sigma} s_{j\sigma} - \sum_{ij\sigma} t_{ij} (\eta^*_{ij} \sigma s_{i\sigma} s_{j\bar{\sigma}} + \eta_{ij} \sigma s^\dagger_{i\sigma} s^\dagger_{j\bar{\sigma}}) + h.c. \tag{5}$$

In the slave boson mean field theory we diagonalize the above Hamiltonian and determine the Hartree-Fock parameters by minimizing the total energy/free energy with the *global constraint* that total number of spinons and chargeons $\sum_i \left(d^\dagger_i d_i + e^\dagger_i e_i + \sum_\sigma s^\dagger_{i\sigma} s_{i\sigma} \right) = N$, number of lattice sites.

To establish the principle of an RVB mechanism of superconductivity without doping at a Mott transition point we quote three general results from our mean field theory: i) $\tilde{H}_{\text{ch}}$, the boson Hamiltonian has a charge gap E_c which vanishes at a $U = U_c$, as we decrease U from a large value - this is the desired Mott transition, ii) the RVB mean field transition temperature $T_c(RVB)$ starts from a value zero at $U = \infty$ and increases continuously as we decrease U and iii) the chargeons develop anomalous expectation value $\langle e_i \rangle = \langle d_i \rangle \neq 0$ below U_c with a $T_c(BE) \neq 0$.

To see the possibility of superconductivity we express the Cooper pair order parameter in terms of the slave boson variables in the conducting phase: $\langle \sum_\sigma \sigma c_{i\sigma} c_{j\bar{\sigma}} \rangle \sim \langle e^\dagger_i \rangle \langle e^\dagger_j \rangle \Delta_{ij} + \langle e_i \rangle \langle d^\dagger_j \rangle \eta_{ij} + \text{otherterms}$. As $\langle e_i \rangle$, $\langle d_i \rangle$

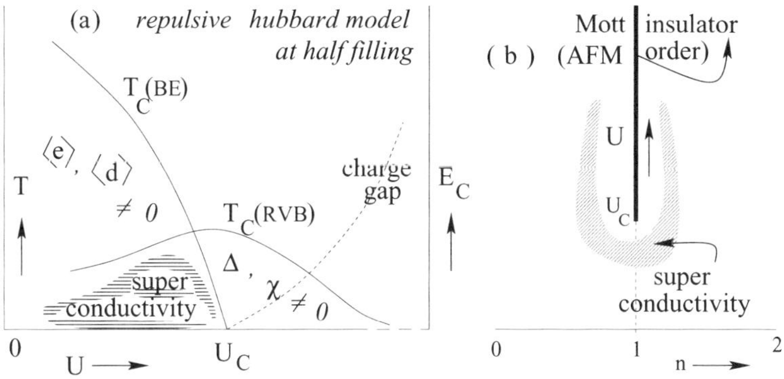

Figure 2. (a) T-U Phase diagram (Schematic). $T_c(BE)$ is the chargeon Bose condensation temperature, $T_c(RVB)$ is the mean field transition temperature of the RVB order parameter and E_c is the charge gap in the Mott insulating state. (b) Schematic Ground state phase diagram in the U-n plane

are nonzero in the conducting phase and Δ_{ij} and η_{ij} continue to remain finite, we have superconductivity within the above mean field theory.

Physically we can understand the above as follows. Coherent electron singlet pair delocalization naturally leads to a finite density of doubly occupied and empty sites present in a phase coherent fashion. In terms of the slave boson picture, presence of real empty and doubly occupied sites in the ground state in a phase coherent fashion corresponds to the bose condensation of chargeons.

Figure (2a) illustrates the region where superconductivity is likely to occur in the T vs U plane. Figure (2b) sketches the ubiquitous superconducting region in the U vs n plane.

Slave boson mean field as applied directly to the Hubbard model in the present paper has to be understood carefully. First, as in the RVB mean field theory applied to Heisenberg Hamiltonian, we get electron pair condensation in the Mott insulating state. This has to be properly interpreted, in view of the presence of charge gap present even in our mean field theory. Use of gauge fluctuations eliminates this fictitious superconducting order in the Mott insulating state, consistent with Elitzur's theorem. Secondly, our mean field theory produces superconductivity even as $U \to 0$. This is an artifact of Hartree-Fock factorization. However, we believe that our results are qualitatively correct close to the Mott transition point. As pointed out earlier, nesting instabilities in the conducting phase can be avoided by appropriate choice of t_{ij}'s, the band parameters.

In this paper we do not go into the detailed energetics of mean field solutions corresponding to different symmetries, for example extended-s and d symmetry of the spinon pair order parameter Δ_{ij} and the charge pair order parameter η_{ij}.

4. Predicting New Superconductors and the Scale of T_c

A practical problem with the 'doping of Mott insulator', an RVB route to high T_c superconductivity is the chemical complications/difficulty in doping Mott insulators. The pressure route does not have this complication.

CuO is the mother compound of the cuprate high T_c family. CuO_2 ribbons form a three dimensional network, each oxygen is shared by two ribbons mutually perpendicular to each other. The strong square planar character from the four oxygens surrounding a Cu isolates out one non degenerate valence d-orbital with one valence electron. This makes CuO an orbitally non-degenerate spin-$\frac{1}{2}$ Mott insulator, a potential candidate for our pressure route to RVB high T_c superconductivity. The frustrated ($\sim 120^0$) superexchange leads to a complex three dimensional magnetic order with a Neel temperature $\sim 200K$. These frustrations should help in stabilizing short range singlet correlations, which will help in singlet Cooper pair delocalization on metallization.

The CuO_2 ribbons give CuO a character of coupled 1d chains. This makes it some what similar to quasi-one dimensional Bechgaard salts, which has a Mott insulator to superconductor transition, via an intermediate metallic antiferromagnetic state as a function of physical or chemical pressure. The intermediate metallic antiferromagnetic state represents a successful competition from nesting instabilities of flat Fermi surfaces arising from the quasi-one dimensional character. Once the quasi-one dimensional character is reduced by pressure nesting of Fermi surface is also reduced and the RVB superconductivity takes over.

If other oxides in perovskite family are any guidance, the metallization should take place in tens of GPa's. Unless a crystallographic transition intervenes and changes the valence electron physics drastically, CuO should undergo a Mott insulator superconductor transition, perhaps with an intermediate antiferromagnetic metallic state. The superconducting T_c should be a finite fraction of the Neel temperature, as is the case with Bechgaard salts or $K_3(NH_3)C_{60}$. Thus an optimistic estimate of T_c will be 50 to 100 K.

Similar statements can be made of the more familiar La_2CuO_4 , insulating YBCO and the $CaCuO_2$, the infinite layer compound or the family of Mott insulating cuprates such as Hg and Tl based insulating cuprates. Infinite layer compound has the advantage of absence of apical oxygen and should be less prone to serious structural modifications in the pressure range of interest to us. The quasi-2d Hubbard model describing the CuO_2 planes does have an appreciable t', making nesting magnetic instabilities weaker. Thus we expect that on metallization a superconducting state to be stabilized with a small or no antiferromagnetic metallic intermediate state. The superconducting T_c could be fairly high $\sim 200K$, as the intrinsic antiferromagnetism scale of the CuO_2 layer is very high $\sim 500K$. The quasi-2d cuprates have a special advantage in the sense we may selectively apply ab-plane pressure in thin films by epitaxial mismatch and ab plane compression.

One way of applying chemical pressure in cuprates is to increase the effective electron band width by increasing the band parameters such as t and t' in the Hubbard model. This can be achieved by replacing oxygens in the CuO_2 planes by either sulfur or selenium, which because of the larger size of the bridging 3p or 4p orbitals increase the band width and at the same time should reduce the charge transfer or Mott-Hubbard gap. So if nature allows us to make a complete or partial replacement of oxygen in the CuO_2 planes one might achieve metalization/superconductivity without doping. Some possible compounds are $La_2CuS_2O_2$, La_2CuS_4 and $CaCuS_2$ or their Se versions.

Superconductivity in doped $BaBi0_3$ was one of the sources of inspiration for the RVB theory of Anderson (2). On closer look this system reveals itself as a Hubbard model with repulsive interaction (15). High pressure is likely to induce a high T_c superconducting state without any external doping.

Acknowledgements

I thank Erio Tosatti for bringing to my attention Mott insulator superconductor transition under pressure in $(NH_3)K_3C_{60}$ family.

References

1. A. Bednorz and A. Muller, Z. Phys. **B 64** 189 (1986)
2. P.W. Anderson, Science **235** 1196 (1987); The theory of high temperature superconductivity (Princeton University Press, NY, 1996); P. W. Anderson, G. Baskaran, T. Hsu and Z. Zou, Phys. Rev. **B 58** 2790 (1987)
3. G. Baskaran, Z. Zou and P.W. Anderson, Sol. State Commm. **63** 973 (1987)
4. G. Baskaran and P.W. Anderson, Phys. Rev. **B37** 580 (1988); G. Baskaran, Physica Scripta, **T27** 53 (1989);
5. S. Kivelson, J. Sethna and D. Rokhsar, Phys. Rev. **B 38** 8865 (1987); G. Kotliar, Phys. Rev. **B 37** 3664 (1988); I. Affleck and J. B. Marston, Phys. Rev. **B 37** 3774 (1988); S. Sorella et al., Phys. Rev. Lett. **88** 117002 (2002); A. Paramekanti, Mohit Randeria and Nandini Trivedi, Phys. Rev. Lett. **87, 217002** (2002)
6. D. Jerome, Science **252** 1509 (1991); C. Bourbonnais, D. Jerome, in Advances in Synthetic Metals (Elsevier, 1999) page 206, Eds. B. Bernier, S. Lefran and G. Bidan; T. Vuletic et al., E. Phys. J **B25** 319 (2002)
7. T. Ishiguro, K. Yamaji and G. Saito, Organic Superconductor (Springer, Berlin 1998)
8. A. Ramirez, Superconductivity Rev. **1** 1 (1994)

9. O. Zhou et al., Phys. Rev. **B 52** 483 (1995)
10. D. Jerome and H.J. Schulz, Adv. Phys. **31** 229 (1982);also reprinted in Adv. Phys. **51** 293 (2002); L. N. Bulaevski, Adv. Phys. **37** 443 (1988)
11. H. Kino and H. Fukuyama, J. Phys. Soc. Jap. **64** 2726 (1995); K. Kanoda, Physica **C282** 820 (1997); R. H. Mackenzie, Science **278** 820 (1997); J. Schmalian, Phys. Rev. Lett. **81** 4232 (1998) K. Kuroki and H. Aoki, Phys. Rev. **B60** 3060 (1999)
12. M. Capone, M. Fabrizio, C. Castellani and E. Tosatti, Science **296** 2364 (2002)
13. Minoru Takahashi, J. Phys. **C 10** 1289 (1977) used Kato's method, T. Kato, Prog. Theor. Phys. **4** 514 (1949)
14. T. Hsu, Phys. Rev. **B 41** 11379 (1990); G. Baskaran , Phys. Rev. **B 64** 092508 (2001)
15. G. Baskaran and P.W. Anderson (1987, unpublished)

EXPLORING THE PHASE DIAGRAM OF THE QUASI-2D ORGANIC SUPERCONDUCTORS κ -(BEDT-TTF)$_2$X

Proceedings of the ARW NATO Workshop Hvar, Croatia, October 2002

MICHAEL LANG
Johann Wolfgang Goethe-University, FOR412
D-60054 Frankfurt am Main, Germany

JENS MÜLLER and FRANK STEGLICH
Max-Planck Institute for Chemical Physics of Solids
D-01187 Dresden, Germany

JOHN SCHLUETER
Material Science Division, Argonne National Laboratory,
Argonne, Illinois 60439, U.S.A.

TAKAHIKO SASAKI
Institute for Materials Research, Tohoku University,
Sendai, Japan

Abstract. By means of high-resolution thermal expansion and specific heat measurements on the title compounds, we explore various aspects of their unified phase diagram. Particular attention has been paid to the interrelation of high-temperature anomalies, i.e. a glass transition associated with the ethylene end groups at $T_g \simeq 50 - 80\,K$ and a density-wave-type transition at $T^* \simeq 40\,K$, with the superconducting state.

Key words: Organic Superconductors, Thermodynamic Properties, Thermal Properties

1. Introduction

The organic charge-transfer salts of the (BEDT-TTF)$_2$X family (BEDT-TTF or simply ET denotes bis(ethylenedithio)-tetrathiafulvalene and X a monovalent anion) represent fascinating model systems for exploring the interplay of strong electron-electron and electron-phonon interactions in reduced dimensions. Of particular interest are the κ-phase (ET)$_2$X salts with various monovalent anions such as X = Cu(NCS)$_2$, Cu[N(CN)$_2$]Br

A.C. Hewson and V. Zlatić (eds.), Concepts in Electron Correlation, 85–93.

and $Cu[N(CN)_2]Cl$. They exhibit unusual normal (N)- and superconducting (SC)-state properties which resemble in some aspects those of the layered high-T_c cuprates. Apparent similarities include the quasi-two dimensional (quasi-2D) electronic structure and the proximity to an antiferromagnetically (AFM) ordered state, cf. Fig. 5. Moreover, the presence of strong AFM spin fluctuations above T_c as inferred from various magnetic and thermal properties have been associated with a magnetically mediated non-s-wave type of SC not only for the cuprates but also for the present organic systems (1). For the latter, however, such a scenario is at variance with the results of numerous other studies, see, e.g. (2) and references cited therein.
In this paper we summarize our recent results of a detailed thermodynamic study employing thermal expansion and specific heat measurements on the above κ-type salts. Particular attention is paid to the anomalous properties at elevated temperatures and their interrelation to the SC state. A full account of the work is given in (2, 3).

2. Classification of anomalies and discussion

Figure 1 compiles the results of the coefficient of thermal expansion measured perpendicular to the conducting planes, $\alpha_\perp$, for the various κ-$(ET)_2X$ salts investigated: X = $Cu(NCS)_2$ (in short κ-$Cu(NCS)_2$), $Cu[N(CN)_2]Br$ (κ-Br) and $Cu[N(CN)_2]Cl$ (κ-Cl). For all compounds, a variety of anomalies can be discerned and classified as (A) step-like changes in α at temperatures T_g around $50-80\,K$, (B) phase-transition-like anomalies at an intermediate temperature T^* around $40\,K$ only for the SC salts and (C) phase-transition anomalies into the SC (κ-$Cu(NCS)_2$, κ-H_8-Br) and AFM (κ-$Cu[N(CN)_2]Cl$) ground states. In what follows, we will discuss all three kinds of anomalies and their implications for the SC state.

(A) Glass-like transitions at $T_g \simeq 50-80\,K$

Figure 2 shows on expanded scales the anomalies in $\alpha_\perp$ for all three compounds near T_g. Both the distinct hysteresis between heating and cooling curves as well as their characteristic cooling-rate dependence (not shown) are clear manifestations of the glassy nature of these anomalies (3). It is well known that in positionally and/or orientationally disordered systems, relaxation processes can lead to glass-like transitions where below the glass temperature T_g, a short-range order characteristic for this temperature becomes frozen, see (3) and references cited therein.

As discussed in detail in ref. (3), the relevant entities involved in the freezing process for the present κ-$(ET)_2X$ salts are the terminal ethylene groups $[(CH)_2]$ of the ET molecules. From these results together with earlier x-ray-diffraction studies (4), we infer that upon cooling from room temperature, the gradual ordering in the relative orientation of the ethylene groups

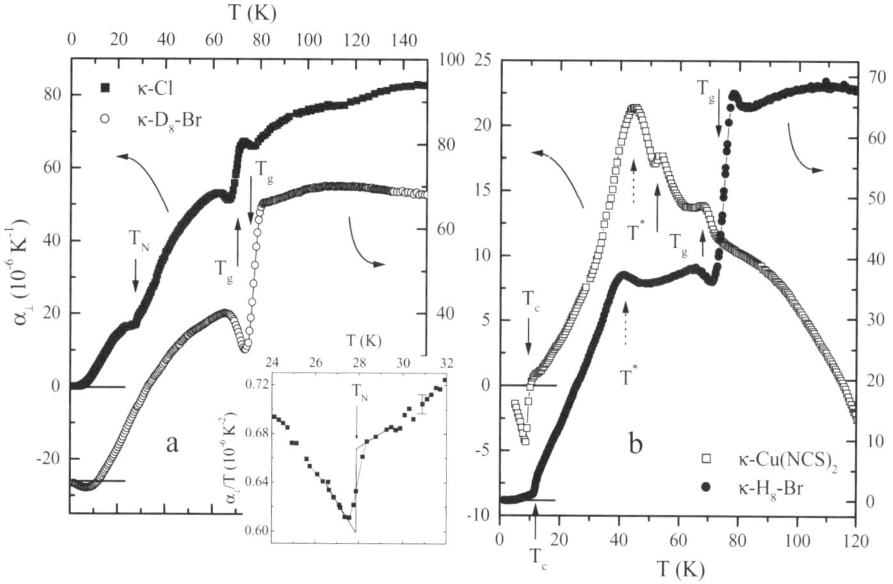

Figure 1. Cross-plane thermal expansion coefficient for various κ-$(ET)_2X$ salts; H_8 (D_8) indicate protonated (deuterated) ethylene endgroups. Inset shows the phase transition into the AFM ground state for κ-Cl.

in one of two possible conformations becomes interrupted at T_g leaving a residual frozen disorder at low temperatures. The amount of disorder depends on the cooling rate employed at T_g and is expected to be at a $10 - 20\,\%$ level. Its implications for the ground-state properties can be discerned by recalling the results of magnetic and transport measurements on protonated (H_8) and deuterated (D_8) κ-Br: rapid cooling through a temperature around $80\,K$ ($\simeq T_g$) was found to cause a strong reduction of the SC volume fraction and may even transform a SC into an insulator (5).

(B) Phase-transition anomalies at $T^* \simeq 40\,K$

For the SC salts, i.e. κ-H_8-Br and κ-$Cu(NCS)_2$, the $\alpha_\perp$ data (Fig. 1b) reveal distinct local maxima at $T^* = 38$ and $45\,K$, respectively. In addition, directional-dependent measurements (not shown) disclose a strongly anisotropic lattice response at T^*. In an attempt to extract the anomalous contributions to the uniaxial coefficients of thermal expansion α_i at T^*, $\delta\alpha_i$, we use a smooth interpolation of the data outside of the α_i anomalies. The so-derived α_{bi} serve as rough estimates of the background expansivities.

In Fig. 3 we display the quantities $\delta\alpha_i(T) = \alpha_i(T) - \alpha_{ib}(T)$ over an extended temperature range which also covers the SC transition. We find pronounced 2nd-order phase-transitions-like anomalies at slightly different temperatures T^* for both compounds. Replacing the somewhat broadened features by idealized sharp jumps employing an "equal-areas" construction

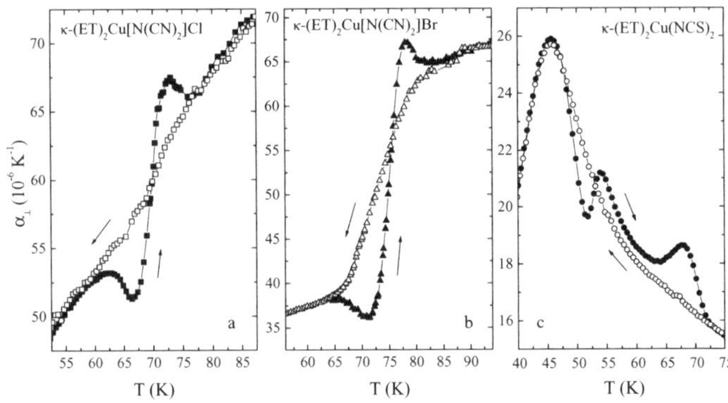

Figure 2. Blow-up of the cross-plane thermal expansion coefficient around the glass-like transition for various κ-$(ET)_2X$ salts. Arrows indicate cooling and heating runs.

in an α/T vs T plot yields transition temperatures $T^* = (41 \pm 2)\,K$ and $(47 \pm 2)\,K$ for the κ-H_8-Br and κ-$Cu(NCS)_2$ salts, respectively. We note that the overall shape of these anomalies, i.e. their width and asymmetry, but not the peak itself, depends somewhat on the interpolation procedure used to determine the background expansivities. An important piece of information enclosed in Fig. 3 is a striking interrelation between the $\alpha(T)$ anomalies at T^* and those at T_c for both compounds: for each axis, a large (small) feature at T^* is accompanied by a large (small) discontinuity at T_c, while the signs of both anomalies are just reversed! According to the Ehrenfest relation

$$\left(\frac{\partial T_0}{\partial p_i}\right)_{p_i \to 0} = V_{mol} \cdot T_0 \cdot \frac{\Delta \alpha_i}{\Delta C}$$

the uniaxial-pressure dependence of a 2nd-order phase transition at T_0 is related to the discontinuities at T_0 in the coefficient of thermal expansion, $\Delta\alpha_i$, and specific heat ΔC. Thus, the data of Fig. 3 imply that for both compounds the uniaxial-pressure coefficients of T^* and those of T_c are strictly anticorrelated in their signs but correlated in their magnitudes. In the case of the κ-$Cu(NCS)_2$ salt, for example, we infer that uniaxial pressure applied perpendicular to the planes causes a substantial shift of T^* to higher temperatures and, at the same time, a strong reduction of T_c. Likewise, uniaxial pressure along the in-plane b-axis neither affects T^* nor does it have any effect on T_c. The anomaly at T^* is particularly interesting because it coincides with the temperature where magnetic, transport, and elastic properties exhibit anomalous behavior: a sharp peak in the spin-lattice relaxation rate $(T_1T)^{-1}$ (7, 6) accompanied by a reduction of the spin susceptibility as determined by Knight-shift (7) or ESR (8) measurements as well as pronounced softening of ultrasonic modes (9, 10). In addition,

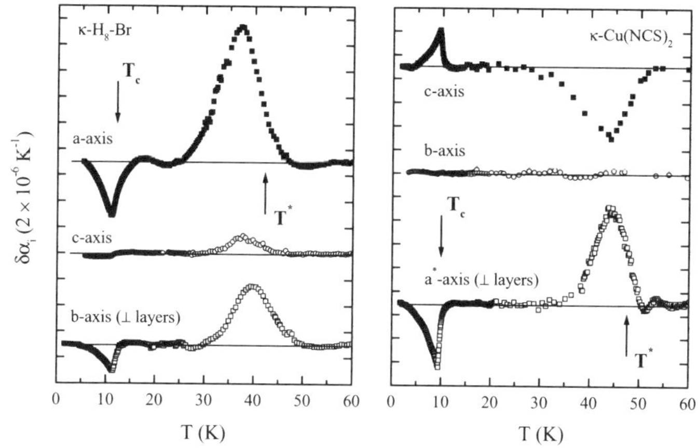

Figure 3. Anomalous contributions to the uniaxial thermal expansion coefficients at T_c and T^* derived by subtracting a smooth background contribution.

our analysis yields for both compounds an increase of T^* under hydrostatic pressure conditions, in accordance with the results of pressure studies (7, 9).

Concerning the nature of the anomaly at T^* we note that it is independent of the crystal structure as κ-Cu(NCS)$_2$ is not isomorphic to κ-Br but it appears to be related to the electronic structure which is very similar for the two SC compounds (4).

Based on the above results, we propose that T^* marks a density-wave instability that involves only the minor 1D parts of the Fermi surface while leaving the major 2D portions unaffected. This implies that cooling through T^* is accompanied by the formation of a **real** gap on a small fraction of the Fermi surface as opposed to a **pseudogap** on the major parts (1). The above interpretation is consistent with recent results of both magnetic and transport measurements on the SC salts yielding the onset of a small but distinct anisotropy below T^* (11).

(C) Transitions into the magnetic and superconducting groundstates

The phase transitions into either the AFM ordered (κ-Cl),(cf. Fig.1a) or SC (κ-H$_8$-Br, κ-Cu(NCS)$_2$), (cf. Figs. 1 and 3) ground state manifest themselves in distinct and strongly anisotropic phase-transition anomalies in α_i. As for the transition at T_N, our directional-dependent measurements (not shown) reveal a finite discontinuity only along the axis perpendicular to the conducting planes ($\Delta\alpha_\perp < 0$), cf. inset Fig. 1a, while $\Delta\alpha_\| \approx 0$ for both in-plane expansion coefficients (3). Applying the Ehrenfest relation, we thus conclude that uniaxial pressure applied perpendicular to the conducting planes causes a reduction of T_N whereas in-plane pressure leaves T_N unaffected. Taking these uniaxial-pressure coefficients together results in a negative pressure dependence of T_N under hydrostatic-pressure conditions in accordance with the results of pressure studies (12). As for the nature of the AFM state, our results seem difficult to reconcile with existing models. For neither the nesting-induced itinerant magnetism (13, 14), nor for a Mott-Hubbard-type magnetic state of localized spins (15, 1) - even when next nearest neighbor couplings are included - is a vanishingly small in-plane pressure coefficient expected.

Applying the same thermodynamic analysis to the directional-dependent $\alpha_i(T)$ anomalies at the SC transition for the κ-$(ET)_2X$ series clearly demonstrates the lack of a simple form of systematics regarding the in-plane pressure coefficients of T_c: While for the κ-H_8-Br salt both in-plane pressure coefficients of T_c are negative, an either zero or even positive in-plane pressure effect is found for κ-$Cu(NCS)_2$ (16, 3). Such a non-universal behavior is in clear contrast to the predictions of the purely 2D electronic models proposed by Kino et al. (17) and Kondo et al. (18) yielding in all cases a suppression of T_c under in-plane stress. On the contrary, our studies reveal that it is the extraordinarily large negative inter-layer pressure coefficient of T_c which is common to all κ-$(ET)_2X$ systems investigated so far including the X = I_3 salt (16). Apparently, it is this component which predominates the reduction of T_c under hydrostatic pressure.

As for the symmetry of the SC state, arguments in favor of an unconventional order parameter with d-wave symmetry for the present κ-$(ET)_2X$ family have been derived from temperature dependent studies and, more recently, from orientational-dependent measurements aiming at a direct determination of the gap anisotropy. Conversely, there are numerous experimental investigations which indicate a SC state characterized by a finite gap. For a discussion on the present status of this controversy, see e.g. (2).

A decisive technique to probe certain aspects of the gap structure – in particular, the question whether gap zeroes exist or not – is provided by specific heat measurements. In case this integral thermodynamic probe were to detect an electronic quasiparticle contribution $C_{es}(T)$ for $T \ll T_c$ that varies exponentially weakly with the temperature, the existence of gap zeroes on parts of the Fermi surface could be definitely ruled out. In Fig. 4 we display the quantity $\Delta C(T) = C(T, B = 0) - C(T, B = 8T)$, where $C(T, 8T)$ is the specific heat in the N-state, measured on a tiny high-quality single crystal of κ-$Cu(NCS)_2$ with a mass of only $0.72\,mg$.

Given a B-independent specific heat above T_c as proved experimentally (19, 20), the analysis of $\Delta C(T)$ has the advantage that the unknown phonon contribution and all other extraneous contributions cancel each other out. Figure 4 demonstrates that $\Delta C(T)$ deviates markedly from a weak-coupling BCS-behavior (broken line) but is in excellent agreement with the so-called

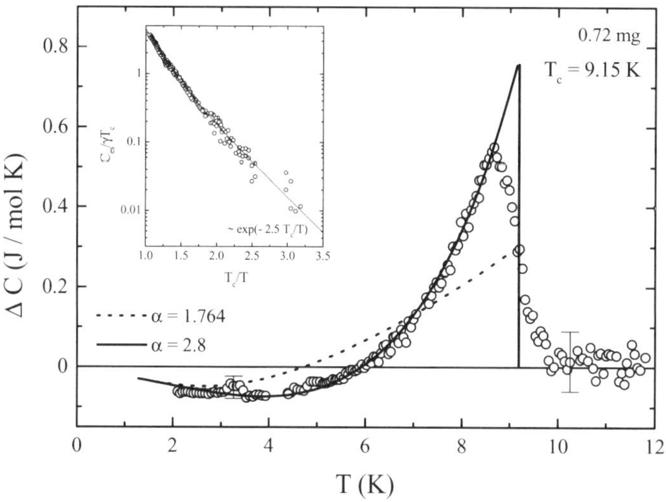

Figure 4. Specific heat difference $\Delta C = C(0T) - C(8T) = C(0T) - C_n$, where C_n denotes the N-state specific heat. Dotted and solid thick lines represent BCS curves for weak and strong coupling, respectively. Inset shows the electronic contribution in a BCS-plot.

α-model, an empirical extension of the BCS formalism to strong-coupling superconductors (21). As the inset of Fig. 4 clearly shows, $C_{es} = \Delta C + \gamma T$, where γ is the Sommerfeld coefficient, reveals an exponentially weak temperature dependence at low temperatures incompatible with a d-wave order parameter, cf. inset of Fig. 4. The same conclusions have been drawn recently from similar measurements on the κ-H_8-Br salt (20).

3. Summary and outlook

The present study discloses important new aspects which impose clear constraints for both experimentalists and theorists attempting to unravel the nature of the states below and above T_c in these κ-$(ET)_2X$ salts. At elevated temperatures, a glass transition has been identified that defines the boundary between an ethylene liquid at $T > T_g$ and a glassy state at $T < T_g$ where a certain amount of disorder in the terminal $[CH_2]_2$ units becomes frozen. With increasing cooling rate through T_g, the level of disorder ($10-20\,\%$) increases and may substantially influence the electronic properties at low temperatures, especially the SC state.

At intermediate temperatures T^*, our studies disclose a second-order phase transition for the metallic/SC salts. We propose that instead of a pseudogap on the major 2D fractions of the Fermi surface, T^* is associated with the formation of a density wave, i.e. a real gap, on the minor 1D parts which competes with superconductivity for stability.

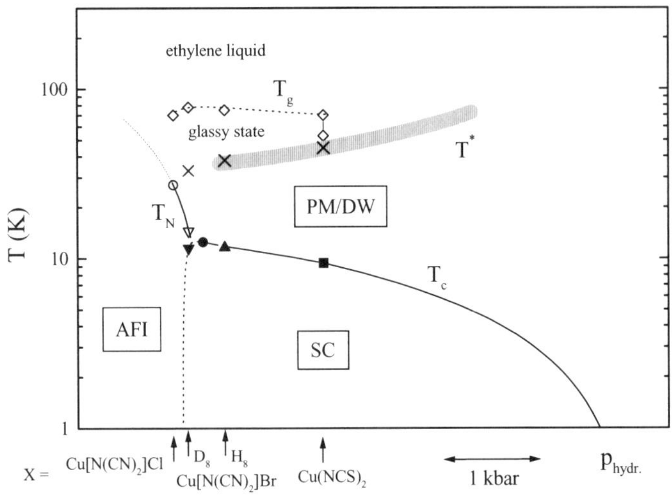

Figure 5. Compilation of thermal expansion anomalies for the various κ-$(ET)_2X$ salts in a pressure-temperature phase diagram. Arrows indicate the location of the different salts at ambient pressure. AFI, PM/DW and SC denotes antiferromagnetic insulator, paramagnetic metal in coexistence with a density-wave state and superconductor, respectively. Solid lines represent the hydrostatic pressure dependencies of T_N and T_c taken from the literature.

As for the response of T_c to pressure, our finding of an extraordinarily large interplane pressure effect but a non-universal behavior in respect to the in-plane pressure coefficients demonstrates that attempts to understand these quasi-2D organic superconductors solely on the basis of 2D electronic models are not appropriate. It is a combination of electronic correlations, electron phonon-interactions as well as interlayer-coupling effects which have to be considered in order to provide a realistic modelling of this family of superconductors.

References

1. A. Kanoda, Hyperfine Interact. **104**, 235 (1997)
2. J. Müller, M. Lang, R. Helfrich, F. Steglich, T. Sasaki, Phys. Rev. B **65** 140509 (2002)
3. J. Müller, M. Lang, F. Steglich, J. A. Schlueter, A. M. Kini, T. Sasaki, Phys. Rev. B **65**, 144521 (2002)
4. A.M. Kini, U. Geiser, H. H. Wang, K. D. Carlson, J. M. Williams, W. K. Kwok, K.G. Vandervoort, J. M. Thompson, D. L. Stupka, D. Jung, M.-H. Whangbo, Inorg. Chem. **29**, 2555 (1990)
5. A. Kawamoto, K. Miyagawa, K. Kanoda, Phys. Rev. B **55**, 14 140 (1997)
6. A. Kawamoto, K. Miyagawa, Y. Nakazawa, K. Kanoda, Phys. Rev. Lett. **74**, 3455 (1995)

7. H. Mayaffre, P. Wzietek, C. Lenoir, D. Jérome, P. Batail, Europhys. Lett. **28**, 205 (1994)
8. V. Kataev, G. Winkel, D. Khomskii, D. Wohlleben, W. Crump, K.F. Tebbe, J. Hahn, Solid State Commun. **83**, 435 (1992)
9. K. Frikach, M. Poirier, M. Castonguay, K.D. Truong, Phys. Rev. B **61**, R6491 (2000)
10. T. Simizu, N. Yoshimoto, M. Nakamura, M. Yoshizawa, Physica B **281&282**, 896 (2000)
11. T. Sasaki, N. Yoneyama, A. Matsuyama, N. Kobayashi, Phys. Rev. B **65**, 060505(R) (2002)
12. S. Lefebvre, P. Wzietek, S. Brown, C. Bourbonnais, D. Jérome, C. Mézière, M. Fourmigué, P. Batail, Phys. Rev. Lett. **85**, 5420 (2000)
13. P. Wzietek, H. Mayaffre, D. Jérome, S. Brazovskii, J. Phys. I **6**, 2011 (1996)
14. M.A. Tanatar, T. Ishiguro, H. Ito, M. Kubota, G. Saito, Phys. Rev. B. **55**, 12 529 (1997)
15. K. Miyagawa, A. Kawamoto, Y. Nakazawa, K. Kanoda, Phys. Rev. Lett. **75**, 1174 (1995)
16. J. Müller, M. Lang, F. Steglich, J. A. Schlueter, A. M. Kini, U. Geiser, J. Mohtasham, R. W. Winter, G. L. Gard, T. Sasaki, N. Toyota, Phys. Rev. B **61**, 11739 (2000)
17. H. Kino and H. Fukuyama, J. Phys. Soc. Jpn. **65**, 2158 (1996)
18. H. Kondo and T. Moriya, J. Phys. Soc. Jpn. **67**, 3695 (1998)
19. B. Andraka, J. S. Kim, G. R. Stewart, K. D. Carlson, H. H. Wang, J. M. Williams, Phys. Rev. B **40**, 11 345 (1989)
20. H. Elsinger, J. Wosnitza, S. Wanka, J. Hagel, D. Schweitzer, W. Strunz, Phys. Rev. Lett. **84**, 6098 (2000)
21. H. Padamsee, J. E. Neighbor, C. A. Shifman, J. Low Temp. Phys. **12**, 387 (1973)

ENHANCEMENT OF SUPERCONDUCTIVITY BY STRONG CORRELATIONS: A MODEL STUDY

Proceedings of the ARW NATO Workshop Hvar, Croatia, October 2002

MICHELE FABRIZIO, ERIO TOSATTI
International School for Advanced Studies (SISSA-ISAS) and Istituto Nazionale per la Fisica della Materia (INFM), Via Beirut 2, 34014 Trieste, Italy and International Center for Theoretical Physics (ICTP), Strada Costiera 11, 34014 Trieste, Italy

MASSIMO CAPONE, CLAUDIO CASTELLANI
Università di Roma "La Sapienza" and Istituto Nazionale per la Fisica della Materia (INFM), Piazzale A. Moro 2, 00185 Roma, Italy.

Abstract.

High-temperature superconductivity in doped Mott insulator cuprates denies the conventional belief that electron repulsion is detrimental to superconductivity. We discuss a parallel situation in the alkali fullerides which, even if lacking the cuprate's spectacular T_c's, are no lesser puzzle. The repulsive electron correlations in fullerides are of comparable strength to the cuprates: yet trivalent fullerides superconduct, and that with an s-wave order parameter. We discuss theoretical studies of a model for alkali fullerides which reveals, in presence of a weak pairing attraction (here of Jahn Teller origin), an unsuspected superconducting pocket near the Mott insulating state. The peculiar pairing mechanism in this model shows that close to an ideal, singlet Mott state, superconducting pairing can manage not just to avoid frustration by strong electron-electron repulsion, but actually to take vast advantage from it. Although derived by solving a very specific tetravalent fulleride model, we argue that this mechanism is in fact common to a wider class of strongly correlated models, including those proposed for cuprates (1).

Key words: High Temperature Superconductivity, Strongly Correlated Systems, Mott Transition, Fullerenes

A.C. Hewson and V. Zlatić (eds.), Concepts in Electron Correlation, 95–113.

1. Introduction

Among the unsolved questions raised by the discovery of high T_c superconductivity in cuprates, a challenging one is why these good superconductors are such bad metals in the normal phase. It is in fact conjectured in many quarters that the main reason for the poor metallic behavior, namely the strong electron-electron correlation, is also the key ingredient to understand the high T_c. That conjecture, even if supported by experimental evidence, is somehow heretical with respect to the Migdal-Eliashberg theory for conventional superconductors, consisting of phonon mediated attraction acting among Landau quasiparticles.

Within Landau Fermi liquid theory, strong repulsive correlations between electrons have two main effects on quasiparticles: first of all they increase the effective mass, $m_* > m$, and secondly they reduce the quasiparticle weight, $Z < 1$, of the single particle excitations. The mass enhancement increases the quasiparticle density of states (DOS) at the chemical potential, ρ_*, with respect to the uncorrelated one, ρ, according to

$$\rho_* = \frac{m_*}{m}\rho.$$

On the other hand, the quasiparticle attraction, parametrized for instance by the coupling constant V_*, is renormalized with respect to the bare value V as

$$V_* = Z^2 \Gamma_V \, V,$$

Γ_V being the vertex correction. In the standard situation of an attraction provided by coupling to phonons, vertex corrections are small, $\Gamma_V \simeq 1$. Therefore the dimensionless parameter λ_*, which controls the strength of the attraction, is renormalized down with respect to the uncorrelated value $\lambda = \rho V$ according to

$$\lambda_* = \rho_* V_* \simeq Z^2 \frac{m_*}{m} \rho V = Z^2 \frac{m_*}{m} \lambda < \lambda,$$

since usually $Z \leq m/m_*$. It seems therefore hardly conceivable within Landau Fermi liquid theory that strong correlations might help superconductivity, except in the fortunate situation where vertex corrections, rather than being negligible, would be large. For instance, if vertex corrections were to compensate for the wave-function renormalization, say $\Gamma_V \sim 1/Z^2$, then

$$\lambda_* \simeq \frac{m_*}{m} \lambda > \lambda,$$

which might indeed produce a T_c enhancement.

This hypothetical possibility is based on the assumption that Landau Fermi liquid theory is valid well below T_c. However, when the mass enhancement pushes $\lambda_* \sim 1$, T_c becomes of the order of the quasiparticle bandwidth and a description of the excitations above T_c in terms of Landau quasiparticles may not be correct. This situation is compatible with the behavior of the cuprates. In fact, while their normal phase shows anomalous non-Fermi-liquid properties, there are spectroscopic evidences that single particle excitations well below T_c can still be represented by Landau-Bogoliubov quasiparticles (2).

There are models for cuprates where the above hypothetical scenario seems to be realized. For instance the t-J model, where strong repulsive correlations cause the single occupancy constraint, and yet superconductivity does arise. The on-site constraint strongly suppresses the coherent electron motion, without affecting the pairing mechanism, because the latter is provided by the intersite spin exchange J. So even if the normal phase of the t-J model will show some exotic properties, for instance spin-charge separation within the RVB phase and similar features (3), the superconducting phase seems still describable in terms of quasiparticles (spinon-holon bound states) subject to a strong attraction J which causes them to pair off, with a loss of kinetic energy given by a coherent bandwidth of similar magnitude (4, 5, 6).

Here we discuss another class of superconductors, the alkali doped fullerenes, or fullerides. While these compounds lack the spectacular T_c's of the cuprates, they nonetheless appear to share some aspects with them. The fullerides are molecular conductors with a narrow bare bandwidth of about $W \simeq 0.5$ eV (for a broad review see Ref. (7)). The relative strength of correlations is comparable with that of cuprates, with a large intramolecular Coulomb repulsion $U \sim 2/3W$ (8, 9), which undoubtedly makes them strongly correlated superconductors, like the cuprates. Unlike the cuprates, which condense with a d-wave order parameter so as to minimize on-site Coulomb repulsion, fullerides are s-wave superconductors. Here an efficient intramolecular pairing mechanism and the orbital degeneracy, are evidently able to induce superconductivity avoiding the relatively strong U. Historically, because of a consensus that the pairing mechanism is a conventional electron-phonon coupling, they never attracted quite the same attention as the cuprates. We actually find instead that this circumstance is an advantage when one wants to understand the truly revolutionary effects of extreme correlations on superconductivity.

C_{60} molecules possess a threefold degenerate t_{1u} LUMO which forms in a crystal lattice three narrow bands. These bands are occupied by the n valence electrons provided by the alkali ions M in M_nC_{60}. According to band structure calculations, alkali doped fullerenes should be metallic for any $0 < n < 6$. On the contrary however, only M_3C_{60} are metallic and superconducting, while for instance M_4C_{60} are non magnetic narrow gap insulators. The phonons which are believed to be mostly involved in superconductivity of M_3C_{60} are the eigth fivefold degenerate molecular vibrations of H_g symmetry, Jahn-Teller coupled to the t_{1u} orbitals (10). The superconducting critical temperature in cubic fullerenes varies from 18 K in K_3C_{60} to 28 K in Rb_3C_{60}, and appears to be a monotonically increasing function of the lattice spacing a. Because this grows parallel to the density of states (11), this increase is qualitatively consistent with standard BCS theory, which is therefore often used to discuss superconductivity in fullerides in spite of very strong correlations (7). In some trivalent compounds however the behavior of T_c versus the lattice spacing a is opposite. Cs_3C_{60}, a bct compound with a larger unit cell volume per C_{60} than either K_3C_{60} and Rb_3C_{60}, shows an increasing T_c with pressure, reaching at 15 Kbar the highest $T_c \sim 40$ K among all fullerides (12). Unless the difference between cubic and noncubic lattices happened to play here a fundamental role (13), one might speculate that T_c actually *decreases* when the volume is too large. An even more compelling suggestion in the same direction is provided by the ammoniated n=3 compound $NH_3K_3C_{60}$, which can be thought of as a very expanded anisotropic K_3C_{60}, and is an S=1/2 antiferromagnetic insulator, certainly of the Mott type. Under pressure it undergoes an insulator-to-metal transition, and superconductivity is recovered with a large $T_c \sim 28$ K at the transition (14), decreasing for increasing pressure. This result is very important. It shows clearly that superconductivity in fullerides does, like in cuprates, occur close to a Mott insulator. A corollary is that ignoring strong correlations cannot in the end be right in the superconducting phase either.

If the behavior of these compounds could be accepted as evidence that T_c varies nonmonotonically with volume, the family of fullerides under pressure would realize the whole T_c versus λ curve – first increase, then decrease – within the narrow span of unit cell volumes per C_{60} ranging from 670 Å^3 in K_3C_{60} at 28 Kbar ($T_c \sim 6$ K) to something of order 800 Å^3 and 830 Å^3 in Cs_3C_{60} at 15 Kbar ($T_c \sim 40$ K) and at residual pressure ($T_c \sim 18$ K), respectively. In other words, in the unit cell volume interval $\Delta V/V \sim 0.2$, fulleneres could explore almost the full region from weak to strong coupling superconductivity. This in itself suggests that a simple BCS-type of approach will not be appropriate. We will argue here that strong correlations are the clue ingredient to explain such a drastic changes in a narrow unit-cell volume interval.

Evidence of non-BCS behavior of superconductivity in trivalent fullerides is widespread. Most strikingly, NMR measurements have shown the existence of a spin gap $\Delta_{spin} \simeq 0.07 \div 0.1$ eV even in metallic K_3C_{60} (15). The spin gap magnitude found in n=3 fullerides is moreover quite comparable with the spin gap of an isolated C_{60}^{3+} molecular ion, suggesting that molecular features do survive in the metallic phase, as expected close to a Mott transition, but not otherwise explainable within band theory.

Turning to tetravalent fullerides such as K_4C_{60} and Rb_4C_{60}, they should, as was previously said, also be metallic according to DFT-LDA calculations. In reality, they appear on the contrary to be on the nonmagnetic insulating side of an insulator-to-metal transition that can be observed under pressure (16). The zero pressure insulating phase could at first sight be explained away (say within Hartree-Fock theory) as a band insulator. By assuming for example a large static Jahn-Teller effect (17), the four electrons would occupy two out of three Jahn Teller split t_{1u} orbitals. The mean field treatment of Coulomb repulsion in Hartee-Fock simply adds to the Jahn-Teller distortion, amplifying its effect and leading in this case to an exceedingly wide insulating gap. However, the gap in the real materials is by far not that large, and in fact about zero in K_4C_{60} and nearly zero in Rb_4C_{60} (16). Moreover, there is no experimentally detectable static Jahn-Teller distortion (18). Finally, and most strikingly, the magnetic response is dominated here too by a 0.1 eV spin gap (19, 16, 15). This spin gapt is totally unaccounted for in the band insulator model provided by Hartree Fock, or for that matter by any other non-strongly correlated theory. All of this indicates that K_4C_{60} and Rb_4C_{60} are not really band-insulators but most likely singlet, nonmagnetic *ideal* Mott insulators, systems possessing no spin entropy and devoid of any breaking of symmetry.

Our earliest analysis of the role of correlations in fullerides started precisely with K_4C_{60}, in an attempt to understand how and why an ideal Mott insulator could be favored instead of the Hartree-Fock band insulator (17). In order to treat strong correlations we resorted to the Dynamical Mean Field Theory (DMFT) that permits an analysis of both Fermi Liquid states and Mott insulating states. We indeed found that the Mott insulator is the most stable phase for realistic parameters. That analysis however also led to unexpected developments (20), which are the subject of these lecture notes. In addition to the Mott state it was found that, by increasing continuously the bandwidth W of our model for K_4C_{60}, the Mott insulator developed a direct and continuous transition not to a simple metal, but to a superconductor, which only subsequently turned into a normal metal. A careful study of this superconducting region, showed that a) the unexpected superconducting phase shares many common features with high T_c superconductors; and that b) the Fermi liquid hypothetical scenario mentioned above seems to describe quite well the onset of superconductivity as well as the approach to the Mott insulating phase. The analysis of that model also allows to identify the condition under which the scenario applies: namely when the vertex correction of the attractive channel compensates for the wave function renormalization, $\Gamma_V \sim 1/Z^2$. This situation can only occur when the pairing attraction involves other degrees of freedom than the charge, because charge fluctuations get completely suppressed close to the Mott transition. In fullerides the pairing attraction is in fact a result of Jahn-Teller coupling, that involves orbital and spin degrees of freedom. The related tendency to stabilize Jahn Teller distorted low spin states for the molecular ion persists for all charge states (10, 21). This is parallel to the situation in the t-J model of cuprates, where the attraction is provided by the spin exchange J, again unrelated to charge.

2. A model for alkali doped fullerenes

After this long introduction, let us describe the model for alkali doped fullerenes we studied. The Hamiltonian is

$$H = \sum_{RR'} \sum_{ij} \sum_{\sigma} t^{ij}_{RR'} c^{\dagger}_{R,i\sigma} c_{R',j\sigma} + \frac{U}{2} \sum_{R} (n_R - n)^2 + H_{Hund}, \tag{1}$$

where $c^{\dagger}_{R,i\sigma}$ is the electron creation operator at site R in the orbital i ($i = 1,2,3$) of the t_{1u} multiplet with spin σ and $n_R = \sum_{i\sigma} c^{\dagger}_{R,i\sigma} c_{R,i\sigma}$ the occupation number at site R. $t^{ij}_{RR'}$ are the hopping matrix elements giving rise to three narrow bands of width $W \simeq 0.5$ eV. For convenience we will assume nearest neighbor hopping to be diagonal in the orbital index. U is the on-site Coulomb repulsion which tends to suppress valence fluctuations around the average electron density $\langle n_R \rangle = n$.

The last term H_{Hund} is the multiplet exchange splitting required by Hund's rules. Within the t_{1u} subspace, we can define angular momentum operators

$$L_{R,i} = \sum_{jk} \sum_{\sigma} c^{\dagger}_{R,j\sigma} \left(\hat{L}_i\right)_{jk} c_{R,k\sigma}, \tag{2}$$

where $\left(\hat{L}_i\right)_{jk} = -\mathrm{i}\varepsilon_{ijk}$, as well as spin operators

$$S_{i,R} = \frac{1}{2} \sum_{\alpha\beta} \sum_{i} c^{\dagger}_{R,i\alpha} \left(\hat{\sigma}_i\right)_{\alpha\beta} c_{R,i\beta}, \tag{3}$$

$\hat{\sigma}_i$ being the Pauli matrices. In terms of these operators the Hund term reads

$$H_{Hund} = J_H \sum_{R} \left(2\mathbf{S}_R \cdot \mathbf{S}_R + \frac{1}{2}\mathbf{L}_R \cdot \mathbf{L}_R\right) + \frac{5}{6}(n_R - 3)^2. \tag{4}$$

The bare J_H is of course positive, favoring high spin states in artificially frozen high symmetry molecular ions. However, there is no such freezing in real C_{60}^{n-}, and the Jahn-Teller (JT) coupling to the H_g molecular vibrations can and does (10, 21) reverse Hund's rule, generally favoring instead low spin and angular momentum configurations. We include approximately this effect by assuming $J_H < 0$. This is formally equivalent to treating the electron-vibron coupling within an antiadiabatic approximation, where it can be shown to renormalize $J_H \rightarrow J_H - 3E_{JT}/4$, E_{JT} being the Jahn-Teller energy gain. The antiadiabatic approximation is more justified than it may seem in strongly correlated fullerides, where the vibron frequencies of order 0.1 eV, is not in fact larger, but rather comparable, with the bare electron hopping $|t|$. Vibrational frequencies are in fact safely larger than the *renormalized* hopping $Z|t|$ close to the Mott state, where $Z \ll 1$. In any case, the neglect of retardation involved in replacing the true JT coupling with a negative J_H does disfavor superconductivity, by preventing high energy screening of the Coulomb interaction U. Therefore, adoption of the $J_H < 0$ approximation in place of the fully retarded JT effect is conservative, and makes physical sense in this problem.

Let us for a moment consider the isolated molecule with $J_H < 0$ and with two valence electrons, or, equivalently, four (having in mind K_4C_{60} type compounds). The ground state has quantum numbers $S = L = 0$. The spin gap to the spin triplet state with quantum numbers $S = L = 1$ is $5|J_H|$. We may therefore fix an approximate J_H magnitude by comparing this value of the spin gap with the exact singlet-triplet gap for the isolated molecular ion. The n=4 C_{60} ion has of course a more favorable Jahn Teller distortion in the S=0 ground state than in the S=1 excited state, and this gives rise to a spin gap $\Delta_{spin} \sim 0.1$ eV. This value is quite comparable with that observed by magnetic resonance experiments in K_4C_{60}. The comparison leads to a crude estimate $J_H \simeq -0.02$ eV, namely $J_H/U \simeq -0.02$, the value we adopt hereafter. While this will constitute a very weak pairing attraction, we shall see that it may still give rise to a remarkably important consequences.

Since all relevant interactions, namely U and J_H, are local, the model described by the Hamiltonian (1) is suitable to be treated by the so-called Dynamical Mean Field Theory (DMFT) (22), which is a very efficient tool to deal with strong local correlations. We studied (1) by DMFT with an average number of electrons (holes) per site $n = 2$ at fixed $J_H/U = -0.02$ but at various ratios of U/W. Before presenting the results, let us analyse the behavior of (1) in the two extreme limits $U/W \ll 1$, weak coupling, and $U/W \gg 1$, strong coupling.

In the weak coupling limit, the Hamiltonian (1) describes a metal with three 1/3-filled degenerate bands. Since $J_H < 0$ is an explicit attraction present in the model, one needs to check the stability of the metal towards Cooper pairing. In the absence of U, J_H would indeed lead to a superconducting phase with an s-wave order parameter

$$\langle \Delta_R \rangle = \sum_{i=1}^{3} \langle c^\dagger_{R,i\uparrow} c^\dagger_{R,i\downarrow} \rangle \neq 0. \qquad (5)$$

However, for our choice of $J_H/U = -0.02$, the dimensionless superconducting coupling $\lambda = 10\rho|J_H|/3 = 0.2\rho U/3$, ρ being the DOS at the Fermi energy per spin and band, is much smaller than the Coulomb pseudopotential $\mu_* = \rho U$. Therefore the weakly correlated metal is stable against superconductivity.

In the strong coupling limit, $U/W \gg 1$, the system is a Mott insulator. Each site is occupied by two electrons which, due to the inverted Hund's rules, form a spin as well as an orbital singlet. Since in this limit the intermolecular superexchange generated by virtual hopping processes is smaller than the on-site $|J_H|$ the molecular singlet state remains stable against the triplet. Therefore the Mott insulator is nonmagnetic and has a gap in both spin and orbital excitations, of the same order of magnitude as the molecular gaps, in agreement with the observed behavior in tetravalent fullerides. This ideal Mott insulating phase – or rather recalling that $J_H < 0$ really mimicks the Jahn-Teller effect, this " Mott-Jahn-Teller" insulating phase (23) – is a kind of on-site version of a valence-bond spin liquid phase.

3. Dynamical Mean Field results

The above discussion shows that, as U/W increases, there should be in our model a transition from a weakly correlated metal to a non-degenerate singlet Mott insulator. However, such a transition cannot be continuous. In fact the only way a metal which satisfies Luttinger's theorem can turn smoothly into an insulator is by flattening the energy vs. momentum dispersion, namely by a diverging effective mass. This would imply that a finite entropy per site is released at the Mott transition, which is not compatible with the insulator being non-degenerate, and hence with zero residual entropy. Therefore either the transition is discontinuous, or else a third phase must intrude between the metal and the Mott insulator. This third phase will most likely break a symmetry of the Hamiltonian so to avoid Luttinger's theorem, and allow the large metallic Fermi surface to shrink continuously as one goes from the metal phase towards the Mott phase.

This is precisely what we find by DMFT: a narrow but well defined superconducting region with s-wave order parameter (5) intrudes between the metal and the Mott insulator (see Fig. 1). Of course for the much larger realistic value of $U/W \simeq 2$, we do find an insulating phase, quite

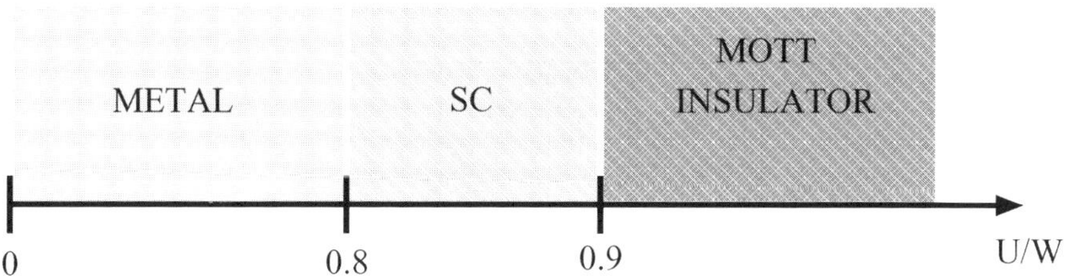

Figure 1. Phase diagram as a function of U/W. SC stands for superconductor

compatible with the behavior of real zero-pressure tetravalent fullerides. It is in fact a very good question whether in view of our results the pressure-induced insulator-metal transition which was demonstrated in these compounds (16) could be re-examined closely, searching for a possible narrow superconducting region.

Theoretically, the existence of a superconducting pocket in our model is a surprise. At weak coupling our model has no superconducting instability, and we would intuitively not expect the s-wave pairing tendency to increase by increasing intra-site electron repulsion. Even more striking are the properties of the superconducting phase.

In Fig. 2 we plot the superconducting single particle gap Δ as function of $|J_H|/W$. The points correspond to our case with $U = 50|J_H|$, where superconductivity occurs only within a small region just before the Mott transition. The gap is extremely large for such a weak attraction. For comparison, we also show the BCS gap calculated in the same model but for $U = 0$: only after multiplication of the latter by 1000 are they comparable.

Therefore, in the presence of a large U, we find that superconductivity appears where it would seem less probable, and with gap values thousand of times larger than at $U = 0$! We thus have found a soluble model (at least as far as DMFT can be regarded as a solution) showing explicitly that strong correlations are not detrimental but may on the contrary enormously enhance superconductivity, which is of course suggestive of high T_c superconductors.

The analogy gets even closer by calculating the spin gap Δ_{spin} extracted by the main peak values of the dynamical spin susceptibility, also drawn in Fig. 2 reduced by a factor 20. The two gaps, the superconducting gap and the spin gap follow each other at the onset of superconductivity, but quite soon they depart, the superconducting gap disappearing in the insulator, and the spin gap smoothly merging with the large one of the singlet Mott insulator. This behavior is reminiscent of the pseudo-gap region in underdoped cuprates.

Finally we note that superconductivity occurs in the small region $\Delta W/W \sim 0.12$ around $W \sim 1.18U$, reaching at this value almost the largest possible

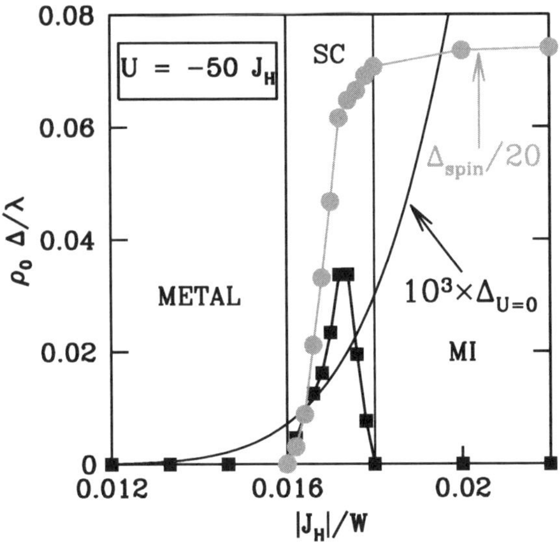

Figure 2. Superconducting gap (square points) and spin gap reduced by a factor 20 (circles) as functions of $|J_H|/W$ at fixed $U = 50|J_H|$. Also drawn is the superconducting gap at $U = 0$, $\Delta_{U=0}$, as function of $|J_H|/W$, multiplied by one thousand.

superconducting gap at the given attraction $10|J_H|/3$. This is compatible with the behavior of T_c vs. unit cell volume in fullerenes, as discussed in the introduction.

Since Fermi liquid theory seems to apply all the way up to the Mott transition in the single band Hubbard model, we can also try to interpret our results within the same framework. Indeed our model (1) has many features which suggest it as a possible condidate for the Fermi liquid scenario for correlation enhanced superconductivity discussed in the introduction. The Hund's rules, even if inverted, are known to be most likely violated in a metal but satisfied in an insulator. Hence we reasonably expect that a strong U will not impede the working of J_H but rather favor it.

Let us assume that the vertex corrections compensate exactly for the wavefunction renormalization so that the renormalized

$$J_{H*} = Z^2 \Gamma_{J_H} J_H \simeq J_H,$$

is practically unchanged. We do expect a reduction of the Coulomb pseudo-potential U_* felt by the quasiparticles, which is equal to U only at weak coupling. The DMFT solution of the single band Hubbard model suggests that a wrong quasiparticle charge configuration, namely a quasiparticle *double occupancy*, costs energy ZU and not U, which also agrees with the behavior of the Gutzwiller wavefunction. Let us assume the same holds in our multiband model, namely the Coulomb pseudopotential renormalizes like $U_* = ZU$. Then the quasiparticle scattering amplitude in the s-wave channel A_* should modify into

$$A_* \simeq U_* + \frac{10}{3} J_{H*} = ZU + \frac{10}{3} J_H = \left(Z - \frac{0.2}{3} \right) U. \tag{6}$$

While at weak coupling, $Z \simeq 1$ the scattering amplitude A_* is repulsive, at sufficiently strong correlation, such that $Z \leq Z_c \simeq 0.067$, the repulsion may turn into attraction and the system becomes unstable towards superconductivity. The conjecture (6) for the quasiparticle scattering amplitude also explains the large values of the gap compared with BCS values. Since the Mott transition occurs for $W \simeq U$, when Z is further reduced down to $\sim Z_c/2$, the effective attraction A_* gets of the same order as the quasiparticle bandwidth, $W_* = ZW \simeq Z_c W/2 \simeq Z_c U/2 = 10|J_H|/6$. In the spirit of Landau Fermi liquid theory, this should correspond to the maximum of the superconducting gap. The actual gap value, $\Delta \sim W_*/10 = |J_H|/6$ linear in the bare attraction, is in fact compatible with that expected at the maximum of the Δ vs. λ curve for a generic unretarded attraction at $U = 0$ (24). The notable difference is that this value is here obtained for a bare bandwidth $W \sim 60|J_H|$, namely for a bare $\lambda \sim 0.056$, where BCS theory would rather predict $\Delta_{BCS} \sim W \exp(-1/\lambda)$ much smaller than the actual value we obtain $\Delta \simeq W\lambda/20$.

3.1. FERMI LIQUID THEORY

So far the Fermi liquid scenario is just a conjecture, which needs to be supported by calculations. DMFT allows to extend the normal metal solution inside the region of superconducting instability by simply preventing gauge-symmetry breaking in the self-consistency equations. This provides a full description of the *normal metal* phase up to the Mott transition, which can be compared with Landau Fermi liquid theory. Our model possesses a spin SU(2) and an orbital O(3) symmetry so that the Landau functional contains 7 parameters: the mass enhancement, which in DMFT is simply $m_*/m = 1/Z$, and six f-parameters. Indeed we can define six density operators in momentum space, the charge density

$$n_k = \sum_{i=1}^{3} \sum_{\sigma} c^\dagger_{k,i\sigma} c_{k,i\sigma},$$

the spin density

$$\sigma_k = \sum_{i=1}^{3} \sum_{\alpha\beta} c^\dagger_{k,i\alpha} \sigma_{\alpha\beta} c_{k,i\beta},$$

the angular momentum density in the symmetric and antisymmetric channels, with components $(a, b = 1, 2, 3)$

$$l_k^a = \sum_{i,j=1}^{3} \sum_{\sigma} c_{k,i\sigma}^\dagger \left(\hat{L}_a\right)_{ij} c_{k,j\sigma},$$

and

$$\left(l^a \sigma^b\right)_k = \sum_{i,j=1}^{3} \sum_{\alpha\beta} c_{k,i\alpha}^\dagger \left(\hat{L}_a\right)_{ij} \hat{\sigma}_{\alpha\beta}^b c_{k,j\beta}.$$

In addition we can introduce five independent bilinears of the angular momentum operators, namely $\hat{W}_1 = \sqrt{3}(\hat{L}_2\hat{L}_3 + \hat{L}_3\hat{L}_2)$, $\hat{W}_2 = \sqrt{3}(\hat{L}_3\hat{L}_1 + \hat{L}_1\hat{L}_3)$, $\hat{W}_3 = \sqrt{3}(\hat{L}_1\hat{L}_2 + \hat{L}_2\hat{L}_1)$, $\hat{W}_4 = \sqrt{3}(\hat{L}_1\hat{L}_1 - \hat{L}_2\hat{L}_2)$, and finally $\hat{W}_5 = (\hat{L}_1\hat{L}_1 + \hat{L}_2\hat{L}_2 - 2\hat{L}_3\hat{L}_3)$, in terms of which we define, with $d = 1, \ldots, 5$,

$$w_k^d = \sum_{i,j=1}^{3} \sum_{\sigma} c_{k,i\sigma}^\dagger \left(\hat{W}_d\right)_{ij} c_{k,j\sigma},$$

as well as its antisymmetric partner $(a = 1, \ldots, 3)$

$$\left(w^d \sigma^a\right)_k = \sum_{i,j=1}^{3} \sum_{\alpha\beta} c_{k,i\alpha}^\dagger \left(\hat{W}_d\right)_{ij} \hat{\sigma}_{\alpha\beta}^a c_{k,j\beta}.$$

Through all the above definitions, the Landau functional reads

$$\begin{aligned} E = & \sum_k \sum_{i=1}^{3} \sum_{\sigma} \epsilon_k c_{k,i\sigma}^\dagger c_{k,i\sigma} + \\ & \frac{1}{V} \sum_{kk'} \left(f^S \, n_k \, n_{k'} + f^A \, \sigma_k \cdot \sigma_{k'}\right) + \\ & \frac{1}{V} \sum_{a=1}^{3} \sum_{kk'} \left[h^S \, l_k^a \, l_{k'}^a + \sum_{b=1}^{3} h^A \left(l^a \sigma^b\right)_k \left(l^a \sigma^b\right)_{k'}\right] + \\ & \frac{1}{V} \sum_{d=1}^{5} \sum_{kk'} \left[g^S \, w_k^d \, w_{k'}^d + \sum_{b=1}^{3} g^A \left(w^d \sigma^b\right)_k \left(w^d \sigma^b\right)_{k'}\right]. \end{aligned} \quad (7)$$

Here ϵ_k are the single-quasiparticle energies with respect to the chemical potential. We introduce the F-parameters $F^{S(A)} = 6\rho f^{S(A)}/Z$, $H^{S(A)} = 4\rho h^{S(A)}/Z$ and $G^{S(A)} = 12\rho g^{S(A)}/Z$, where ρ is the bare DOS, through which the susceptibilities χ to each field which couples to one of the above densities are given by the standard expression

$$\frac{\chi}{\chi^{(0)}} = \frac{1}{Z} \frac{1}{1+F}. \quad (8)$$

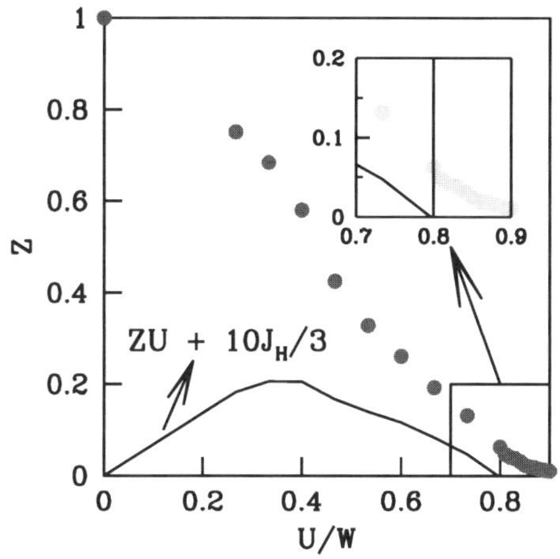

Figure 3. Quasiparticle residue for the metallic solution. The inset shows the enlarged region where superconductivity appears, which is signalled by a vertical line at $U \simeq 0.8W$. The line is the conjectured A_* of Eq. (6) which indeed crosses zero at $U \simeq 0.8W$ when $Z \sim 0.067$.

Here $F = F^{S(A)}, H^{S(A)}, G^{S(A)}$ depending on the form of the external field. By calculating within DMFT the quasiparticle residue Z as well as all the six susceptibilities, we obtain all F-parameters.

Fig. 3 shows the decrease of Z in the metallic solution on approaching the Mott transition (MIT). Superconductivity sets in at Z very close to our conjectured value $Z_c = 0.67$.

The charge compressibility, κ, is a decreasing function of U, in agreement with the system approaching the incompressible insulator, see Fig. 4. Through Eq. (8) this implies that F^S is always positive and diverges faster than $1/Z$ at the MIT, actually like $1/Z^2$. All other susceptibilities, (see for example Fig. 4 where the spin susceptibility is shown), initially increase due to Stoner enhancement, then turn around at $U/W \sim 0.7$, and finally vanish at the Mott transition, consistent with the spin and orbital gaps. This implies that F^A, $H^{S(A)}$ and $G^{S(A)}$ all start negative and decrease roughly like $-U/W$, as can be perturbatively checked, until $U \sim ZW$, when they turn upward, cross the zero and finally diverge as $+1/Z^2$ at the MIT.

The Landau F-parameters are related to the quasiparticle scattering amplitudes strictly only at low frequencies and momentum transfers, hence it is not rigorously justified to use them to estimate the scattering amplitudes in the Cooper channel, primarily a high momentum transfer process. Yet the F-parameters have been often used to that purpose, for instance to justify the triplet p-wave pairing in He^3. In our case, this approximation

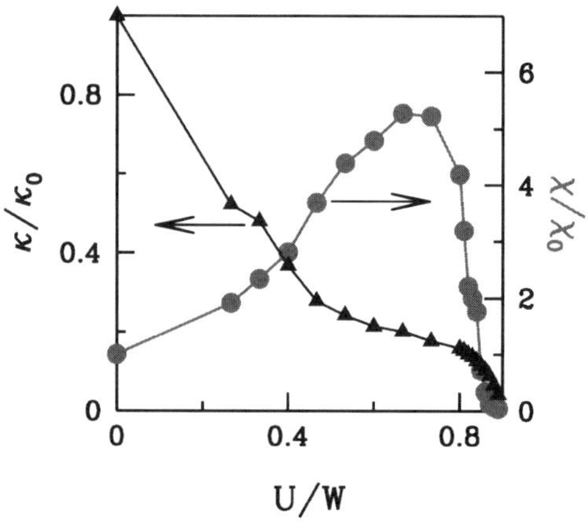

Figure 4. Charge compressibility κ and spin susceptibility χ normalized to the uncorrelated values κ_0 and χ_0, respectively. All other susceptibilities behave qualitatively like χ.

leads to the following estimate of the quasiparicle scattering amplitude in the s-wave channel of Eq. (5):

$$A = \frac{Z}{12\rho}\left[\frac{F^S}{1+F^S} - 3\frac{F^A}{1+F^A} - 3\frac{H^S}{1+H^S} + 9\frac{H^A}{1+H^A} + 5\frac{G^S}{1+G^S} - 15\frac{G^A}{1+G^A}\right]. \tag{9}$$

Fig. 5 shows the scattering amplitude A as calculated through the above formula.

At weak coupling, $A = U + 10J_H/3 > 0$ is repulsive and increases with U/W until at $U/W \sim 0.5$ it turns downwards and crosses zero signalling a superconducting instability: one which we actually find when allowing gauge-symmetry breaking. As the MIT is approached, all F-parameters diverge hence

$$A \to -\frac{Z}{2\rho},$$

saturates to a value about equal to half the quasiparticle bandwidth $W_* = ZW$. For comparison we have also drawn in Fig. 4 our conjectured A_* of Eq. (6). Up to the metal-to-superconductor transition, A_* seems to agree quite well with A, although close to the MIT A_* would remain constant while A vanishes as Z. If the estimate of A through the F-parameters is at least qualitatively correct, this behavior would imply that the attraction among the quasiparticles cannot exceed their bandwidth.

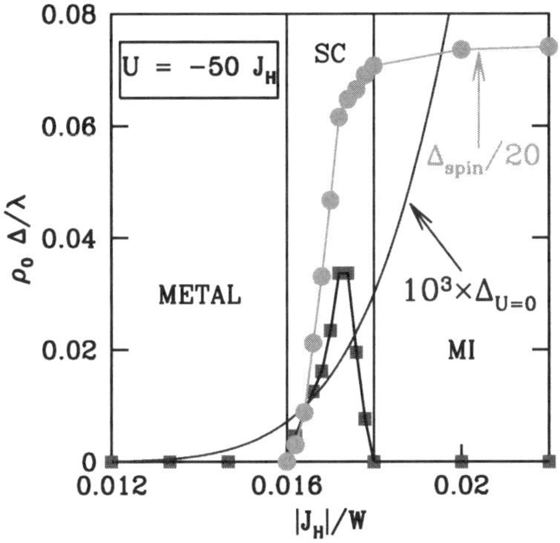

Figure 5. Quasiparticle scattering amplitude A in units of the quasiparticle bandwidth ZW as calculated through Eq. (9) (circles) and as conjectured in Eq. (6) (line).

Although the expression (9) seem to correctly reproduce the onset of superconductivity, its validity needs further justification. As Nozières pointed out to us (25), the same scheme applied to the single band Hubbard model would lead to an incorrect result. In the single band Hubbard model, the s-wave quasiparticle scattering amplitude within the same approximation is

$$A = \frac{Z}{2\rho}\left[\frac{F^S}{1+F^S} - 3\frac{F^A}{1+F^A}\right]. \tag{10}$$

Close to the MIT, both F-parameters diverge, $F^S \sim 1/Z^2$ and $F^A \sim 1/Z$, respectively. Hence one would still predict $A \sim -ZW$ near the MIT, namely an instability towards an s-wave superconductor, which seems to be very unlikely. However the single band Hubbard model is pathological. While the uniform magnetic susceptibility is finite across the MIT, implying $F^A \sim 1/Z$, the local susceptibility diverges like $1/Z$. Therefore the quasiparticle interaction in the spin channel must behave in a completely different way depending whether the momentum transfer is zero or is finite, which correspond within DMFT to either uniform or local response, respectively. Since the typical momentum transfer involved in superconductivity is finite rather than zero, one would be tempted to use in the expression (10) for A the analogous of the F-parameters as naively extracted by local susceptibilities

$$\frac{\chi_{local}}{\chi_{local}^{(0)}} = \frac{1}{Z}\frac{1}{1+F_{local}}.$$

This would still imply $F^S_{local} \sim 1/Z^2$, but instead $F^A_{local} = const. < 0$, namely a repulsive $A > 0$ at the MIT. This is obvioulsy a more realistic prediction.

In our model there is in fact no such ambiguity since the uniform and the local susceptibilities behave exactly in the same way, hence both procedures lead to the same conclusion $A \sim -ZW/2$ near the MIT. Moreover the use of *local* Landau parameters might be justified within DMFT where the lattice problem is mapped onto an Anderson impurity model where the coupling to the conduction bath is determined self-consistently by imposing that the conduction-electron local Green's function coincides with the impurity one. At self-consistency one can indeed imagine that a Fermi liquid theory for the effective Anderson impurity model should be valid.

We conclude this overview of the Fermi liquid properties of the Hamiltonian (1) close to the Mott transition, by studying another quantity which sheds some light on the real physical behavior. For our choice of a very small $J_H/U = -0.02$, superconductivity arises when the Mott-Hubbard bands are already well separated from the narrow quasiparticle resonance. This suggests that the effective Anderson impurity model has already flowed in the Kondo regime, where a two-component description might be used (26). Let us consider the probability $P(n)$ for a site to be occupied by n electrons in the ground state. In Fig. 6 we draw $P(n)$ for the superconductor (panels B to D) and for the nearby Mott insulator (panel A), which are quite similar. In spite of the large superconducting gap, there is no evidence of preformed pairs or bipolarons as underlined by the strong steady peaking of $P(n)$ at $n = 2$. We assume the *two-component* form of $P(n)$

$$P(n) = Z\,P^{(SC)}_{qp}(n) + (1-Z)\;P_{ins}(n),$$

where $P_{ins}(n)$ is the probability distribution in the nearby Mott insulator. Since Z, $P(n)$ and $P_{ins}(n)$ are all known, this allows to extract the quasiparicle probability distribution $P^{(SC)}_{qp}(n)$, which is drawn in Fig. 6A'–6D'. It shows oscillations between even and odd n's, as expected for a superconductor, with no relevant variations as a function of U/W upon approaching the Mott transition ($U/W \simeq 0.9$), in agreement with an intermediate coupling regime as implied by $A \sim -W_*$. To check the consistency of the *two-component* form, we also extracted the $P^{(METAL)}_{qp}(n)$ from the (metastable) metallic solution obtained by preventing gauge-symmetry breaking. The results are very similar to the probability distribution of the free Fermi

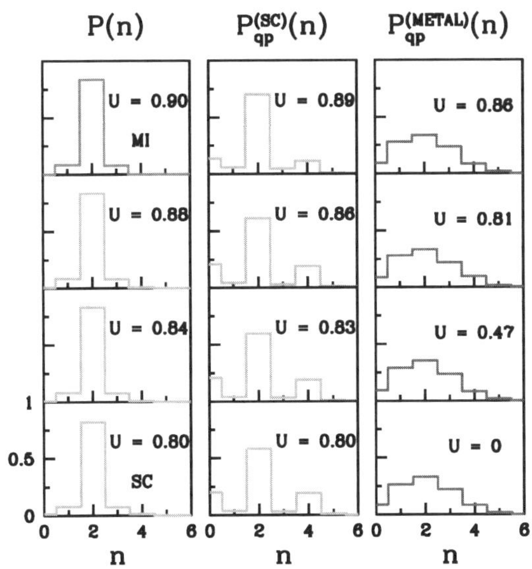

Figure 6. Occupation probabilities for the particles, $P(n)$, and for the quasiparticles in the superconducting and in the metallic phases, $P_{qp}^{(SC)}(n)$ and $P_{qp}^{(METAL)}(n)$, respectively.

gas, $U = J_H = 0$, drawn in Fig. 6D". The agreement is quite remarkable, as $P_{qp}^{(METAL)}(n)$ is but a tiny fraction of order Z of $P(n)$ and there is no free parameter.

These results strongly support to a *two-component* description where fermion-like quasiparticles of small weight Z get strongly paired while floating in a prevailing Mott insulator background. That background slows down quasiparticle motion while taking away their Coulomb repulsion but not their pair attraction. As a remark we notice that the behavior of $P(n)$ is also consistent with our conjecture that the quasiparticle repulsion renormalizes as $U_* = ZU$.

4. Conclusions

We have shown in a simplified model for alkali doped fullerides that strong correlations may enhance superconductivity just close to the Mott transition. The detailed analysis of the results indicates that such effect appears when the pairing occurs in a channel not involving the charge degrees of freedom which are most affected by the repulsion. Then superconductivity is automatically pushed to an intermediate coupling regime where the quasiparticle bandwidth, strongly suppressed by correlations, gets of the order of the attraction, which stays unrenormalized.

In our model (1) this newer pairing mechanism, able not only to avoid U but to take advantage from U, is provided by an inverted Hund's rule mimicking the Jahn-Teller coupling to molecular vibrations. However the general mechanism is quite general and seems valid in other cases too, as for instance in $t - J$ models for cuprates.

Acknowledgements

This work was sponsored through MIUR COFIN, by INFM/G, and by the EU, through contract ERBFMRXCT970155 (FULPROP). We are grateful to P. Nozieres and to G. Santoro for illuminating discussions.

References

1. Based on lectures presented at Hvar, and at Varenna (2002).
2. A. Kaminski *et al.*, Phys. Rev. Lett. **84**, 1788 (2000).
3. P. W. Anderson, Science **235**, 1196 (1987).
4. G. Kotliar and J. Liu, Phys. Rev. B **38**, 5142 (1988).
5. S. Sorella *et al.*, Phys. Rev. Lett. **88**, 117002 (2002).
6. A. Paramehanti, M. Randeria, and N. Trivedi, Phys. Rev. Lett. **87**, 217002 (2001).
7. O. Gunnarsson, Rev. Mod. Phys. **69**, 575 (1997).
8. R.L. Martin and J.P. Ritchie, Phys. Rev. B **48** 4845 (1993);
9. R.W. Lof, M.A. van Veenendaal, B. Koopmans, H.T. Jankman, G.A. Sawatzky, Phys. Rev. Lett. **68**, 3924 (1992).
10. N. Manini, E. Tosatti, and A. Auerbach, Phys. Rev. B **49**, 13008 (1994)
11. O. Zhou *et al.*, Science **255**, 833 (1992).
12. T.T.M. Palstra *et al.*, Solid State Commun. **93**, 327 (1995).
13. N. Manini, G.E. Santoro, A. Dal Corso, and E. Tosatti, Phys. Rev. B **66**, 115107 (2002)
14. O. Zhou *et al.*, Phys. Rev. B **52**, 483 (1995); S. Margadonna, K. Prassides, H. Simoda, Y. Iwasa, and M. Mezouar, Europhys. Lett. **56**, 61 (2001).
15. V. Brouet, H. Alloul, T.N. Le, S. Garaj, and L. Forro, Phys. Rev. Lett. **86**, 4680 (2001).
16. R. Kerkoud *et al.*, J. Phys. Chem. Solids **57**, 143 (1996).
17. M. Capone, M. Fabrizio, P. Giannozzi, and E. Tosatti, Phys. Rev. B **62**, 7619 (2000).
18. C.A. Kuntscher, G.M. Bendele, P.W. Stephens, Phys. Rev. B **55**, R3366 (1997).
19. G. Zimmer *et al.*, in *Physics and Chemistry of Fullerenes and Derivatives*, edited by H. Kuzmany, O. Fink, M. Mehring, and S. Roth (World Scientific, Singapore, 1995).
20. M. Capone, M. Fabrizio, and E. Tosatti, Phys. Rev. Lett. **86**, 5361 (2001); M. Capone, M. Fabrizio, C. Castellani, and E. Tosatti, Science **269**, 2364 (2002).
21. M. Lueders *et. al.*, Phil. Mag. B 82, 1611 (2002).
22. A. Georges, G. Kotliar, W. Krauth, and M.J. Rozenberg, Rev. Mod. Phys. **68**, 13 (1996).
23. M. Fabrizio and E. Tosatti, Phys. Rev. B 55, 13465 (1997).
24. R. Micnas, J. Ranninger, and S. Robaszkiewicz, Rev. Mod. Phys. **62**, 113 (1990).

25. P. Nozières, private communication and preprint (2002, to appear in Jour. of Stat. Phys.).
26. G. Kotliar, Eur. Phys. J. B **11**, 27 (1999), and references therein.

INELASTIC LIGHT SCATTERING AND THE CORRELATED METAL-INSULATOR TRANSITION

Proceedings of the ARW NATO Workshop Hvar, Croatia, October 2002

J.K. FREERICKS
Department of Physics, Georgetown University, Washington, DC 20057, U.S.A.

T.P. DEVEREAUX
Department of Physics, University of Waterloo, Canada

R. BULLA
Theoretische Physik III, Elektronische Korrelationen und Magnetismus, Institut für Physik, Universität Augsburg, D-86135, Augsburg, Germany

Abstract. Inelastic light scattering is an important probe of the two-particle charge excitations in a correlated material. We show how to determine the nonresonant response to inelastic light scattering exactly in the limit of large spatial dimensions by employing dynamical mean field theory. We examine the optical photon case of Raman scattering and the X-ray photon scattering case which exchanges both energy and momentum with the charge excitations. A number of formal details that have not appeared elsewhere are included here.

Key words: Metal-Insulator Transition, Inelastic Light Scattering, Dynamical Mean-Field Theory, Infinite Spatial Dimensions, Exact Solution, Hubbard Model, Falicov-Kimball Model, Electronic Raman Scattering, Charge Dynamics

1. Introduction

In correlated materials, two-particle properties, such as charge fluctuations, may be quite different from single-particle properties (such as the interacting density of states measured in photoemission). In particular, one expects significant renormalization effects of the charge excitations near the correlated metal-insulator transition. There are a number of direct experimental probes of the charge excitations in a material. The most common probe is

A.C. Hewson and V. Zlatić (eds.), Concepts in Electron Correlation, 115–122.

an optical conductivity measurement, which is either performed directly in a transmission geometry in the THz range, or involves a Kramers-Kronig analysis of reflectivity data for higher frequencies. Another important experimental probe involves the inelastic scattering of light from the charge excitations. If optical photons are used, the process is called electronic Raman scattering, and only energy is exchanged with the correlated charge excitations. If X-rays are employed, then both energy and momentum are exchanged with the charge excitations.

The theoretical description of inelastic light scattering is complicated, and has only been solved recently in the limit of large dimensions (1, 2, 3). This solution involves the exact evaluation of the two-particle diagrams for nonresonant inelastic light scattering. Nonresonant scattering implies that there is no feedback effect from the energy of the incident photon, and only the transferred photon energy (and momentum) enters into the scattering response function. While it is well known that many resonant scattering effects exist, and indeed, resonant effects are necessary for some experiments to even be possible, the main qualitative features of the inelastic scattering process are contained in the nonresonant formalism. Furthermore, the computations are significantly more tractable.

2. Formal Development

We employ dynamical mean-field theory (DMFT) to solve the inelastic light scattering problem. In DMFT, the electronic self energy is local, and only the local piece of the irreducible vertex function enters into any physical response function. We describe below how to formulate the inelastic light scattering problem on a hypercubic lattice in $d \to \infty$ dimensions. It turns out that much of the formalism is independent of the explicit form of the Hamiltonian, but the interactions must all be local. For concreteness, we consider the Hubbard Hamiltonian (4)

$$\mathcal{H} = -\frac{t^*}{2\sqrt{d}} \sum_{\langle i,j\rangle,\sigma} (c^\dagger_{i\sigma} c_{j\sigma} + c^\dagger_{j\sigma} c_{i\sigma}) + U \sum_i c^\dagger_{i\uparrow} c_{i\uparrow} c^\dagger_{i\downarrow} c_{i\downarrow}, \tag{1}$$

with rescaled hopping integral t^* (5) (between nearest neighbors—the summation is over nearest-neighbor pairs) and screened Coulomb integral U. The $c^\dagger_{i\sigma}$ ($c_{i\sigma}$) operators create (destroy) an electron with spin σ at site i. The noninteracting bandstructure is $\epsilon(\mathbf{k}) = -\lim_{d\to\infty} \sum_{i=1}^{d} t^* \cos(k_i)/\sqrt{d}$ and the noninteracting DOS becomes $\rho(\epsilon) = \exp(-\epsilon^2)/t^*\sqrt{\pi}$. The inelastic light scattering response function is given by an effective density-density correlation function $S(\mathbf{q},\nu) = -\frac{1}{\pi}[1+n(\nu)]\mathrm{Im}\chi(\mathbf{q},\nu)$ for transferred momentum

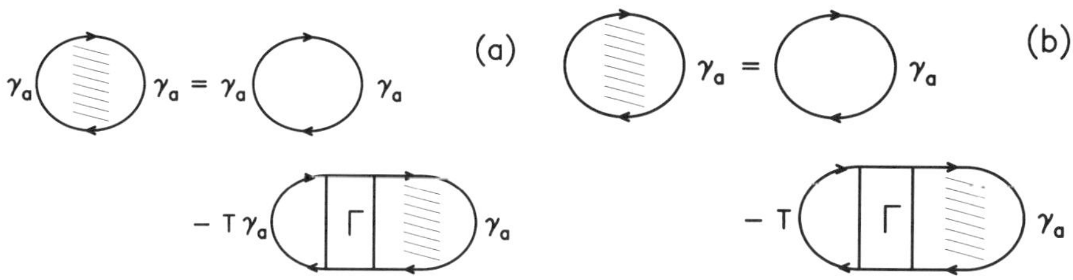

Figure 1. Coupled Dyson equations for the inelastic light scattering density-density correlation functions described by the scattering amplitude γ_a. Panel (a) depicts the Dyson equation for the interacting correlation function, while panel (b) is the supplemental equation needed to solve for the correlation function (the difference in the two equations is the number of γ_a factors). The symbol Γ stands for the local dynamical irreducible charge vertex. In situations where there are no charge vertex corrections (like B_{1g} scattering along the zone-diagonal), the correlation function is simply given by the first (bare-bubble) diagram on the right hand side of panel (a).

$\mathbf{q}$ and energy ν with

$$\chi(\mathbf{q},\nu) = \langle[\tilde{\rho}(\mathbf{q}),\tilde{\rho}(-\mathbf{q})]\rangle_{(\nu)} \tag{2}$$

formed with an "effective" density operator given by

$$\tilde{\rho}(\mathbf{q}) = \sum_{\mathbf{k},\sigma}\gamma_a(\mathbf{k})c^\dagger_\sigma(\mathbf{k}+\mathbf{q/2})c_\sigma(\mathbf{k}-\mathbf{q/2}), \tag{3}$$

where the $c^\dagger_\sigma(\mathbf{k})$ and $c_\sigma(\mathbf{k})$ operators create or destroy an electron with spin σ and momentum $\mathbf{k}$ and $n(\nu) = 1/[1-\exp(\nu/T)]$ is the Bose factor. The strength of the scattering γ_a is determined by the curvature of the band as

$$\gamma_a(\mathbf{k}) = \sum_{\alpha,\beta} e^s_\alpha \frac{\partial^2\epsilon_{\mathbf{k}}}{\partial k_\alpha \partial k_\beta} e^i_\beta. \tag{4}$$

Here $\mathbf{e^{i,s}}$ denote the incident, scattered photon polarization vectors, respectively, and we have chosen units $k_B = c = \hbar = t^* = 1$ and have set the hypercubic lattice constant equal to 1. We can classify the scattering amplitudes by point group symmetry operations. If we choose $\mathbf{e}^i = (1,1,1,...)$ and $\mathbf{e}^s = (1,-1,1,-1,...)$, then we have the B_{1g} sector, while $\mathbf{e}^i = \mathbf{e}^s = (1,1,1,...)$ projects out the A_{1g} sector since the B_{2g} component is identically zero in our model due to the inclusion of only nearest-neighbor hopping (1). Hence $\gamma_{A_{1g}}(\mathbf{q}) = -\epsilon(\mathbf{q})$ and $\gamma_{B_{1g}}(\mathbf{q}) = t^*\sum_{j=1}^\infty \cos\mathbf{q}_j(-1)^j/\sqrt{d}$.

The Dyson equation for the density-density correlation function takes the form given in Figure 1. Note that there are two coupled equations illustrated in Figures 1 (a) and (b); these equations differ by the number of γ_a factors in them. The irreducible vertex function Γ is the dynamical charge vertex which is known explicitly only for the Falicov-Kimball model (6). If the scattering amplitude γ does not have a projection onto the full symmetry of the lattice, then there are no vertex corrections from the local dynamical charge vertex (7). This is the only case that can be analyzed for the Hubbard model.

Let's begin our discussion on the imaginary axis in the B_{1g} sector. If we restrict ourselves to the zone diagonal, then $\mathbf{q} = (q, q, q, ..., q)$. Examining the diagrams in Figure 1 (a), we see that there is a bare response plus a vertex correction term. The vertex correction term, however, must vanish, because the leftmost piece of that diagram is

$$\begin{aligned} 0 &= -T \sum_n \sum_{\mathbf{k}} \lim_{d\to\infty} \sum_{j=1}^{d} \frac{\cos(k_j)(-1)^j}{\sqrt{d}} \frac{1}{i\omega_n + \mu - \Sigma(i\omega_n) - \epsilon(\mathbf{k} - \frac{1}{2}\mathbf{q})} \\ &\times \frac{1}{i\omega_n + i\nu_l + \mu - \Sigma(i\omega_n + i\nu_l) - \epsilon(\mathbf{k} + \frac{1}{2}\mathbf{q})} \end{aligned} \tag{5}$$

for "energy" transfer $i\nu_l = 2i\pi Tl$ and momentum transfer $\mathbf{q}$ [in Eq. (5) $i\omega_n = i\pi T(2n+1)$ is the fermionic Matsubara frequency and μ is the chemical potential]. It vanishes, because each term indexed by j is equal in magnitude, but opposite in sign, so the overall summation is equal to zero. Hence the B_{1g} response on the zone diagonal (including the Raman response) is given by the bare bubble.

The bare bubble on the zone diagonal is simple to calculate directly (we shift $\mathbf{k} \to \mathbf{k} + \mathbf{q}/2$):

$$\begin{aligned} \chi_0(\mathbf{q}, i\nu_l) &= -T \sum_n \sum_{\mathbf{k}} \lim_{d\to\infty} \sum_{i,j=1}^{d} \frac{\cos(k_i + \frac{1}{2}q)\cos(k_j + \frac{1}{2}q)(-1)^{i+j}}{d} \\ &\times \frac{1}{i\omega_n + \mu - \Sigma(i\omega_n) - \epsilon(\mathbf{k})} \\ &\times \frac{1}{i\omega_n + i\nu_l + \mu - \Sigma(i\omega_n + i\nu_l) - \epsilon(\mathbf{k} + \mathbf{q})}. \end{aligned} \tag{6}$$

Now, the terms with $i \neq j$ are all equal in magnitude, but there are as many positive as negative, so they vanish—only the terms with $i = j$ survive. If we assume that n and $n + l$ are both larger than 0, and define $Z_n = i\omega_n + \mu - \Sigma(i\omega_n)$, then we can rewrite the fractions as integrals of exponentials

$$\begin{aligned} \chi_0(\mathbf{q}, i\nu_l) &= T \sum_n \sum_{\mathbf{k}} \lim_{d\to\infty} \sum_{j=1}^{d} \frac{\cos^2(k_j + \frac{1}{2}q)}{d} \\ \times &\int_0^\infty d\lambda \int_0^\infty d\lambda' \exp[i\lambda Z_n + i\lambda' Z_{n+l} - i\lambda\epsilon(\mathbf{k}) - i\lambda'\epsilon(\mathbf{k}+\mathbf{q})]. \end{aligned} \tag{7}$$

To evaluate this integral, we first expand the functions $\epsilon(\mathbf{k})$ and $\epsilon(\mathbf{k} + \mathbf{q})$ in terms of the Cartesian momentum components. Then it is obvious that each j term is equal in magnitude, so the sum over j is trivial. The next step is to expand each exponential factor that has a $1/\sqrt{d}$ prefactor in a Taylor series expansion, and keep the lowest nonvanishing terms in the multiple integrals (8). The multiple integral over momentum then becomes

$$\begin{aligned}\lim_{d\to\infty} & \int dk_1 \int dk_2 ... \int dk_d \cos^2(k_1 + \frac{q}{2}) \\ & \prod_{j=1}^{d} \Big(1 + \frac{i}{\sqrt{d}}\lambda \cos k_j + \frac{i}{\sqrt{d}}\lambda \cos(k_j + q) - \frac{\lambda^2}{2d}\cos^2 k_j \\ - & \frac{\lambda\lambda'}{d}\cos k_j \cos(k_j + q) - \frac{\lambda'^2}{2d}\cos^2(k_j + q) + ...\Big). \end{aligned} \tag{8}$$

Here we have chosen the extra prefactor to lie in the 1 direction. Each integral over k_j, except the first, is equal to each other and equal to $1 - (\lambda^2 + 2\lambda\lambda' X(\mathbf{q}) + \lambda'^2)/4d + ...$. The first integral becomes $\frac{1}{2}[1 - (\lambda^2 + 2\lambda\lambda' X(\mathbf{q}) + \lambda'^2)/4d - (\lambda\lambda' + \lambda'^2 \cos q)\sin^2 q/8d + ...]$. Here we use the notation $X(\mathbf{q}) = \lim_{d\to\infty} \sum_{j=1}^{d} \cos q_j/d$ ($= \cos q$ for a zone-diagonal momentum). The next step is to rewrite each factor as an exponential, and then take the infinite product. The result for the integral over $\mathbf{k}$ is

$$\frac{1}{2}\exp\left[-\frac{\lambda^2 + 2\lambda\lambda' X(\mathbf{q}) + \lambda'^2}{4}\right] \tag{9}$$

since the terms proportional to $\sin^2 q$ coming from the 1-direction are just a $1/d$ correction. The end result for the susceptibility is then

$$\chi_0(\mathbf{q}, i\nu_l) = -\frac{1}{2}\int d\epsilon \rho(\epsilon) \frac{1}{Z_n - \epsilon}\frac{1}{\sqrt{1 - X^2}} F_\infty\left(\frac{Z_{n+l} - X\epsilon}{\sqrt{1 - X^2}}\right), \tag{10}$$

with F_∞ the Hilbert transformation of the DOS: $F_\infty(z) = \int d\epsilon \rho(\epsilon)/(z - \epsilon)$.

The analytic continuation of this expression is straightforward, and produces the final result for the B_{1g} response

$$\begin{aligned}\chi_{B_{1g}}(\mathbf{q}, \nu) &= \frac{i}{4\pi}\int_{-\infty}^{\infty} d\omega \{ f(\omega)\chi_0(\omega; X, \nu) - f(\omega + \nu)\chi_0^*(\omega; X, \nu) \\ &- [f(\omega) - f(\omega + \nu)]\tilde{\chi}_0(\omega; X, \nu)\} \end{aligned} \tag{11}$$

with

$$\begin{aligned}\chi_0(\omega; X, \nu) &= -\int_{-\infty}^{\infty} d\epsilon \rho(\epsilon) \frac{1}{\omega + \mu - \Sigma(\omega) - \epsilon}\frac{1}{\sqrt{1 - X^2}} \\ &\times F_\infty\left(\frac{\omega + \nu + \mu - \Sigma(\omega + \nu) - X\epsilon}{\sqrt{1 - X^2}}\right), \end{aligned} \tag{12}$$

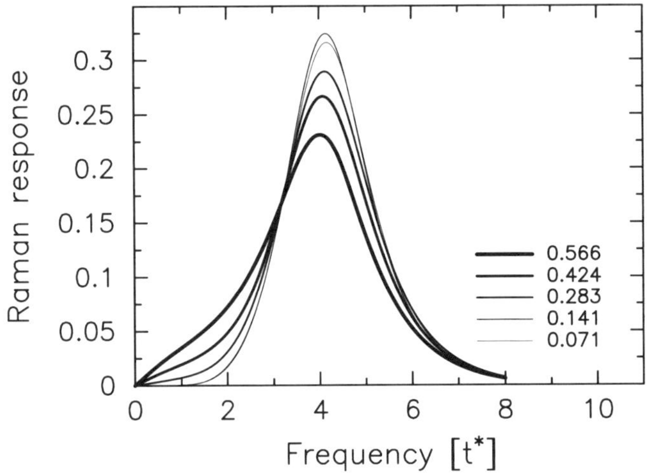

Figure 2. Nonresonant B_{1g} Raman response ($X = 1$) for different temperatures at $U = 4.24$ and half filling. The numbers in the legends label the temperature.

and

$$\tilde{\chi}_0(\omega; X, \nu) = -\int_{-\infty}^{\infty} d\epsilon \rho(\epsilon) \frac{1}{\omega + \mu - \Sigma^*(\omega) - \epsilon} \frac{1}{\sqrt{1 - X^2}} \times F_\infty \left(\frac{\omega + \nu + \mu - \Sigma(\omega + \nu) - X\epsilon}{\sqrt{1 - X^2}} \right). \tag{13}$$

Here $f(\omega) = 1/[1 + \exp(\omega/T)]$ is the Fermi factor.

A similar, but more complicated analysis can be performed for the A_{1g} response, or the B_{1g} response off of the zone diagonal, but we don't have enough space to report those results here, and they cannot be analyzed numerically for the Hubbard model.

3. Results

We employ a numerical renormalization group analysis to determine the self energy and Green's function of the Hubbard model on the real axis (9). We begin by showing Raman scattering results at half filling for $U = 4.24$ and a variety of temperatures in Figure 2. At this value of U, the system is a correlated insulator for all temperatures. The results display all of the behavior seen on correlated insulators like FeSi (10), or SmB_6 (11), or the high-temperature superconductors (12). In particular, we see a charge-transfer peak at high-energy and the onset of low-energy spectral weight at a low (but nonzero) temperature. The curves also cross at the so-called isosbestic point (near $\nu \approx 3.2$).

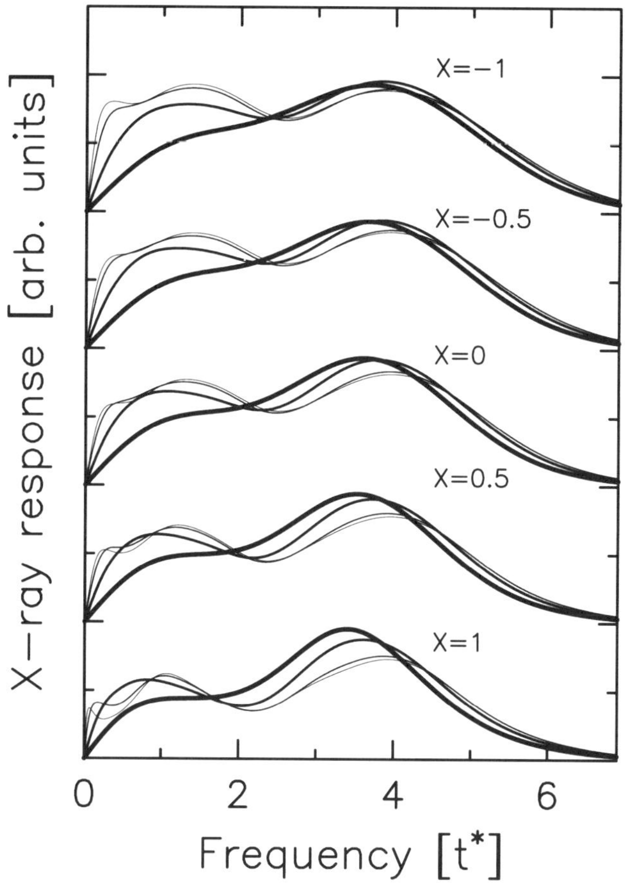

Figure 3. Nonresonant B_{1g} inelastic X-ray response for different temperatures at $U = 3.54$ and $\rho_e = 0.9$. Five values of the transferred photon momentum are plotted, each shifted by an appropriate amount, and running from the zone center ($X = 1$) to the zone boundary ($X = -1$) along the zone diagonal. The temperature decreases with decreasing thickness of the lines and ranges from 0.503 to 0.114 to 0.042 to 0.026.

We show the inelastic X-ray scattering at four different temperatures for a slightly smaller value of U and at $n = 0.9$ in Figure 3. Here the behavior is quite different because the system is metallic for all temperatures. In particular, we see a Fermi-liquid peak form and evolve toward zero frequency as T is lowered for momentum transfer near the zone diagonal. But for finite momentum transfer, the peak never fully evolves. In addition, there is significant "mid-IR" spectral weight occuring at energies below the charge-transfer peak but not corresponding to the Fermi peak.

4. Conclusions

We have presented a number of new results for the inelastic scattering of light with correlated materials. On the insulating side of the metal-insulator transition, Raman scattering results agree well with experiments that have been performed on a wide variety of different materials. We find a number of interesting features for correlated metals as well, and it would be interesting to experimentally measure both Raman scattering and inelastic X-ray scattering for these materials. We expect this behavior might be able to be seen in a variety of different heavy-fermion compounds.

Acknowledgements

J.K.F. acknowledges support of the National Science Foundation under grants DMR-9973225 and DMR-0210717. T.P.D. acknowledges support from the National Research Council of Canada. R.B. acknowledges support by the Deutsche Forschungsgemeinschaft, through Sonderforschungsbereich 484.

References

1. J.K. Freericks and T.P. Devereaux, J. Cond. Phys. (Ukraine) **4**, 149 (2001); Phys. Rev. B **64**, 125110 (2001).
2. J.K. Freericks, T.P. Devereaux, and R. Bulla, Acta Phys. Pol. B **32**, 3219 (2001); Phys. Rev. B **64**, 233114 (2001); unpublished.
3. T.P. Devereaux, G. E. D. McCormack, and J.K. Freericks, *submitted to* Phys. Rev. Lett. (cond-mat/0208049).
4. J.C. Hubbard, Proc. Royal Soc. London, Ser. A **276**, 238 (1963).
5. W. Metzner and D. Vollhardt, Phys. Rev. Lett. **62**, 324 (1989).
6. A. M. Shvaika, Physica C **341–348**, 177 (2000); J. K. Freericks and P. Miller, Phys. Rev. B **62**, 10022 (2000); A. M. Shvaika, J. Phys. Stud. **5**, 349 (2002).
7. A. Khurana, Phys. Rev. Lett., **64**, 1990 (1990).
8. E. Müller-Hartmann, Int. J. Phys. B **3**, 2169 (1989).
9. R. Bulla, Phys. Rev. Lett. **83**, 136 (1999); R. Bulla, T. A. Costi, and D. Vollhardt, Phys. Rev. B **64**, 045103 (2001).
10. P. Nyhus, S.L. Cooper, and Z. Fisk, Phys. Rev. B **51**, R15626 (1995).
11. P. Nyhus, S.L. Cooper, Z. Fisk, and J. Sarrao, Phys. Rev. B **52**, R14308 (1995); Phys. Rev. B **55**, 12488 (1997).
12. X. K. Chen, J.G. Naeini, K.C. Hewitt, J.C. Irwin, R. Liang, W.N. Hardy, Phys. Rev. B **56**, R513 (1997); J.G. Naeini, X.K. Chen, J.C. Irwin, M. Okuya, T. Kimura, and K. Kishio, Phys. Rev. B **59**, 9642 (1999); M. Rübhausen, O.A. Hammerstein, A. Bock, U. Merkt, C.T. Rieck, P. Guptasarma, D.G. Hinks, M.V. Klein, Phys. Rev. Lett. **82**, 5349 (1999); S. Sugai and T. Hosokawa, Phys. Rev. Lett. **95**, 1112 (2000); M. Opel, R. Nemetschek, C. Hoffmann, R. Philipp, P.F. Müller, R. Hackl, I. Tütto, A. Erb, B. Revaz, E. Walker, H. Berger, and L. Forró, Phys. Rev. B **61**, 9752 (2000); F. Venturini, M. Opel, T. P. Devereaux, J. K. Freericks, I. Tüttő, B. Revaz, E. Walker, H. Berger, L. Forró and R. Hackl, Phys. Rev. Lett. **89**, 107003 (2002).

ORBITAL PHYSICS VERSUS SPIN PHYSICS: THE ORBITAL-HUBBARD MODEL

Proceedings of the ARW NATO Workshop Hvar, Croatia, October 2002

LOUIS FELIX FEINER
Institute for Theoretical Physics, Utrecht University, Leuvenlaan 4, NL-3584 CC Utrecht, The Netherlands
Philips Research Laboratories, Prof. Holstlaan 4, NL-5656 AA Eindhoven, The Netherlands

ANDRZEJ M. OLEŚ
Marian Smoluchowski Institute of Physics, Jagellonian University, Reymonta 4, PL-30059 Kraków, Poland
Max-Planck-Institut für Festkörperforschung, Heisenbergstrasse 1, D-70569 Stuttgart, Germany

Abstract. To elucidate the similarities and differences between orbital and spin degrees of freedom, we analyze an orbital-Hubbard model with two orbital flavors, corresponding to pseudospin 1/2, and contrast its behavior with that of the familiar (spin-1/2) Hubbard model. The orbital-Hubbard model describes a partly filled spin-polarized e_g band on a cubic lattice, as occurs in ferromagnetic manganites.

We demonstrate that the absence of SU(2) invariance in orbital space has important implications — superexchange contributes in all orbital ordered states, the Nagaoka theorem does not apply, and the kinetic energy is enhanced as compared with the spin case. As a result orbital-ordered states are destabilized by doping, and instead a strongly correlated *orbital liquid* with disordered orbitals is realized.

Key words: Orbital-Hubbard Model, Hubbard Model, Superexchange, Orbital Liquid, Orbital Ordering, SU(2) Symmetry, Nagaoka Theorem, Gutzwiller Factor

Recently there has been renewed interest in orbital degrees of freedom in Mott insulators (1). In undoped perovskites such as $KCuF_3$, $LaMnO_3$, $LaTiO_3$, and $LaVO_3$, strong on-site Coulomb repulsion U eliminates charge fluctuations and replaces them by effective low-energy interactions of super-exchange (SE) type. When the electrons occupy partly-filled degenerate e_g or t_{2g} orbitals, it is necessary to consider the orbital degrees of freedom on equal footing with the electron spins (2, 3). The SE interactions are there-

A.C. Hewson and V. Zlatić (eds.), Concepts in Electron Correlation, 123–131.

fore strongly frustrated — the quantum fluctuations are then enhanced, and might even lead to a *spin liquid* state in $LiNiO_2$ (4). Another possibility, that an *orbital liquid* (OL) is stabilized and coexists with long-range spin order, was pointed out recently for t_{2g} systems (5). In contrast, the SE in undoped e_g systems, $KCuF_3$ (2) and $LaMnO_3$ (6), favors alternating orbital (AO) order which coexists with the antiferromagnetic spin (AS) order.

In this paper we study a generic model of correlated electrons in a spin-polarized e_g band, with two orbital flavors described by a pseudospin 1/2 in the orbital Hilbert space, and consider its relation to the (spin) Hubbard model for electrons with spin $s = 1/2$. We investigate: (i) in what respect long-range order in *orbital systems* is different from that in *spin systems*, and (ii) whether the orbitals are *ordered* or rather form a *disordered* OL. These questions are both of fundamental nature and of immediate interest for understanding the metallic ferromagnetic manganites (3, 7). Our purpose here is to elucidate the physical mechanisms which operate in the e_g band by contrasting them with those known in spin systems.

So we consider spinless e_g electrons [in a ferromagnetic spin (FS) state] on a cubic lattice with kinetic energy

$$H_t = -t \sum_{\alpha} \sum_{\langle ij \rangle \| \alpha} c^{\dagger}_{i\zeta_\alpha} c_{j\zeta_\alpha}, \tag{1}$$

where hopping t between sites i and j occurs only for the directional orbitals $|\zeta_\alpha\rangle$ oriented along the bond $\langle ij \rangle$, i.e., $|\zeta_\alpha\rangle \propto 3x^2 - r^2$, $3y^2 - r^2$, and $3z^2 - r^2$, when $\langle ij \rangle$ is along the cubic axis $\alpha = a$, b, and c, respectively. To describe the local electron interactions one needs to choose an orthogonal basis for the two orbital flavors. One possibility is $|x\rangle \equiv x^2 - y^2$ and $|z\rangle \equiv (3z^2 - r^2)/\sqrt{3}$, called *real orbitals*, whereupon the local Coulomb interaction becomes $U n_{ix} n_{iz}$. The disadvantage is that the expression for the kinetic energy then takes a different form for each axis (8). We thus prefer to use instead the basis of *complex orbitals* $|+\rangle = \frac{1}{\sqrt{2}}(|z\rangle - i|x\rangle)$ and $|-\rangle = \frac{1}{\sqrt{2}}(|z\rangle + i|x\rangle)$, corresponding to "up" and "down" pseudospin flavors, and write the e_g orbital-Hubbard model in the form

$$\mathcal{H} = -\frac{1}{2} \sum_{\alpha} \sum_{\langle ij \rangle \| \alpha} \Big[t\Big(c^{\dagger}_{i+} c_{j+} + c^{\dagger}_{i-} c_{j-}\Big) + \gamma t\Big(e^{-i\chi_\alpha} c^{\dagger}_{i+} c_{j-} + e^{+i\chi_\alpha} c^{\dagger}_{i-} c_{j+}\Big)\Big] + U \sum_i n_{i+} n_{i-}, \tag{2}$$

with $\chi_{a,b} = \pm 2\pi/3$, $\chi_c = 0$, and $\gamma = 1$. The phase factors $e^{\pm i\chi_\alpha}$ are characteristic of the orbital problem – the orbitals have an actual shape in real space so that each hopping process depends on the bond direction.

The representation in Eq. (2) has several advantages: (i) It displays manifestly the cubic symmetry, as the transformation $c_{i\pm}^\dagger \to c_{i\pm}^\dagger e^{\pm 2i\pi/3}$, which corresponds to cyclic permutation of the cubic axes, leaves the Hamiltonian Eq. (2) invariant. (ii) It exhibits clearly the difference between the spin case and the orbital casc, and in fact allowed us to introduce the parameter γ by which one can turn the e_g-band orbital-Hubbard model ($\gamma = 1$) into what looks formally like a spin-Hubbard model ($\gamma = 0$). (iii) It shows explicitly that rotational SU(2) symmetry for the pseudospins is absent (2). The total pseudospin operator $\mathcal{T}^z = \sum_i T_i^z$, with $T_i^z = \frac{1}{2}(n_{i+} - n_{i-})$, is conserved only at $\gamma = 0$ (i.e., $[\mathcal{T}^z, \mathcal{H}] = 0$), while the terms $\propto \gamma$ commute instead with the staggered pseudospin operator $\mathcal{T}_{\mathbf{Q}}^z = \sum_i \exp(i\mathbf{Q} \cdot \mathbf{R}_i) T_i^z$, where $\mathbf{Q} = (\pi, \pi, \pi)$.

Because the electrons interact by the local Coulomb interaction U, they are prone to instabilities towards orbital order, which we compare with magnetic instabilities in the spin case (9). At half-filling ($n = 1$) the simplest possibility to reduce the interaction energy $\propto U$ would be to polarize the system completely into *ferro orbital* (FO) states, $|\Phi_{\rm FO}\rangle = \prod_i c_i^\dagger(\psi, \theta)|0\rangle$ (uniform order (10)). As in the spin case, another possibility is AO order, $|\Phi_{\rm AO}\rangle = \prod_{i \in A} c_i^\dagger(\psi_A, \theta_A) \prod_{j \in B} c_j^\dagger(\psi_B, \theta_B)|0\rangle$, i.e., with orbitals alternating between two sublattices A and B. If the band is partly filled ($n < 1$), such states must involve a coherent mixture of occupied and empty sites.

It is an important feature of the $|+\rangle$-polarized (FO+) and $|+\rangle/|-\rangle$-staggered (AO$\pm$) *complex states*, that they retain cubic symmetry (11). By contrast, the FO and AO *real states*, such as $|x\rangle$-polarized (FOx), $|z\rangle$-polarized (FOz), or $(|x\rangle + |z\rangle)/(|x\rangle - |z\rangle)$-staggered (AOxz), with either quasi-one-dimensional (FOz and AOxz) or two-dimensional (FOx) dispersion, break cubic symmetry (and are thus favored in lower dimensional systems, e.g. FOx in a 2D square lattice (12)). This nonequivalence between real and complex states is a manifestation of the *broken* SU(2) symmetry. We focus here on the orbital-ordered states with complex orbitals (13), which occur in Hartree-Fock (HF) approximation for large enough U (11).

In the case of the FO+ state, the electron bands split above a critical value of U, and lead (14) to a finite order parameter $T^z = \langle T_i^z \rangle \neq 0$, but the mechanism of the instability is different from that known in the spin case. At $\gamma = 0$ one recovers the Stoner criterion $U_o N(E_F) = 1$ for the existence of FS order, with the FS states becoming saturated only at larger U [see Fig. 1(a)]. By contrast, at $\gamma > 0$ the FO states appear as a global property of the band rather than as a Fermi surface instability [Fig. 1(b)]. For large U ($\gtrsim 6t$) a gap opens, only the lower band $\varepsilon_{\mathbf{k}}^{\rm FO} = -tA_{\mathbf{k}} + U(\frac{1}{2}n - T^z E_{\mathbf{k}})$

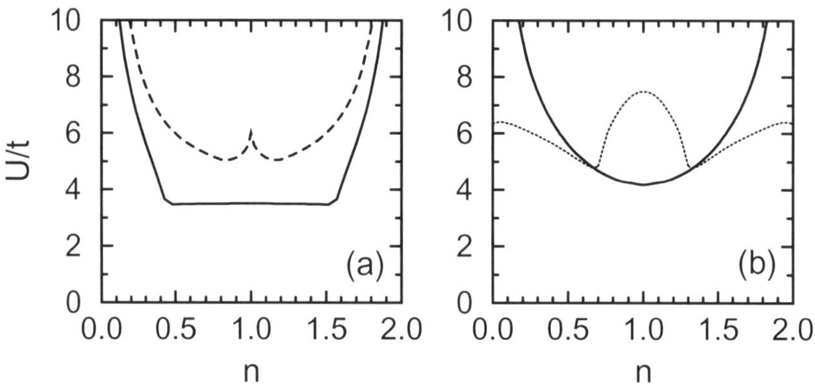

Figure 1. HF instabilities towards FO states (full lines) for: (a) $\gamma = 0$, and (b) $\gamma = 1$ [dotted line shows $N^{-1}(\omega)$]. Saturated FS states occur only in (a) above the dashed line.

is occupied, and

$$T^z = \frac{1}{2}\frac{1}{N}\sum_{\mathbf{k}}\frac{1}{E_{\mathbf{k}}}, \qquad E_{\mathbf{k}} = \Big[1 + \Big(\frac{\gamma t}{UT^z}\Big)^2 B_{\mathbf{k}}^2\Big]^{1/2}, \tag{3}$$

where $A_{\mathbf{k}} = c_x + c_y + c_z$, $B_{\mathbf{k}}^2 = c_x^2 + c_y^2 + c_z^2 - (c_xc_y + c_yc_z + c_zc_x)$, with $c_x = \cos k_x$, *etcetera*, and the sum is over the occupied part of the Brillouin zone (BZ). Unlike in the spin case, $T^z = n/2$ only at $U = \infty$, since the saturated FO+ state is not an eigenstate of $\mathcal{H}$. Thus the FO+ state resembles the AS phase in the spin-Hubbard model.

In the AO phase one has four subbands in the reduced BZ. For large U only the lowest two with dispersion $\varepsilon_{\mathbf{k},\pm}^{\rm AO} = \pm\gamma t B_{\mathbf{k}} + U(\frac{1}{2}n - T^z F_{\mathbf{k}})$ are occupied, and the order parameter, $T^z = \langle T_{i\in A}^z\rangle = -\langle T_{j\in B}^z\rangle$, is given by

$$T^z = \frac{1}{2}\frac{1}{N}\sum_{\mathbf{k}}\frac{1}{F_{\mathbf{k}}}, \qquad F_{\mathbf{k}} = \Big[1 + \Big(\frac{t}{UT^z}\Big)^2 A_{\mathbf{k}}^2\Big]^{1/2}, \tag{4}$$

rather similar to the FO case Eqs. (3), but with the interchange $A_{\mathbf{k}} \leftrightarrow \pm\gamma B_{\mathbf{k}}$. The reason is readily recognized from Eq. (2): for FO order, the diagonal hopping $\propto c_{i\pm}^\dagger c_{j\pm}$ that gives $A_{\mathbf{k}}$ is order-preserving, while the off-diagonal terms $\propto \gamma$ are order-perturbing and reduce T^z. For AO order this is reversed: the off-diagonal hopping $\propto c_{i\pm}^\dagger c_{j\mp}$ that gives $B_{\mathbf{k}}$ is compatible with the order, while the diagonal one disturbs it. The similarity between the FO and AO states at $\gamma \simeq 1$ becomes even more transparent at large U, where, at $x = 1 - n > 0$,

$$T_{\rm FO}^z = \frac{1}{2}\Big\{(1-x) - \frac{3}{(1-x)^2}\Big(\frac{\gamma t}{U}\Big)^2\Big\}, \tag{5}$$

$$T_{\rm AO}^z = \frac{1}{2}\Big\{(1-x) - \frac{3-2x}{(1-x)^2}\Big(\frac{t}{U}\Big)^2\Big\}. \tag{6}$$

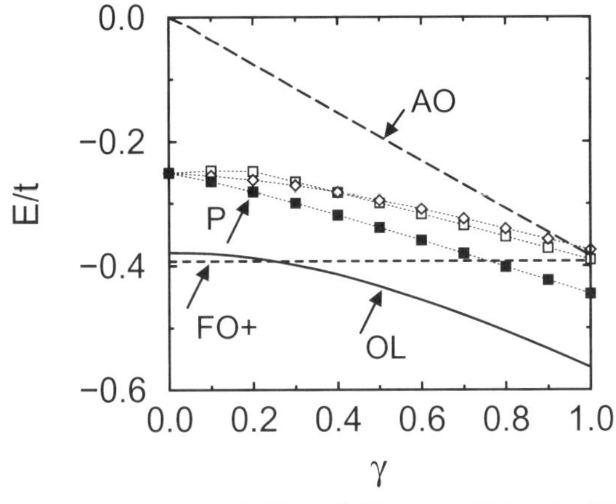

Figure 2. Energies per site at $n = 0.75$ and $U = \infty$: E in the KR approach for OL (solid line), FO+ (dashed line), AO$\pm$ state (long-dashed line), and $E_\square$ for: the ground state (filled squares) and excited states (empty symbols) of a single plaquette (P).

Note that a SE contribution $\propto (\gamma t)^2/U$ appears also in the FO+ state, because the off-diagonal hopping permits virtual charge fluctuations. In the genuine orbital case ($\gamma = 1$) the reduction of the order parameter by SE is the same for FO and AO at $x = 0$, but at $x \gtrsim 0$ it is slightly larger for the FO phase, and surprisingly the energy per site of the FO phase is *lower* than that of the AO phase. Thus near half-filling FO+ order is more stable than AO$\pm$ order at any U, because the FO phase not only gains more kinetic energy $\propto -3tx$ than the AO phase $\propto -2t\gamma x$ due to the difference in band edge between $-tA_{\mathbf{k}}$ and $-\gamma t B_{\mathbf{k}}$, but also has lower SE energy (15). Instead, AO$\pm$ order dominates at larger doping $x \gtrsim 0.27$ (11, 16), as a consequence of its peculiar density of states with large weight close to the band edges. Note that this is *opposite* to the spin case ($\gamma = 0$), where the Néel (AS) state has lower energy near $n = 1$ and the FS state takes over above a critical doping $x_c \simeq t/2U$.

Another striking difference is that the Nagaoka theorem, stating that at $U = \infty$ a single hole in the half-filled Hubbard model gives a spin-polarized (FS) ground state (17), *does not hold* for the orbital-Hubbard model at $\gamma > 0$, because it requires conservation of $\mathcal{T}^z$ [see, e.g., the proof in Ref. (17)]. Thus polarized states are harder to stabilize, as is illustrated explicitly by an exact calculation for a four-site plaquette with three electrons. Whereas at $\gamma = 0$ the ground state is fourfold degenerate, corresponding to maximum spin $S = \frac{3}{2}$ as required by the Nagaoka theorem, at $\gamma > 0$ it splits into a nondegenerate ground state and three excited states (Fig. 2), none of which can be classified by a pseudospin quantum number. One notes the large energy gain in the ground state (from $E_\square = -0.25t$ at $\gamma = 0$ down to $E_\square \simeq -0.44t$ at $\gamma = 1$) when the orbitals get disordered and full advantage is taken of the pseudospin non-conserving hopping.

We will now argue that indeed orbital (FO or AO) order is not robust at $\gamma > 0$ and gets replaced by a disordered (OL) phase, if one goes *beyond* the HF approximation and includes electron correlation effects in the disordered phase as well. We consider specifically the $U = \infty$ limit, where the OL competes with fully saturated FO (5) and AO (6) states. Needing a reliable variational method to calculate the correlation energy, we adapted the slave-boson approach introduced by Kotliar and Ruckenstein (KR) for the spin-Hubbard model (19) to the orbital case. The introduction of slave bosons ($b_{i+}^\dagger$ and $b_{i-}^\dagger$ for occupied, $e_i^\dagger$ for empty sites) and pseudofermions ($f_{i+}^\dagger$ and $f_{i-}^\dagger$) was done for the *complex* $\{|+\rangle, |-\rangle\}$ orbitals, which is essential to arrive at a gauge (cubic) invariant formulation (20). Treating the bosons in mean-field approximation yields

$$\begin{aligned}\mathcal{H}_{U=\infty} &= -\sum_{i\lambda}\mu_{i\lambda}n_{i\lambda} - \frac{1}{2}t\sum_{\langle ij\rangle}\Big[q_+ f_{i+}^\dagger f_{j+} + q_- f_{i-}^\dagger f_{j-} \\ &+ \gamma\sqrt{q_+q_-}\Big(e^{+i\chi_\alpha} f_{i+}^\dagger f_{j-} + e^{-i\chi_\alpha} f_{i-}^\dagger f_{j+}\Big)\Big], \qquad (7)\end{aligned}$$

with $n_{i\lambda} = f_{i\lambda}^\dagger f_{i\lambda}$, $\sqrt{q_\pm} = [x/(1-\langle n_{i\pm}\rangle)]^{1/2}$, and the local constraints (19) implemented by means of Lagrange multipliers $\{\mu_{i+}, \mu_{i-}\}$. The self-consistent solution corresponds to an *OL state* with $\langle n_{i+}\rangle = \langle n_{i-}\rangle = \frac{1}{2}(1-x)$ and Gutzwiller renormalization factor $q(x) = q_\pm(x) = 2x/(1+x)$ (21). The fermion bands, $\varepsilon_{\mathbf{k},\pm}^{\mathrm{OL}} = -tq(x)[A_{\mathbf{k}} \pm \gamma B_{\mathbf{k}}]$, interpolate correctly between the uncorrelated ($x \simeq 1$) and Mott insulator ($x = 0$) limits. Since the OL state is incoherent (it is described by the density matrix $\hat{\rho}_i = \frac{1}{2}\mathbf{1}_i$ at every site), it is SU(2) symmetric: random complex or random real orbitals are equivalent, and indeed the same *correlated disordered* OL state is obtained using real orbitals (22).

The ordered states can be obtained too within the present KR slave-boson formalism by a suitable choice of the Lagrange multipliers, e.g. $\mu_+ = 0$, $\mu_- = -\infty$ in Eq. (7) gives the FO+ state. Such states do not experience any band narrowing, as double occupancy is eliminated at $U = \infty$, and the correlation energy vanishes (23). As a result, only the $\varepsilon_{\mathbf{k}}^{\mathrm{FO}} = -tA_{\mathbf{k}}$ band ($\varepsilon_{\mathbf{k},\pm}^{\mathrm{AO}} = \pm\gamma tB_{\mathbf{k}}$ bands) is (are) partly filled in the FO+ (AO$\pm$) state. Therefore, the bands in the OL state represent formally a (renormalized) superposition of the FO and AO bands.

It is instructive to consider the variation with γ of the total energy E of ordered and disordered states at fixed doping (Fig. 2). In the spin model ($\gamma = 0$) the FS phase has somewhat lower energy than the disordered state close to half-filling (24). When γ is increased, E_{FO} does not change, whereas $E_{\mathrm{AO}\pm}$ decreases from zero $\propto \gamma$, but at $x = 0.25$ still does not surpass the FO+ state for $\gamma = 1$. However, in spite of the band narrowing $\propto q(x)$,

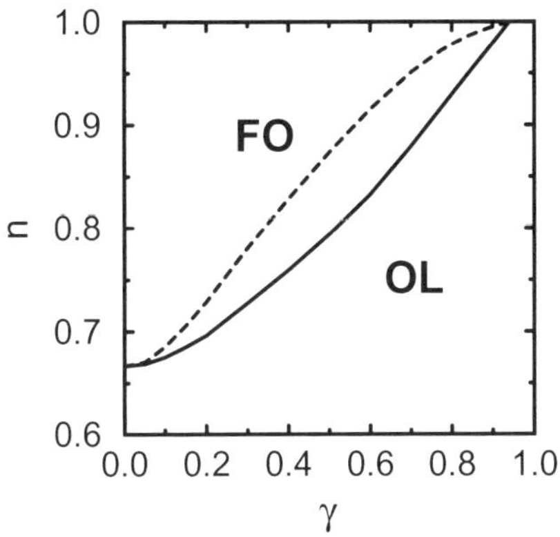

Figure 3. Phase diagram as function of γ: OL versus real FOx(z) (full line) and complex FO+ (dashed line) states at $U = \infty$.

considerably more (kinetic) energy is gained in the OL state, because both hopping channels contribute (25).

Finally, we compare at $U = \infty$ the energies of all states, both complex and real, varying n and γ. One finds that AO states are never stable, while FO states are stable only at small γ (Fig. 3). At $\gamma = 0$ the FO+ and FOx(z) states are degenerate, but at any $\gamma > 0$ the real phases have lower energy, with FOz (FOx) being more stable at $n < 0.71$ ($n > 0.71$). The range of FO order shrinks gradually with increasing γ, and *above* $\gamma \simeq 0.94$ *the OL phase is stable in the entire range of* n. At finite U the kinetic energy will become more dominant and favor disorder even more, except near $n \simeq 1$ where SE stabilizes real-orbital AO order (2, 3, 4, 6, 16). We thus argue that for the e_g orbital-Hubbard model ($\gamma = 1$) doping induces a crossover to the OL state *at any* U, supporting earlier conjectures that such a disordered state is realized (21, 26). Indeed, the disordered OL state provides a natural explanation why the magnons in the ferromagnetic metallic state are isotropic. As we have shown elsewhere (27), the stiffness constant (determined mainly by the double exchange) is proportional to the Gutzwiller band narrowing factor $q(x)$ and increases with hole doping x — thus it measures the kinetic energy of strongly correlated e_g electrons released by doping.

In conclusion, the Nagaoka theorem does not apply to the orbital-Hubbard model of correlated e_g electrons, and ordered states are harder to realize than in the spin case. This is manifest in the inverted stability of the ordered phases with complex orbitals, with ferro (staggered) orbital order favored at small (large) doping. Yet, the exciting suggestion that such complex-orbital ordered states could be stable at finite doping (11) is not confirmed, because of the *inherent tendency of* e_g *systems towards orbital disorder* due to the enhancement of the kinetic energy when SU(2) symmetry is absent.

Acknowledgements

We thank P. Horsch, G. Khaliullin, D. I. Khomskii, P. Wölfle, and particularly K. Rościszewski for valuable discussions. This work was supported by the Committee of Scientific Research (KBN) Project No. 5 P03B 055 20.

References

1. Y. Tokura and N. Nagaosa, Science **288**, 462 (2000).
2. K. I. Kugel and D. I. Khomskii, Usp. Fiz. Nauk **136**, 621 (1982) [Sov. Phys. Usp. **25**, 231 (1982)]; A. M. Oleś, L. F. Feiner, and J. Zaanen, Phys. Rev. B **61**, 6257 (2000).
3. A. M. Oleś, Acta Phys. Polon. B **32**, 3303 (2001).
4. L. F. Feiner, A. M. Oleś, and J. Zaanen, Phys. Rev. Lett. **78**, 2799 (1997).
5. G. Khaliullin and S. Maekawa, Phys. Rev. Lett. **85**, 3950 (2000); G. Khaliullin, P. Horsch, and A. M. Oleś, ibid. **86**, 3879 (2001).
6. L. F. Feiner and A. M. Oleś, Phys. Rev. B **59**, 3295 (1999).
7. A. P. Ramirez, J. Phys.: Condens. Matter **9**, 8171 (1997).
8. A. Takahashi and H. Shiba, Eur. Phys. J. B **5**, 413 (1998); J. van den Brink and D. I. Khomskii, Phys. Rev. Lett. **82**, 1016 (1999).
9. P. Fazekas, *Lecture Notes on Electron Correlation and Magnetism* (World Scientific, Singapore, 1999).
10. Electron creation operators $c_i^\dagger(\psi_i, \theta_i)$ depend on two angles $\{\psi_i, \theta_i\}$ and create e_g electrons in orbital coherent states $|\Omega_i\rangle = \cos(\psi_i/2)e^{-i\theta_i}|i+\rangle + \sin(\psi_i/2)e^{+i\theta_i}|i-\rangle$.
11. A. Takahashi and H. Shiba, J. Phys. Soc. Jpn. **69**, 3328 (2000); J. van den Brink and D. I. Khomskii, Phys. Rev. B **63**, 140416 (2001).
12. F. Mack and P. Horsch, Phys. Rev. Lett. **82**, 3160 (1999).
13. Such states would however cost Jahn-Teller energy which we neglect here, see: R. Englman, *The Jahn-Teller Effect in Molecules and Crystals* (Wiley, London, 1972); Y. Motome and M. Imada, Phys. Rev. B **60**, 7921 (1999).
14. In the absence of SU(2) symmetry only the decoupling $Un_{i+}n_{i-} \simeq U(\langle n_{i+}\rangle n_{i-} + n_{i+}\langle n_{i-}\rangle - \langle n_{i+}\rangle\langle n_{i-}\rangle)$ gives complex states with finite $T = \langle T_i^z\rangle$.
15. The reduction of T in the FOx (FOz) state is only half of that in the FO+ state. As the SE energy gain is halved as well, these states are unstable at finite U for $n \simeq 1$. For the real states in HF: $Un_{i+}n_{i-} \simeq -U(\langle T_i^+\rangle c_{i-}^\dagger c_{i+} + c_{i+}^\dagger c_{i-}\langle T_i^-\rangle - \langle T_i^+\rangle\langle T_i^-\rangle)$.
16. S. Maezono and N. Nagaosa, Phys. Rev. B **62**, 11 576 (2000).
17. Y. Nagaoka, Phys. Rev. **147**, 392 (1966).
18. Q. Yuan, T. Yamamoto, and P. Thalmeier, Phys. Rev. B **62**, 12 696 (2000).
19. G. Kotliar and A. E. Ruckenstein, Phys. Rev. Lett. **57**, 1362 (1986).
20. Like the SU(2) invariant formulation for spins: R. Frésard and P. Wölfle, Int. J. Mod. Phys. B **6**, 685 (1992).
21. The nonvariational slave-fermion approximation gives a different band renormalization $\propto x$ [S. Ishihara, M. Yamanaka, and N. Nagaosa, Phys. Rev. B **56**, 686 (1997)], and underestimates the stability of the OL phase.

22. The cubic invariant KR approach for real states consists in substituting $b_{i\pm}^{\dagger} \mapsto (b_{iz}^{\dagger} \pm i b_{ix}^{\dagger})/\sqrt{2}$, and treating the amplitudes $\langle b_{iz} \rangle$ and $\langle b_{ix} \rangle$ in mean field.
23. Equivalent results are therefore obtained for the orbital ordered states by a single slave-fermion approach.
24. At $\gamma = 0$ the present method is believed to give an upper bound for the stability of FS states;(9) they are stable below $x \simeq 0.33$ in the cubic lattice, very close indeed to $x \simeq 0.32$ found for a single spin-flip in the Gutzwiller wave function [B. S. Shastry, H. R. Krishnamurthy, and P. W. Anderson, Phys. Rev. B **41**, 2375 (1990)].
25. At $\gamma = 1$ and $n = 0.75$ the energy gain in the OL state comes close to that in the exact ground state for a plaquette (Fig. 2).

26. R. Kilian and G. Khaliullin, Phys. Rev. B **58**, R11 841 (1998).
27. A. M. Oleś and L. F. Feiner, Phys. Rev. B **65**, 052414 (2002).

QUASIPARTICLES IN PHOTOEMISSION SPECTRA OF MANGANITES

Proceedings of the ARW NATO Workshop Hvar, Croatia, October 2002

J. BAŁA and ANDRZEJ M. OLEŚ
Marian Smoluchowski Institute of Physics, Jagellonian University, Reymonta 4, PL-30059 Kraków, Poland
Max-Planck-Institut für Festkörperforschung, Heisenbergstrasse 1, D-70569 Stuttgart, Germany

Abstract. We compare the spectral functions of a hole moving in the orbital-ordered undoped $LaMnO_3$, obtained using a self-consistent Born approximation. The quasiparticle (QP) dispersion and spectral weight depend critically on the type of orbital ordering. If a hole scatters on orbital excitations, the QP dispersion on the orbiton energy scale is modified by orbital polarization and by electron-lattice coupling. A lower QP dispersion and quantum decoherence are obtained due to the inter-planar hole-magnon scattering in the A-AF phase which disproves the classical concept of a ferromagnetic polaron.

Key words: Manganites, Orbital Ordering, Born Approximation, Spectral Functions, Orbital Polaron, Quasiparticle Dispersion, Hole Scattering, Orbiton

1. Orbital t-J model for manganites

It is well known from the cuprates that a hole propagating in an antiferromagnetic (AF) CuO_2 plane of a high temperature superconductor scatters on magnon excitations and thus its spectral properties are completely different from those of a free hole. Stronger, the propagation is only possible due to the processes in which a single magnon is either excited or absorbed, and without them (Ising model), a hole would be confined by a string potential (1). Here we show that although richer models are needed to describe a propagation of a hole in $LaMnO_3$, a parent compound for 'colossal magnetoresistance', the essential feature of such propagation, dressing of a hole by low-energy excitations (2), remains the same.

$LaMnO_3$ is an AF Mott insulator due to strong on-site Coulomb repulsion $\propto U$, with ferromagnetic (FM) (a, b) planes which stagger along the c-axis in the so-called A-AF phase. This kind of order is stabilized by

A.C. Hewson and V. Zlatić (eds.), Concepts in Electron Correlation, 133–141.

the anisotropic superexchange H_J due to the charge excitations for Mn^{3+}–Mn^{3+} pairs with singly occupied degenerate e_g orbitals, described within the spin-orbital model (3). If the electron correlations could be ignored, the hole propagation on the energy scale $\propto t$ (t is the hopping element between $|z\rangle$ orbitals along the c axis) would follow from the tight-binding Hamiltonian H_t. However, the double-exchange hinders the motion along the c-axis (4), and a hole can move mainly within a single FM (a, b) plane. In the present paper we analyze possible scenarios of hole dynamics when it couples either to orbital excitations (5), or phonons (6), or to magnons (7).

First we consider a two-dimensional (2D) model for a single FM (a, b) plane when the oxygen atoms are frozen in their optimal positions dictated by the Jahn-Teller (JT) effect. As $U \gg t$, one has to use the strong-coupling limit represented by the following extended *orbital t-J model* (8),

$$\mathcal{H} = H_t + H_J + H_\Delta + H_{\rm JT} + E_l. \tag{1}$$

It includes, in addition to H_t and H_J which give the orbital (in analogy to spin) t-J model (5), also the polarization of orbitals around a hole (H_Δ) which induces the splitting of e_g orbitals next to the hole (9),

$$H_\Delta = -\Delta \sum_\gamma \sum_{\langle ij\rangle \parallel \gamma} n_i \tau_j^\xi, \tag{2}$$

with $\gamma = a, b, c$ labelling cubic axes, $\tau_j^{a(b)} = -\frac{1}{2}\left(T_j^z \mp \sqrt{3}T_j^x\right)$ and $\tau_j^c = T_j^z$ given by two pseudospin operators $\{T_j^x, T_j^z\}$, and the JT interaction,

$$H_{\rm JT} = -2\sqrt{6}\lambda \sum_i \left(Q_{i,2}T_i^x + Q_{i,3}T_i^z\right), \tag{3}$$

using the notation of Kanamori (10). The last term in Eq. (1) stands for the elastic energy of the distorted lattice (E_l) (11).

The orbital ordering within the (a, b) planes is stabilized even by the superexchange interaction alone (3, 12), and one finds alternating occupied e_g orbitals, as observed experimentally (13),

$$|i\mu\rangle = \cos\left(\tfrac{\pi}{4} - \phi\right)|iz\rangle \pm \sin\left(\tfrac{\pi}{4} - \phi\right)|ix\rangle, \tag{4}$$

where + (−) refers to $i \in A$ ($i \in B$) orbital sublattice, and $|ix\rangle$ ($|iz\rangle$) stands for the local basis orbital $|x^2 - y^2\rangle$ ($|3z^2 - r^2\rangle$) at site i, respectively. In contrast to the spin t-J model, a single hole can now propagate coherently without disturbing the orbital order (5).

The lattice distortion is described by three independent parameters: δ_x (δ_z) – uniform deformation along the a and b (c) cubic directions, and u – oxygen displacement along the Mn–O–Mn bond in the (a, b) plane; all

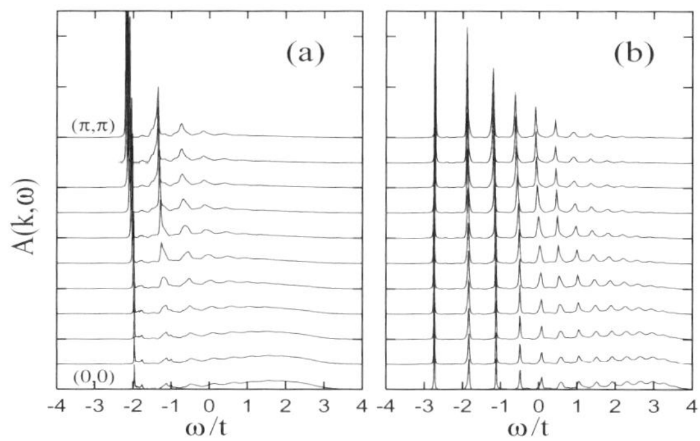

Figure 1. The hole spectral functions $A(\mathbf{k},\omega)$ for the orbital ordering given by $\phi=0$ in the absence of the electron-phonon interaction ($\lambda=0$) for: (a) $\Delta=0$, and (b) $\Delta=t$, as obtained along the $(0,0)-(\pi,\pi)$ direction in the 2D Brillouin zone.

distortions are in the units of the Mn–Mn distance ($d=1$) (11). The JT term (3) acts then at both sublattices as the staggered field (14),

$$H_{\rm JT}=-2\lambda\left[(\delta_x-\delta_z)\sin 2\phi-2\sqrt{3}u\cos 2\phi\right]\Big(\sum_{i\in A}\tilde{T}_i^z-\sum_{i\in B}\tilde{T}_i^z\Big), \quad (5)$$

with the pseudospin operators $\tilde{T}_i^z$ referring to the ground state, Eq. (4). In the linear orbital-wave (LOW) theory (12) the effective Hamiltonian represents a coupled hole-orbiton problem in momentum space,

$$\begin{aligned}\mathcal{H}_{\rm LOW} &= \sum_{\mathbf{k}}\varepsilon_{\mathbf{k}}(\phi)h_{\mathbf{k}}^{\dagger}h_{\mathbf{k}}+\sum_{\mathbf{q}}\omega_{\mathbf{q}}(\phi)\alpha_{\mathbf{q}}^{\dagger}\alpha_{\mathbf{q}}\\ &+ \tfrac{1}{\sqrt{N}}\sum_{\mathbf{k},\mathbf{q}}\left\{h_{\mathbf{k}-\mathbf{q}}^{\dagger}h_{\mathbf{k}}\left[M_{\mathbf{k},\mathbf{q}}\alpha_{\mathbf{q}}^{\dagger}+N_{\mathbf{k},\mathbf{q}}\alpha_{\mathbf{q}+\mathbf{Q}}^{\dagger}\right]+\text{H.c.}\right\}, \quad (6)\end{aligned}$$

with N being the number of lattice sites and two hole-orbiton vertices,

$$M_{\mathbf{k},\mathbf{q}} = 2t\cos(2\phi)\,(u_{\mathbf{q}}\gamma_{\mathbf{k}-\mathbf{q}}+v_{\mathbf{q}}\gamma_{\mathbf{k}})+\Delta\cos(2\phi)\,(\gamma_{\mathbf{q}}-1)\,(u_{\mathbf{q}}+v_{\mathbf{q}}), \quad (7)$$

$$N_{\mathbf{k},\mathbf{q}} = -\sqrt{3}t\,(u_{\mathbf{q}+\mathbf{Q}}\eta_{\mathbf{k}-\mathbf{q}}+v_{\mathbf{q}+\mathbf{Q}}\eta_{\mathbf{k}})-\sqrt{3}\Delta\sin(2\phi)\eta_{\mathbf{q}}\,(u_{\mathbf{q}+\mathbf{Q}}+v_{\mathbf{q}+\mathbf{Q}}), \quad (8)$$

where $\mathbf{Q}=(\pi,\pi)$, $\gamma_{\mathbf{q}}=\frac{1}{2}(\cos q_x+\cos q_y)$, $\eta_{\mathbf{q}}=\frac{1}{2}(\cos q_x-\cos q_y)$, and $\{u_{\mathbf{q}},v_{\mathbf{q}}\}$ follow from the respective Bogoliubov transformation for the orbital waves. The free hole, $\varepsilon_{\mathbf{k}}(\phi)=4t_{ab}^{\phi}\gamma_{\mathbf{k}}$ with $t_{ab}^{\phi}=t(1-2\sin 2\phi)$, and the orbiton, $\omega_{\mathbf{q}}(\phi)=3J\{A(\lambda)[A(\lambda)+\frac{1}{3}(2\cos 4\phi-1)\gamma_{\mathbf{q}}]\}^{1/2}$, dispersion depend on the type of orbital ordering ϕ. Here $A(\lambda)=1+2\lambda^2/(JK)$, with K being the Mn–O spring constant. The Green's function, $G(\mathbf{k},\omega)=[\omega-\varepsilon_{\mathbf{k}}(\phi)-\Sigma(\mathbf{k},\omega)]^{-1}$, is determined in the self-consistent Born approx-

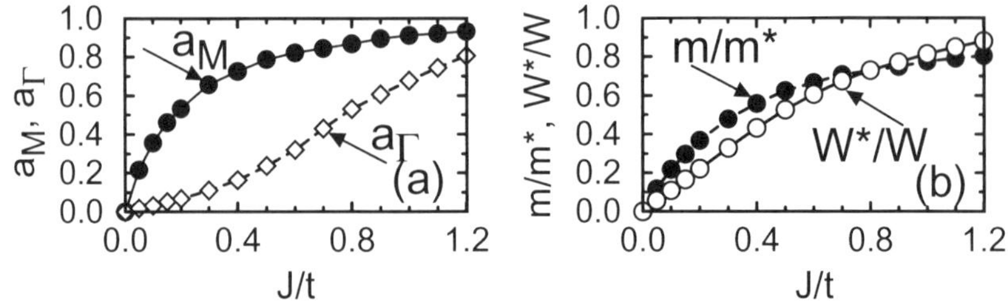

Figure 2. QP's in the orbital t-J model ($\Delta = \lambda = 0$) as functions of J/t for $\phi = 0$ (5): (a) spectral weights at $\mathbf{k} = (\pi, \pi)$ (a_M, full circles) and $\mathbf{k} = (0, 0)$ (a_Γ, empty circles); (b) inverse effective mass m/m^* and the bandwidth W^*/W (full and empty circles).

imation (15) by the hole self-energy,

$$\Sigma(\mathbf{k}, \omega) = \sum_{\mathbf{q}} \left\{ M^2_{\mathbf{k},\mathbf{q}} G[\mathbf{k} - \mathbf{q}, \omega - \omega_{\mathbf{q}}(\phi)] + N^2_{\mathbf{k},\mathbf{q}} G[\mathbf{k} - \mathbf{q}, \omega - \omega_{\mathbf{q}+\mathbf{Q}}(\phi)] \right\}, \tag{9}$$

to obtain the spectral functions, $A(\mathbf{k}, \omega) = -\frac{1}{\pi}\text{Im}G(\mathbf{k}, \omega + i0^+)$.

In the numerical calculations we use the parameters: $K = 200$ eV, the Mn–Mn effective hopping $t = 0.4$ eV, and the orbital superexchange $J \simeq 50$ meV, as estimated using the basic electronic parameters for $LaMnO_3$ (3). Consider first the simplest case with $\lambda = \Delta = 0$. One finds a large redistribution of spectral weight with respect to the free bands, a more drastic effect than for the hole-phonon interaction (6). The spectral function $A(\mathbf{k}, \omega)$ consists of a quasiparticle (QP) at low energy, with a minimum at the $M = (\pi, \pi)$ point, and a broad incoherent part [Fig. 1(a)], similar to the spin case (1). In the realistic regime of $J/t \simeq 0.12$, the QP spectral weight is large at the M (a_M), but almost negligible at the Γ point (a_Γ) [Fig. 2(a)].

The orbital polarization $\propto \Delta$ enhances the hole localization and again modifies the spectral properties. For instance, for the $(|x\rangle \pm |z\rangle)/\sqrt{2}$ orbital ordering ($\phi = 0$), the polarization changes the spectrum into the ladder-like one with practically vanishing QP dispersion [Fig. 1(b)]. This leads to a large effective mass m^*/m of a hole due to its strong confinement. Moreover, the binding energy of the hole at $\mathbf{k} = (\pi, \pi)$ (the bottom of the QP band) increases by energy $\delta E \approx 0.5t \ll 2\Delta$ due to the polarization, suggesting that the rearrangement of e_g orbitals around the hole is small. In contrast, the effective mass m^*/m is much less enhanced when $\Delta = 0$, and decreases with increasing J [Fig. 2(b)]. At the same time the QP bandwidth W^*/W is about $2.2J$ at $J/t < 0.4$, and approaches t at large $J/t > 1$.

The changes of the QP dispersion with Δ and λ become more transparent by comparing the QP's obtained for two orbital orderings shown in Fig. 3. The shape and position of the QP band strongly depends on both

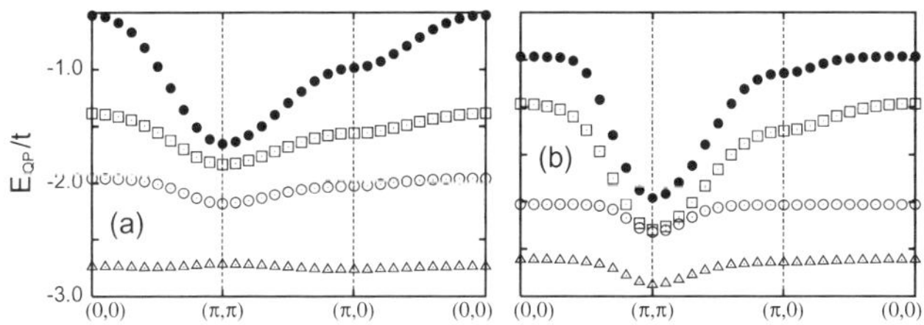

Figure 3. Dispersion of the QP band for: (a) $\phi = 0$, and (b) $\phi = -\pi/12$, with $\Delta = \lambda = 0$ (○), $\Delta = 0$, $\lambda = 10t$ (●), $\Delta = t$, $\lambda = 0$ (△), and $\Delta = t$, $\lambda = 10t$ (□).

orbital polarization (Δ), and on the JT coupling (λ). With polarized e_g orbitals around a hole ($\Delta = t$) and no JT interaction ($\lambda = 0$), the QP band calculated for $\phi = 0$ ($\phi = -\pi/12$) is narrowed by a factor of 50 (20), respectively, in comparison with the free hole dispersion and moves to much *lower* energies. The increasing JT interaction leads to a rigid orbital ordering and consequently to the hole motion with almost no scattering off orbital waves (8, 16). As a result, the binding energy gets small, and the hole undresses, resulting in large dispersion set by the hopping element t.

2. Dynamical electron-phonon coupling

To describe the *dynamical* JT effect we have to quantize the oscillations of the oxygen atoms around their mean-field positions ($\propto u$) and derive the dynamical part of the JT Hamiltonian of the following form (17),

$$\begin{aligned} H_{\rm JT}^{dyn} &= K\sum_i \sum_{\xi=1,2,3} \tilde{Q}_{i,\xi}^2 + \frac{1}{2M}\sum_i \sum_{\xi=1,2,3} P_{i,\xi}^2 \\ &+ 2\sqrt{6}\lambda \sum_i \tilde{T}_i^x \left(\sin 2\phi\, \tilde{Q}_{i,2} - \cos 2\phi\, \tilde{Q}_{i,3}\right), \end{aligned} \tag{10}$$

where M stands for the mass of an oxygen ion, with the JT phonon modes:

$$\tilde{Q}_{i,2} = \frac{1}{\sqrt{2}}\left(u_{ix} - u_{iy}\right), \qquad \tilde{Q}_{i,3} = \frac{1}{\sqrt{6}}(2u_{iz} - u_{ix} - u_{iy}), \tag{11}$$

where $u_{i\nu}$ (with $\nu = x,\, y,\, z$) is the oxygen vibration along the respective Mn–O bond and $P_{i,\xi}$ is the conjugate momentum vector corresponding to the lattice distortion $\tilde{Q}_{i,\xi}$. The breathing mode, $\tilde{Q}_{i,1} = \frac{1}{\sqrt{3}}(u_{ix} + u_{iy} + u_{iz})$, does not couple to the orbital excitations. In the LOW order the total effective Hamiltonian [derived from Eqs. (1) and (10) at $\Delta = 0$] represents a coupled hole-orbiton-phonon problem in momentum space,

$$\mathcal{H}_{eff} = \sum_{\mathbf{k}} \varepsilon_{\mathbf{k}}(\phi) h_{\mathbf{k}}^\dagger h_{\mathbf{k}} + \sum_{\mathbf{q}} \sum_{\xi=1,2} \Omega_{\mathbf{q}}^{(\xi)}(\phi) \beta_{\mathbf{q},\xi}^\dagger \beta_{\mathbf{q},\xi} + \omega_0 \sum_{\mathbf{q},\mu=1,3} B_{\mathbf{q},\mu}^\dagger B_{\mathbf{q},\mu}$$

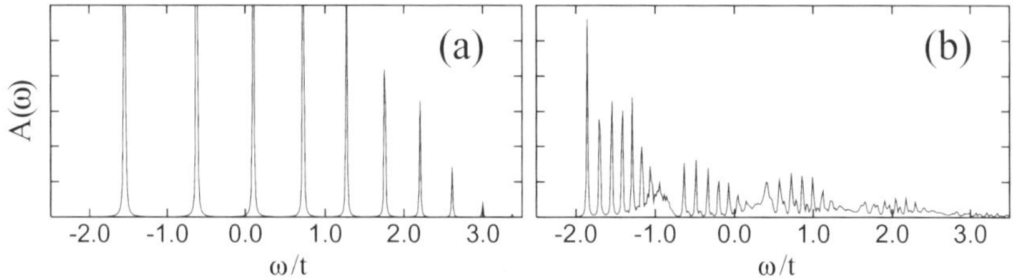

Figure 4. The local hole spectral functions $A(\omega)$ obtained for the $|x^2-z^2\rangle/|y^2-z^2\rangle$ orbital ordering ($\phi=\pi/12$) of $LaMnO_3$ with: (a) $\lambda/t=0$ and (b) $\lambda/t=8$.

$$+\sum_{\mathbf{k},\mathbf{q}}\Big\{h^\dagger_{\mathbf{k}-\mathbf{q}}h_{\mathbf{k}}\Big[M_0\left(\sqrt{2}B^\dagger_{\mathbf{q},1}+B^\dagger_{\mathbf{q},3}\right)$$
$$+\sum_{\xi=1,2}\left(M^{(\xi)}_{\mathbf{k},\mathbf{q}}\beta^\dagger_{\mathbf{q},\xi}+N^{(\xi)}_{\mathbf{k},\mathbf{q}}\beta^\dagger_{\mathbf{q}+\mathbf{Q},\xi}\right)\Big]+\text{H.c.}\Big\}, \qquad (12)$$

with the energy of the phonon mode $\omega_0=\sqrt{2K/M}$ and the hole-phonon vertex $M_0=\sqrt{3/N}\lambda/(2KM)^{1/4}$. $\Omega^{(\xi)}_{\mathbf{q}}(\phi)$ and $M^{(\xi)}_{\mathbf{k},\mathbf{q}}$ are the energies of mixed electron-phonon excitations and the respective vertices, which depend on λ and ϕ (for more details see Ref. (17)). Moreover, $B_{\mathbf{q},\mu}$ are phonon operators representing modes which do not couple to orbitons, while $\beta_{\mathbf{q},\xi}$ represent mixed orbiton-phonon excitations (18).

In the presence of both orbital ordering and lattice distortions, a hole can scatter off both orbitons and phonons, already renormalized by each other. The simplest situation occurs for the $|x^2-z^2\rangle/|y^2-z^2\rangle$ orbital ordering with $\varepsilon_{\mathbf{k}}(\phi)=0$. In this case, similar to a hole in a quantum antiferromagnet (1), a hole can only propagate making mixed orbital-phonon excitations on its path. Such mixed excitations have already been observed in FM $LaMnO_{3+\delta}$ (19). As the orbitons and phonons having no momentum dependence in this case, one finds **k**-independent spectra and the infinite hole effective mass. In the absence the JT effect, the model reduces to the t-J^z model (1) with ladder-like spectral functions shown in Fig. 4(a). When the orbiton-lattice coupling $\propto\lambda$ increases, the QP peak moves to lower energies and the spectrum looses gradually its coherence; for large values of $\lambda\simeq 8t$ it separates into a number of subspectra consisting of a series of lattice vibrations [Fig. 4(b)]. Each hopping of the hole (leading to an excitation of an orbiton) is accompanied by a few vibrational quanta. This spectral function well exemplifies the structure of the electron dressing after a sudden photoemission process. For other orbital orderings the vibrational peaks merge into a single broad maximum next to the QP peak.

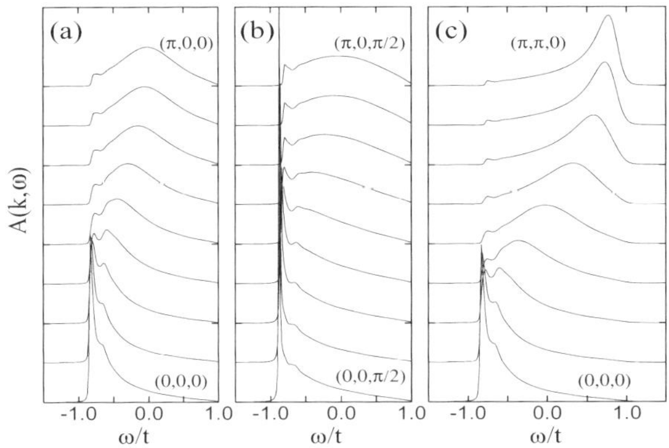

Figure 5. The hole spectral functions obtained in the hole-magnon 3D model for the A-AF phase with $\phi = \pi/24$ along high symmetry lines in the 3D Brillouin zone (7).

3. Scattering on magnons in the A-AF phase and discussion

In this section we consider the limit of rigid orbital ordering ($\lambda \gg t$) for $\Delta = 0$. Then the model given by Eq. (1) leads to free propagation of an e_g hole in the (a, b) plane. However, the A-AF order leads to hole scattering on spin excitations when it hops out off FM planes. Assuming that the orbital ordering repeats itself in (a, b) planes, the hopping part of our three-dimensional (3D) t-J model , $\mathcal{H} = H_t + H_J$, reads (7),

$$H_t = t^{\phi}_{ab} \sum_{\langle ij \rangle_{ab},\sigma} \tilde{c}^{\dagger}_{i\sigma}\tilde{c}_{j\sigma} - t^{\phi}_{c} \sum_{\langle ij \rangle_{c},\sigma} \tilde{c}^{\dagger}_{i\sigma}\tilde{c}_{j\sigma} + \text{H.c.}, \tag{13}$$

where $\tilde{c}^{\dagger}_{i\sigma} = c^{\dagger}_{i\sigma}(1 - n_{i\bar{\sigma}})$. Here $\langle ij \rangle_{ab}$ and $\langle ij \rangle_{c}$ represent the nearest-neighbors in the (a, b) plane and along c direction, and $t^{\phi}_{c} = \frac{1}{2}t(1 + \sin 2\phi)$. The interactions between spins $S = 2$ of Mn^{3+} ions are given by:

$$H_J = -J_{ab} \sum_{\langle ij \rangle_{ab}} \mathbf{S}_i \cdot \mathbf{S}_j + J_c \sum_{\langle ij \rangle_{c}} \mathbf{S}_i \cdot \mathbf{S}_j, \tag{14}$$

with the superexchange constants, $J_{ab} > 0$ and $J_c > 0$.

The scattering off magnons strongly depends on the orbital ordering and can considerably change the hole propagation when $t^{\phi}_{c} > t^{\phi}_{ab}$. Here, we present the spectral functions for the orbital ordering given by $\phi = \pi/24$, where $t^{\phi}_{c}/t^{\phi}_{ab} \simeq 5.2$ and the hole-magnon coupling is strong (see Fig. 5). We find weak QP states at small momenta with incoherent structureless continuum developing above them, which absorbs most of the spectral weight. At larger momenta the spectra widen with thresholdlike peaks developing at $\omega \simeq -0.8t$. Consequently, the whole spectrum bears again little resemblance with the free-hole dispersion. On the contrary, for the $|3x^2 - r^2\rangle/|3y^2 - r^2\rangle$

orbital ordering, where $t_c^\phi/t_{ab}^\phi = 0.5$ the hole hardly feels the AF ordering in the c direction and moves almost coherently in the (a, b) plane (7).

Summarizing, we have studied a single hole propagation in undoped $LaMnO_3$ with orbital and magnetic ordering. The hole scatters strongly on low-energy excitations: phonons, mixed orbiton-phonon, or magnons. The orbital polarization around a carrier leads to its additional confinement. Our results predict a large redistribution of spectral weight with respect to the bands found in local density approximation (LDA) or in LDA+U, and can have important implications on the angular resolved photoemission spectroscopy (ARPES). In all cases we found low-energy QP's, but the minimum of the QP band falls at $\mathbf{k} = (\pi, \pi, k_z)$ when the hole-orbiton scattering dominates, while it is at $\mathbf{k} = (0, 0, \pi/2)$ for the hole-magnon scattering. Which of the considered processes dominates and which dispersion is realized might be answered only by the ARPES experiments. In fact, the orbitons are gapped, and were found recently at energies larger than $\sim$ 150 meV (20). The magnons are gapless, and have much lower energies $\sim$ 30 meV, so the hole scattering on them might dominate, but only if the orbital ordering is close to $|x^2 - z^2\rangle/|y^2 - z^2\rangle$ ($\phi = \pi/12$) rather than to $|3x^2 - r^2\rangle/|3y^2 - r^2\rangle$ ($\phi = -\pi/12$). Thus, the results of future ARPES experiments could also help to establish which type of orbital ordering is realized in $LaMnO_3$, the issue difficult to resolve by resonant x-ray scattering (13). We also argue that the spectral features measured in the highly doped $La_{1.2}Sr_{1.8}Mn_2O_7$ (21) should change substantially when magnetic order changes to the A-AF phase under doping.

This work was supported by the Polish State Committee of Scientific Research (KBN), Project No. 5 P03B 055 20.

References

1. G. Martínez and P. Horsch, Phys. Rev. B **44**, 317 (1991).
2. J. Zaanen and A. M. Oleś, Phys. Rev. B **48**, 7197 (1993).
3. L. F. Feiner and A. M. Oleś, Phys. Rev. B **59**, 3295 (1999).
4. C. Zener, Phys. Rev. **82**, 403 (1951).
5. J. van den Brink, P. Horsch, and A. M. Oleś, Phys. Rev. Lett. **85**, 5174 (2000).
6. V. Perebeinos and P. B. Allen, Phys. Rev. Lett. **85**, 5178 (2000).
7. J. Bała, G. A. Sawatzky, A. M. Oleś, and A. Macridin, Phys. Rev. Lett. **87**, 067204 (2001).
8. J. Bała, A. M. Oleś, and P. Horsch, Phys. Rev. B **65**, 134420 (2002).
9. R. Kilian and G. Khaliullin, Phys. Rev. B **60**, 13458 (1999).
10. J. Kanamori, J. Appl. Phys. **31**, 14S (1960).
11. A. J. Millis, Phys. Rev. B **53** 8434 (1996).
12. J. van den Brink, P. Horsch, F. Mack, and A. M. Oleś, Phys. Rev. B **59**, 6795 (1999).
13. Y. Murakami *et al.*, Phys. Rev. Lett. **81**, 582 (1998).
14. J. Bała and A. M. Oleś, Phys. Rev. B **62**, R6085 (2000).
15. C. L. Kane, P. A. Lee, and N. Read, Phys. Rev. B **39**, 6880 (1989).

16. W.-G. Yin, H.-Q. Lin, and C.-D. Gong, Phys. Rev. Lett. **87**, 047204 (2001).
17. J. Bała, A. M. Oleś, and G. A. Sawatzky, Phys. Rev. B **65**, 184414 (2002).
18. J. van den Brink, Phys. Rev. Lett. **87**, 217202 (2001).
19. Y. G. Pashkevich *et al.*, unpublished (2002).
20. E. Saitoh *et al.*, Nature **410**, 180 (2001).
21. D. S. Dessau *et al.*, Phys. Rev. Lett. **81**, 192 (1998).

METALS NEAR A ZERO-TEMPERATURE MAGNETIC INSTABILITY

Proceedings of the ARW NATO Workshop Hvar, Croatia, October 2002

H. V. LÖHNEYSEN
[1] *Physikalisches Institut, Universität Karlsruhe, D-76128 Karlsruhe, Germany and*
Forschungszentrum Karlsruhe, Institut für Festkörperphysik, D-76021 Karlsruhe, Germany

Abstract. Zero-temperature magnetic phase transitions exhibit an abundance of nearly critical magnetic fluctuations leading to unusual low-temperature properties that have become known as non-Fermi liquid (NFL) behavior. For the prototypical heavy-fermion compound $CeCu_{6-x}Au_x$ this transition may be tuned by either Au concentration, hydrostatic pressure, or magnetic field. In the system $UCu_{5-x}Pd_x$, strong disorder affects the NFL properties. The d-electron Laves phase $ZrZn_2$ was recently found to display superconductivity in coexistence with itinerant ferromagnetism.

Key words: Heavy Fermions, Non-Fermi Liquid, Antiferromagnetic Order, Quantum Phase Transition, Ferromagnetic Superconductor, Ce-Cu-Au, Zr-Zn, U-Cu-Pd

1. Introduction

Metals with strong electronic correlations are often close to a magnetic instability. In a number of these systems, the transition temperature between a paramagnet and a magnetically ordered phase can be tuned to absolute zero by some externally controlled parameter such as chemical composition, pressure, magnetic field, or charge carrier concentration. This offers the possibility to induce a $T = 0$ magnetic-nonmagnetic quantum phase transition (QPT). In the vicinity of this transition, non-Fermi-liquid (NFL) behavior (1) may occur in thermodynamic and transport properties: the linear specific-heat coefficient $\gamma = C/T$ acquires an unusual temperature dependence, often $\gamma \sim -\ln(T/T_0)$, and the T-dependent part of the electrical resistivity $\Delta\rho = \rho - \rho_0$ where ρ_0 is the residual resistivity, often varies as $\Delta\rho \sim T^m$ with $m < 2$, in contrast to the Fermi-liquid (FL) predictions $\gamma =$ const and $m = 2$.

A.C. Hewson and V. Zlatić (eds.), Concepts in Electron Correlation, 143–153.

It is generally believed that the NFL behavior observed in a number of heavy-fermion systems (HFS) at the magnetic-nonmagnetic QPT arises from a proliferation of magnetic excitations (2, 3, 4). If the transition is continuous, it is driven by quantum fluctuations instead of thermal fluctuations in finite-T transitions.

$CeCu_{6-x}Au_x$ provides one of the best studied examples of NFL behavior at a QPT where macroscopic (thermodynamic and transport properties (5, 6, 8)) as well as microscopic measurements (elastic (7) and inelastic neutron scattering (9, 10, 11)) have been performed. This system presents very unusual spin dynamics which will be briefly reviewed. We will also discuss how hydrostatic pressure or magnetic field are operative in tuning the system through a QPT. Another NFL system that has been studied early on (12) is $UCu_{5-x}Pd_x$. Since this system is strongly disordered, it presents an example for a Griffiths phase scenario (13) near a magnetic instability (14, 15). Very recently, the coexistence of ferromagnetism and superconductivity was demonstrated for UGe_2 for pressures close to the critical pressure where ferromagnetism is suppressed (16). In addition, the prototype weak itinerant ferromagnet $ZrZn_2$, long considered to be a prime candidate for p-wave superconductivity (17), was finally found to be superconducting in sufficiently pure samples (18). The purpose of this article is to review briefly the salient features of these three different systems, all being at the verge of a magnetic instability.

2. Non-Fermi liquid behavior and magnetic fluctuations in $CeCu_{6-x}Au_x$

Pure $CeCu_6$ shows no long-range magnetic order down to very low T due to the quenching of Ce $4f$ magnetic moments by the Kondo effect (19, 20), with some evidence for magnetic ordering (either electronic or nuclear) occurring at a few mK (21, 22, 23). With $\gamma = 1.6\,\mathrm{J/moleK^2}$ it is one of the "heaviest" HFS. $CeCu_6$ exhibits a pronounced magnetic anisotropy with the magnetization ratios along the three axes $M_c : M_a : M_b \approx 10 : 2 : 1$ at low T (20). We use the orthorhombic notation, in spite of the (small) monoclinic distortion.

Already at relatively high T, i.e. around 1 K, does $CeCu_6$ exhibit intersite antiferromagnetic fluctuations as observed with inelastic neutron scattering (INS) by peaks in the dynamic structure factor $S(\mathbf{q}, \omega)$ for energy transfer $\hbar\omega = 0.3\,\mathrm{meV}$ at $\mathbf{Q} = (1\ 0\ 0)$ and $(0\ 1{\pm}0.15\ 0)$ (24, 25). The rather large widths of these peaks correspond to correlation lengths extending roughly only to the nearest Ce neighbors. Recently, additional features in the a^*c^* plane at an energy transfer of 0.1 meV were found (26).

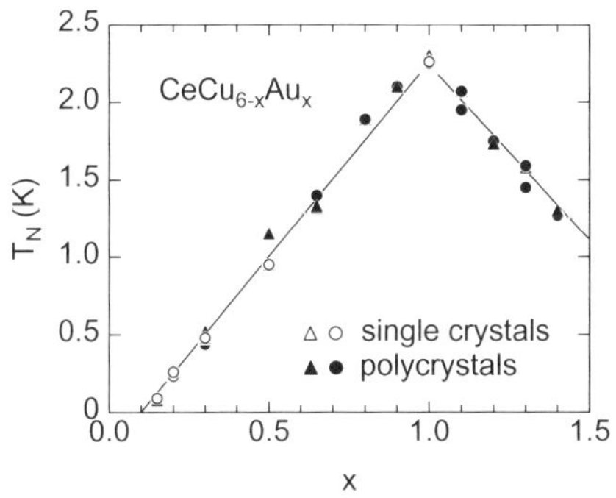

Figure 1. Néel temperature T_{N} of $CeCu_{6-x}Au_x$ versus Au concentration x.

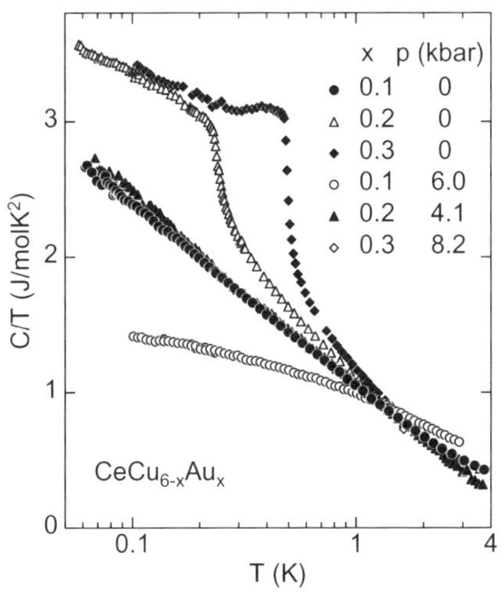

Figure 2. Specific heat C of $CeCu_{6-x}Au_x$ ($x = 0.1, 0.2, 0.3$) plotted as C/T versus log T at ambient pressure and at the critical pressure ($x = 0.2$: $p_c \approx 4$ kbar, 0.3: $p_c \approx 8$ kbar). For $x = 0.1$ which is critical at ambient pressure, data for 6.0 kbar are shown.

Upon alloying with Au, the $CeCu_6$ lattice expands (27), thus weakening the hybridization between conduction electrons and Ce $4f$ electrons. Hence the conduction-electron$-4f$-electron exchange constant J decreases, leading to a stabilization of localized magnetic moments which can now interact via the RKKY interaction, with ensuing antiferromagnetic order (28). Fig. 1 shows the Néel temperature T_{N} of $CeCu_{6-x}Au_x$ vs x. The magnetic structure of $CeCu_{6-x}Au_x$ ($0.15 \leq x \leq 1$) as determined with elastic neutron scattering (7, 29, 30) shows magnetic Bragg peaks in the a^*c^* plane (see Fig. 3 below).

The onset of magnetic order in $CeCu_{6-x}Au_x$ is observed as a kink in C/T, cf. Fig. 2. For the critical concentration $x_c = 0.1$ the linear specific-heat coefficient depends logarithmically on T, $C/T = a\ln(T_0/T)$ with $a = 0.58\,\mathrm{J/molK^2}$ and $T_0 = 6.2\,\mathrm{K}$, between 0.06 and 2.5 K, instead of being constant as for a Fermi liquid and roughly obeyed for $CeCu_6$.

The abundance of low-energy magnetic excitations when T_N is just tuned to zero, has been suggested early on to cause the NFL behavior at the magnetic instability (5). However, the $-\ln T$ dependence of C/T and the unusual linear T dependence of ρ (not shown), also signaling NFL behavior in $CeCu_{6-x}Au_x$ at the magnetic instability, are at variance with spin-fluctuation theories for three-dimensional (3D) itinerant fermion systems which predict (3, 4) $C/T = \gamma_0 - \beta\sqrt{T}$ and $\Delta\rho \sim T^{3/2}$ for antiferromagnets. On the other hand, 2D critical fluctuations coupled to quasiparticles with 3D dynamics do indeed lead to the observed behavior $C/T \sim -\ln T$, $\Delta\rho \sim T$, and a linear dependence of T_N or x on p (31).

A detailed INS investigation at the critical concentration $x = 0.1$ (10) showed that the critical fluctuations as measured with an energy transfer $\hbar\omega = 0.10\,\mathrm{meV}$ extend into the a^*c^* plane. Hence the dynamical structure factor $S(\mathbf{q},\hbar\omega = 0.10\,\mathrm{meV})$, where $\hbar\mathbf{q}$ is the transferred momentum, has the form of rods as indicated by the shaded regions in Fig. 3. A quasi-1D dynamic feature in reciprocal space corresponds to quasi-2D fluctuations in real space. The 3D ordering peaks for $x = 0.15, 0.2$ and 0.3 fall on the rods for $x = 0.1$ which therefore can be viewed as precursors of 3D ordering (7). From the width of the rods in reciprocal space, the prefactor a of the logarithmic C/T dependence could be calculated to within a factor of two of the experimental value (10).

The spin fluctuations also develop specific dynamics at $x = 0.1$ (9). The scattering function $S(\mathbf{q},E,T)$ or the susceptibility $\chi'' = S\cdot(1-\exp(-E/k_BT))$ exhibit E/T scaling (where $E = \hbar\omega$) in the critical $\mathbf{q}$ region, e.g. at $\mathbf{Q}_c = (0.8, 0, 0)$, which can be expressed by

$$\chi''(\mathbf{Q}_c, E, T) = T^{-\alpha} g(E/k_BT) \qquad (1)$$

with $\alpha = 0.75$ (9). The anomalous non-Lorentzian response ($\alpha \neq 1$) does not change for other $\mathbf{q}$ away from the critical region (11). For all $\mathbf{q}$ the susceptibility can be expressed as

$$\chi^{-1}(\mathbf{q}, E, T) = c^{-1}(f(\mathbf{q}) + (-iE + aT)^{\alpha}). \qquad (2)$$

In particular, the T dependence of the static uniform susceptibility $\chi(\mathbf{q} = 0, E = 0)$ for $x = 0.1$ yields the exponent $\alpha \approx 0.8$ to a high degree of accuracy. The simple form of Eq.(2) separates static spatial correlations from the specific temporal correlations, the latter being independent of $\mathbf{q}$. These local fluctuations at the quantum critical point have received considerable theoretical attention, although a detailed model is not available

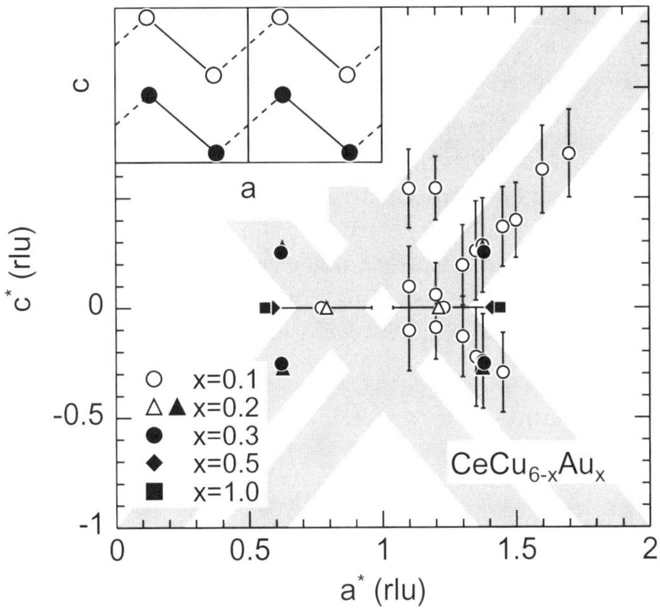

Figure 3. Position of the dynamic correlations ($x = 0.1, \hbar\omega = 0.1\,\text{meV}, T < 100\,\text{mK}$) and magnetic Bragg peaks ($0.15 \leq x \leq 1.0$) in the a^*c^* plane in $CeCu_{6-x}Au_x$. Closed symbols for $x = 0.2$ represent short-range order peaks. The vertical and horizontal bars indicate the Lorentzian linewidths for $x = 0.1$. The four shaded rods are related by the orthorhombic symmetry (we ignore the small monoclinic distortion). The inset shows a schematic projection of the $CeCu_{6-x}Au_x$ structure onto the ac plane where only the Ce atoms are shown. The bars in reciprocal space correspond to planes in real space spanned by b and the lines in the inset.

yet (32, 33). The evolution of the ordered moment with increasing $x > x_c$ (7), may provide a valuable input to test the different models.

The onset of magnetic order in $CeCu_{6-x}Au_x$ is attributed to a weakening of J because of the increase of the molar volume upon alloying with Au as mentioned above. This is confirmed by the observation that T_N of $CeCu_{6-x}Au_x$ decreases roughly linearly under hydrostatic pressure p (6, 34). At the critical pressure p_c where $T_N \to 0$, i.e. $p_c \approx 4\,\text{kbar}$ for $x = 0.2$ and $\approx 8\,\text{kbar}$ for $x = 0.3$, C/T exhibits NFL behavior, i.e., $C/T \sim -\ln T$, with the same coefficients a and T_0 as for $x = 0.1$ at $p = 0$ (see Fig. 2). Consequently, $p = 6\,\text{kbar}$ drives the ambient-pressure NFL alloy $x = 0.1$ into the FL regime.

An inducement of NFL behavior by a magnetic field in the related system $CeCu_{6-x}Ag_x$ was reported previously (35, 36). The low-T properties at the critical field are compatible with the conventional 3D antiferromagnetic spin-fluctuation scenario (4). A detailed comparison of the specific heat and electrical resistivity of $CeCu_{5.8}Au_{0.2}$ revealed distinctive differences depending on whether the QPT is tuned by B or p, presenting strong evidence for pronounced differences in the fluctuation spectra (37). The pressure-tuning results suggest that the strongly anisotropic fluctuation spectrum observed for $x = 0.1$ at ambient pressure, which can be modeled by quasi-2D fluctuations as discussed above, prevails. One may expect that likewise the unexpected energy-temperature scaling of the dynamic susceptibility survives at the QPT under pressure. On the other hand, magnetic field appears to drive the system towards a more isotropic 3D fluctuation spectrum. Indeed, preliminary experiments suggest that the dynamical susceptibility at the field-induced QCP exhibits a Lorentzian behavior ($\alpha = 1$).

3. Phase transitions and non-Fermi-liquid behavior in $UCu_{5-x}Pd_x$ at low temperatures

The antiferromagnetic order of pure UCu_5 below $T_N = 16\,K$ is suppressed quickly upon Cu substitution by Pd, reaching $T_N = 0$ around $x = 1$ (12, 38) (see Fig. 4 for the magnetic phase diagram). Beyond $x = 1.5$, a spin-glass state is found up to $x = 2.3$ which is the stability limit for the cubic $AuBe_5$ structure. Several different mechanisms have been proposed to account for the observed NFL behavior (12) in $UCu_{5-x}Pd_x$. One possibility is the proximity to magnetic order, in particular for $x = 1$ and 1.5. On the other hand, the strong structural disorder has been suggested to lead to a distribution $P(T_K)$ of Kondo temperatures T_K (39) mimicking NFL behavior. Inelastic neutron scattering data on $UCu_{5-x}Pd_x$, again for $x = 1$ and 1.5, revealed an unusual frequency temperature-scaling of the dynamic magnetic susceptibility, i.e., $\chi''(\omega, T) \cdot T^{1/3} \sim (T/\omega)^{1/3} f(\omega/T)$ (40). The independence of $\chi''(\omega, T)$ of the transferred neutron momentum suggests a single-ion mechanism for this scaling behavior which was observed at intermediate $T > 10\,K$ and $\hbar\omega > 1\,meV$. Note that the above scaling for $\chi''(\omega, T)$ is similar to that in $CeCu_{5.9}Au_{0.1}$. Finally, in an attempt to combine disorder and magnetic instability, the unusual C/T behavior and the similarly diverging static magnetic susceptibility $\chi(T)$ were interpreted (14) in terms of a Griffiths phase scenario which was proposed to lead to weak power-law divergences, i.e. $C/T \sim \chi \sim T^{-1+\lambda}$ (13).

The proximity to a magnetic instability in $UCu_{5-x}Pd_x$ for $x = 1$ was directly demonstrated through the observation of a spin-glass-like maximum in the *ac* susceptibility χ^{ac} and a maximum in C/T around 0.2 K (41).

Figs. 5(a) and (b) show C/T and χ^{ac} for $UCu_{5-x}Pd_x$ vs. T on log-log plots (15). The high-temperature data can be reasonably well described by the algebraic T dependences suggested by the Griffiths phase model (13), with a fitting range centered around 1 K in order to exclude the low-T spin-glass order not explicitly treated in the model. The spin-glass

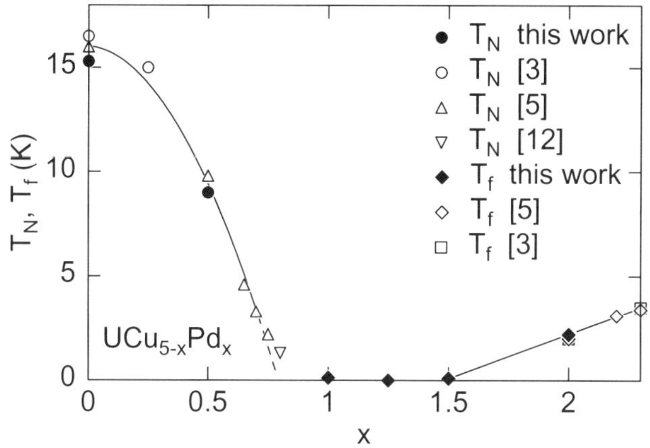

Figure 4. Magnetic phase diagram T vs. x of $UCu_{5-x}Pd_x$. T_N indicates the Néel temperature, T_f the spin-glass freezing temperature.

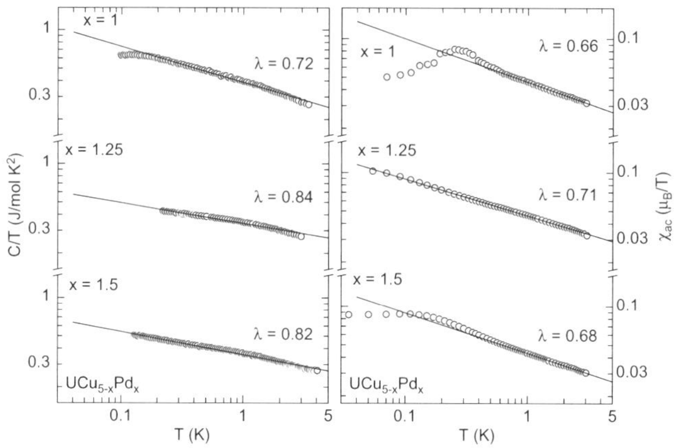

Figure 5. (a) $\ln C/T$ vs. $\ln T$ for $UCu_{5-x}Pd_x$ with $x = 1, 1.25, 1.5$; solid lines are fits of the form $C/T = T^{-1+\lambda}$. (b) $\ln \chi_{ac}$ vs. $\ln T$ for $x = 1, 1.25, 1.5$; solid lines are fits of the form $\chi = T^{-1+\lambda}$.

ordering is clearly visible as a maximum in $\chi^{ac}(T)$ for $x = 1$ and 1.5. The λ values obtained from the specific heat and the linear susceptibility are in reasonable agreement with each other, with systematically larger values for C/T. Interestingly, the λ values are largest for $x = 1.25$, i.e., for the concentration furthest away from the magnetic instabilities. Note that $\lambda = 1$ corresponds to Fermi-liquid behavior.

A further prediction of the Griffiths phase scenario concerns the nonlinear susceptibility which should have the form $\chi_{nl} \propto T^{-3+\lambda}$ (13). Approximate data of χ_{ne} above the spin-glass freezing temperature are consistent with this prediction (15).

4. Superconductivity and ferromagnetism in $ZrZn_2$

The compound $ZrZn_2$ is ferromagnetic (42) despite being made from non-magnetic and even superconducting elements. The magnetic properties are believed to derive primarily from the Zr 4*d* orbitals that have a significant direct overlap (43). Ferromagnetism develops below the Curie temperature $T_{FM} = 28.5\,K$ with an ordered moment $\mu_s = 0.17\mu_B$ per formula unit. $ZrZn_2$ has a large electronic heat capacity at low temperatures, $C/T \approx 47\,mJ/molK^2$, signaling the presence of many low-energy magnetic excitations in addition to spin waves (44). The low T_{FM} and small ordered moment make $ZrZn_2$ unique among stoichiometric ferromagnetic metals and indicate that the compound is close to a ferromagnetic QPT. This proximity has led to numerous proposals that $ZrZn_2$ might be a superconductor (17, 46).

For the highest-quality sample with a low-T residual resistivity of $\rho_0 = 0.62\,\mu\Omega cm$, we observe a rapid drop in the electrical resistivity $\rho(T)$ below $T_{SC} = 0.29\,K$ at ambient pressure (inset of Fig. 6), suggesting an incomplete transition to a zero-resistance state. Application of a field of 0.2 T suppresses the drop as would be expected for a superconducting transition. In addition, a clear diamagnetic signal in the *ac* susceptibility associated with superconducting screening is observed below T_{SC}. For the lowest excitation amplitudes, χ^{ac} approaches -0.65 as $T \to 0$, comparable with the ideal value of -1 (18). This diamagnetic signal rides on top of a large ferromagnetic background.

It has been known for a long time that ferromagnetism in $ZrZn_2$ is rapidly suppressed under pressure p (47). Fig. 6 summarizes the effect of pressure on T_{FM} and T_{SC}. p suppresses both ferromagnetism and superconductivity above a critical pressure of $p_c = 21\,kbar$. In view of its sensitivity to the quality of the sample, the superconductivity in $ZrZn_2$ is likely to be unconventional. The fact that superconductivity in $ZrZn_2$ only occurs in the presence of ferromagnetism where T_{SC} is relative insensitive against pressure, and is hence promoted by the ferromagnetic state, may arise naturally in scenarios where the Cooper pairs are in a parallel-spin (triplet) state, which is already favored in the ferromagnetic state. Such behavior could well be universal for itinerant ferromagnets in the limit of small Curie temperature and long electron mean free path. Further work has to establish the microscopic relation between ferromagnetism and superconductivity and to investigate if and how the ferromagnet is affected by the onset of superconductivity.

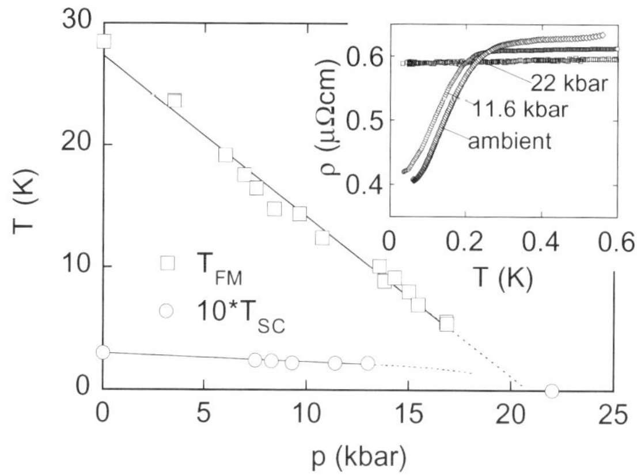

Figure 6. Pressure dependence of the ferromagnetic temperature T_{FM} and of T_{SC}. A few typical $\rho(T)$ curves are shown in the inset. Note that T_{SC} for clarity is magnified by a factor of ten.

Acknowledgements

This work reviewed here was done in collaboration with C. Pfleiderer, A. Schröder, O. Stockert, and R. Vollmer We thank our coworkers and colleagues for help and discussions: F. Huster, A. Neubert, T. Pietrus, M. Sieck, U. Tutsch, M. Waffenschmidt and B. Will; G. Aeppli, N. R. Bernhoeft, R. Chau, P. Coleman, J. Flouquet, S. M. Hayden, A. D. Huxley, M. Loewenhaupt, G. G. Lonzarich, M. B. Maple, G. Müller-Vogt, N. Pyka, A. Rosch and P. Wölfle. We would like to acknowledge support of the European Science Foundation within the program on Fermi-liquid instabilities in correlated metals (FERLIN), and the Deutsche Forschungsgemeinschaft (DFG).

References

1. See, e.g.: Proceedings of the Conference on Non-Fermi-Liquid Behavior in Metals, Santa Barbara (1996), published as special issue of J. Phys.: Cond. Matt., **8**, 9689, (1996).
2. J. A. Hertz, Phys. Rev. B, **14**, 1165, (1976).
3. A. J. Millis, Phys. Rev. B, **48**, 7293, (1993).
4. T. Moriya, T. Takimoto, J. Phys. Soc. Jpn., **64**, 960, and refs. therein, (1995).
5. H. v. Löhneysen et al., Phys. Rev. Lett., **72**, 3262, (1994).
6. B. Bogenberger, H. v. Löhneysen, Phys. Rev. Lett., **74**, 1016, (1995).
7. H. v. Löhneysen et al., Eur. Phys. J. B, **5**, 447, and refs. therein, (1998).

8. H. v. Löhneysen, J. Magn. Magn. Mat., **200**, 532, (1999).
9. A. Schröder, G. Aeppli, E. Bucher, R. Ramazashvili, P. Coleman, Phys. Rev. Lett., **80**, 5623, (1998).
10. O. Stockert, H. v. Löhneysen, A. Rosch, N. Pyka, M. Loewenhaupt, Phys. Rev. Lett., **80**, 5627, (1998).
11. A. Schröder et al., Nature, **407**, 351, (2000).
12. B. Andraka and G. R. Stewart, Phys. Rev. B, **47**, 3208, (1993).
13. A. M. Castro, G. Castilla, B. A. Jones, Phys. Rev. Lett., **81**, 3531, (1998).
14. M. C. de Andrade et al., Phys. Rev. Lett., **81**, 5620, (1999).
15. R. Vollmer, T. Pietrus, H. v. Löhneysen, R. Chau, M. B. Maple, Phys. Rev. B, **61**, 1218, (2000).
16. S. Saxena et al., Nature, **406**, 587. (2000).
17. D. Fay, J. Appel, Phys. Rev. B, **22**, 3173, (1980).
18. C. Pfleiderer et al., Nature, **412**, 58; ibid **412**, 660, (2001).
19. Y. Onuki, T. Komatsubara, J. Magn. Magn. Mat., **63 & 64**, 281, (1987).
20. A. Amato et al., J. Low Temp. Phys., **68**, 371, (1987).
21. E. A. Schuberth, J. Schupp, R. Freese, K. Andres, Phys. Rev. B, **51**, 12892, (1995).
22. L. Pollack et al., Phys. Rev. B, **52**, R15707, (1995).
23. H. Tsujii et al., Phys. Rev. Lett., **84**, 5407, (2000).
24. G. Aeppli et al., Phys. Rev. Lett., **57**, 122, (1986).
25. J. Rossat-Mignod et al., J. Magn. Magn. Mat., **76 & 77**, 376, (1988).
26. O. Stockert et al., to be published.
27. T. Pietrus, B. Bogenberger, S. Mock, M. Sieck, H. v. Löhneysen., Physica B, **206 & 207**, 317, (1995).
28. A. Germann, A. K. Nigam, J. Dutzi, A. Schröder, H. v. Löhneysen, J. Physique Coll., **49**, C8-755, (1988).
29. A. Schröder et al., Physica B, **199 & 200**, 47, (1994).
30. H. Okumura et al., J. Magn. Magn. Mat., **177-181**, 405, (1998).
31. A. Rosch, A. Schröeder, O. Stockert, H. v. Löhneysen., Phys. Rev. Lett., **79**, 150, (1997).
32. Q. Si, S. Raballo, K. Ingersent, J. L. Smith, Nature, **413**, 804, (2001).
33. P. Coleman, C. Pepin, Q. Si, R. Ramazashvili, J. Phys.: Condens. Matt., **13**, R723, (2001).
34. A. Germann, H. v. Löhneysen, Europhys. Lett., **9**, 367, (1989).
35. K. Heuser, E.-W. Scheidt, T. Schreiner, G. R. Stewart, Phys. Rev. B, **57**, R4198, (1998).
36. K. Heuser, E.-W. Scheidt, T. Schreiner, G. R. Stewart, Phys. Rev. B, **58**, R15959, (1998).
37. H. v. Löhneysen, C. Pfleiderer, T. Pietrus, O. Stockert, B. Will, Phys. Rev. B, **63**, 134411, (2001).
38. R. Chau and M. B. Maple, J. Phys.: Condens. Matt., **8**, 9939, (1996).
39. O. O. Bernal, D. E. McLaughlin, H. G. Lukefahr, B. Andraka, Phys. Rev. Lett., **75**, 2023, (1995).
40. M. C. Aronson et al., Phys. Rev. Lett., **75**, 725, (1995).
41. R. Vollmer et al., Physica B, **230-232**, 603, (1997).
42. T. B. Matthias, R. M. Bozorth, Phys. Rev., **109**, 604, (1958).
43. T. Jarlborg, A. J. Freeman, D. D. Kölling, J. Magn. Magn. Mat., **23**, 291, (1981).
44. N. R. Bernhoeft et al., Phys. Scr., **38**, 191, (1988).
45. A. P. J. van Deusen et al., J. Magn. Magn. Mat., **54**, 1113, (1986).
46. K. B. Blagoev, J. R. Engelbrecht, K. S. Bedell, Phys. Rev. Lett., **82**, 133, (1998).

47. T. F. Smith, J. A. Mydosh, E. P. Wohlfahrt, Phys. Rev. Lett., **27**, 1732, (1971).
48. E. B. Sonin, I. Felner, Phys. Rev. B, **57**, R14000, (1998).
49. P. Fulde, R. A. Ferrell, Phys. Rev. A, **135**, 550, (1964).
50. A. I. Larkin, Y. N. Ovchinnikov, Sov. Phys. JETP, **20**, 762, (1975).

HEAVILY DOPED SEMICONDUCTORS: MAGNETIC MOMENTS, ELECTRON-ELECTRON INTERACTIONS AND THE METAL-INSULATOR TRANSITION

Proceedings of the ARW NATO Workshop Hvar, Croatia, October 2002

H. V. LÖHNEYSEN
Physikalisches Institut, Universität Karlsruhe, D-76128 Karlsruhe, Germany and
Forschungszentrum Karlsruhe, Institut für Festkörperphysik, D-76021 Karlsruhe, Germany

Abstract. Electron localization in a metal, ultimately leading to a metal-insulator transition (MIT), can occur because of disorder (Anderson transition) or electron-electron interactions (Mott-Hubbard transition). Both effects play a role in heavily doped semiconductors which have become prototype systems for the study of MIT. In this review we focus on phosphorus-doped Si. The MIT in Si:P can be tuned by varying the P concentration or – for barely insulating samples – by application of uniaxial stress S. The continuous stress tuning allows the observation of dynamic scaling of $\sigma(T, S)$ and hence a reliable determination of the critical exponent μ of the extrapolated zero-temperature conductivity $\sigma(0) \sim | S - S_c |^{\mu}$, i.e. $\mu = 1$, and of the dynamical exponent $z = 3$.

Key words: Doped Semiconductors, Metal-Insulator Transition, Electron-Interactions, Anderson Transition, Hubbard Splitting, Localized Magnetic Moments, Si:P, Thermo(electric) Power, Critical Behaviour, Stress Tuning

1. Introduction

Heavily doped semiconductors have become prototype systems for the study of a metal-insulator transition (MIT) driven by the combined effects of disorder (Anderson transition) and electron-electron interactions (Mott-Hubbard transition). In a strict sense, the MIT viewed as a transition from extended to localized states at the Fermi level, occurs only at $T = 0$ because at finite temperature T thermally activated carrier transport may occur. Hence a metal is defined by a finite *dc* conductivity $\sigma(T)$ for $T \rightarrow 0$, while

A.C. Hewson and V. Zlatić (eds.), Concepts in Electron Correlation, 155–167.

for an insulator $\sigma(0) = 0$. The determination of the critical exponent μ describing how $\sigma(0)$ vanishes at $T = 0$ as a function of a control parameter t at a critical value t_c, i.e. $\sigma(0) \sim\mid t - t_c \mid^{\mu}$, has been a subject of considerable controversy. However, using both carrier concentration N and uniaxial stress S to tune the MIT, identifying the critical region of t where critical behavior is expected to occur, and employing a dynamic scaling analysis, a consistent picture is beginning to emerge.

In this review, we will discuss a few aspects of electron-electron interactions and the critical behavior at the MIT focusing on three-dimensional P-doped Si. In section 2 we treat the local magnetic moments on the metallic side and evidence for a Kondo effect will be presented. The evidence for a Hubbard gap will be discussed in section 3. In section 4 we present results on the critical behavior of the conductivity at the MIT. In particular, we will discuss dynamic scaling properties. A more comprehensive review can be found elsewhere (1).

2. Metallic Si:P: formation of local magnetic moments and Kondo effect

It has been known for a long time that a small fraction of localized moments derived from the spin-degenerate groundstate of the hydrogen-like donor or acceptor wave function on the insulating side, survive on the metallic side of the MIT. These localized magnetic moments have been mostly detected with magnetization (2, 3), magnetic resonance (4, 5) and specific-heat measurements (6, 7). The dependence of the concentration of localized moments on the P concentration N has been mapped out systematically from specific-heat measurements for Si:P (8, 9) and more recently also for Si:(P,B) (10).

The data for Si:P are shown in figure 1 with the overall behavior of Si:(P,B) being quite similar. The concentration of local moments has been determined (i) from zero-field excess specific heat $\Delta C = C - \gamma T - \beta T^3$, i.e. conduction electron and phonon contributions subtracted from the measured specific heat C, via the entropy $S = \int(\Delta C/T)\mathrm{d}T = N_S \ln 2$, and (ii) from the Schottky anomaly observed in applied magnetic fields $B = 1.5$ and 5.7 T, yielding N_{Sch}. In both cases, $S = 1/2$ is assumed for the spin. N_S and N_{Sch} vary smoothly through the critical concentration $N_c = 3.52 \cdot 10^{18}\mathrm{cm}^{-3}$ and localized moments survive in the metallic state due to the disorder and on-site Coulomb repulsion as will be discussed in detail below.

The random distribution of dopant atoms, which has been demonstrated convincingly by scanning tunneling microscopy (12), leads to a wide distribution of nn distances and, hence, nn exchange couplings. This wide

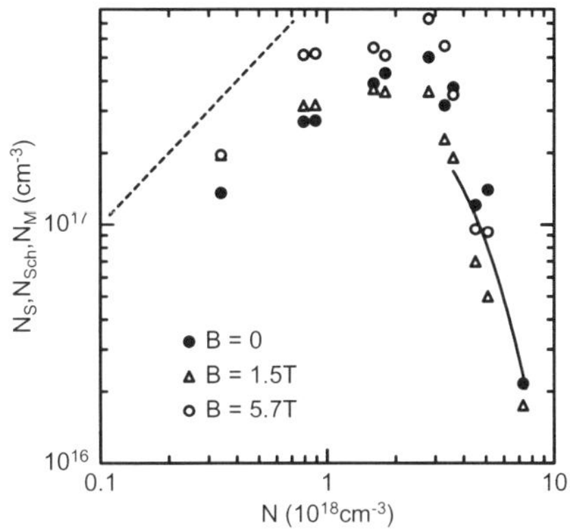

Figure 1. Concentration of magnetic moments N_S determined experimentally in zero field and N_{Sch} in finite magnetic fields 1.5 and 5.7 T versus doping concentration. Dashed line indicates $N_{Sch} = N$ expected in the dilute insulating limit. Solid line gives the concentration of local moments N_M in the metallic phase as calculated theoretically (9).

distribution of exchange couplings led Bhatt and Lee (11) to a description of the magnetic properties of doped semiconductors in terms of a hierarchical coupling of antiferromagnetic pairs of localized moments. In this model, only free spins which are not coupled to pairs at a given T (with density $N_{free}(T)$) contribute to the specific heat at that temperature with

$$C_{loc} = k_B T \cdot \ln 2 \cdot \mathrm{d}N_{free}(T)/\mathrm{d}T \tag{1}$$

while the magnetic susceptibility is given by

$$\chi_{loc} = \mu_0 \mu_B^2 N_{free}(T)/k_B T. \tag{2}$$

In order to test the validity of the Bhatt-Lee model one can calculate C_{loc} from χ_{loc} with the above equations and compare it to ΔC as measured directly (8, 9). The overall good agreement found on the insulating side (2, 13) indicates the validity of the Bhatt-Lee model. However, with increasing $N \geq 3.3 \cdot 10^{18}\mathrm{cm}^{-3}$, i.e. on the metallic side of the MIT, the agreement quickly deteriorates (13), indicating that the Bhatt-Lee model is unsuited for the metallic state.

The origin of localized moments in the metallic state has been investigated in a number of approaches (14, 15, 16, 17, 9), most of them starting with an Anderson-Hubbard model which is the simplest model Hamiltonian featuring the essential elements of disorder, through random hopping integrals t_{ij}, and on-site Coulomb interaction U:

$$H = \sum_{i\sigma} (\varepsilon_i - \mu) n_{i\sigma} + \sum_{ij\sigma} t_{ij} c_{i\sigma}^{+} c_{j\sigma} + U \sum_i n_{i\uparrow} n_{i\downarrow} \tag{3}$$

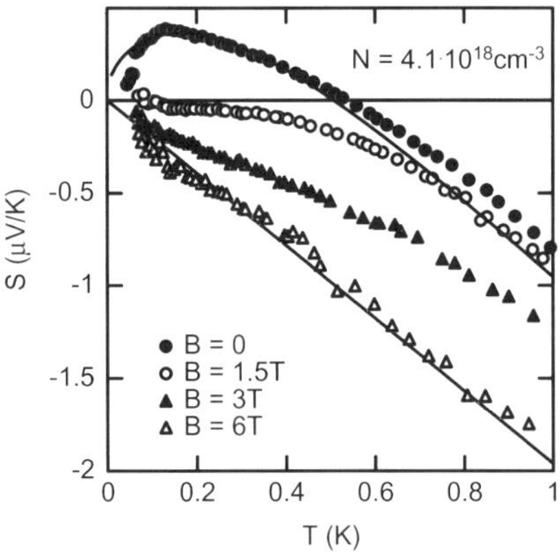

Figure 2. Thermoelectric power S vs. temperature T for different magnetic fields B for a metallic Si:P. Solid line is the best fit of a Kondo thermoelectric power with $T_{\mathrm{K}} = 0.8\,\mathrm{K}$ to the zero-field data (18).

$c_{i\sigma}(c^{+}_{i\sigma})$ is the annihilation (creation) operator for an electron with spin projection σ in the $1s(A_1)$ groundstate of the donor atom at site i, and $n_{i\sigma} = c^{+}_{i\sigma}c_{i\sigma}$. It is usually assumed that the five excited states generated by valley-orbit splitting are sufficiently far away (experimentally $\sim 10\,\mathrm{meV}$, see below) that a single-band model is applicable. The host semiconductor properties enter only through the fact that the positions i are random sites of the Si lattice and that the effective mass and interaction are renormalized. The various approaches differ in the type of mean-field model employed. In the approach taken by Langenfeld and Wölfle (17) an isolated moment ("impurity") forming in an effective homogeneous medium is considered. The calculated concentration N_M of local moments is in good agreement with the experimentally determined values N_S and N_{Sch} on the metallic side (Fig. 1).

The coupling of the localized moments to the itinerant electrons by an effective exchange interaction J_{eff} gives rise to the Kondo effect. Because of disorder, both the density of states at the Fermi level $N(E_F)$ and the effective exchange interaction $J_{eff} \approx \bar{t}^2/U$ where $\bar{t}$ is the average hopping amplitude from the local-moment site to the neighboring sites, will fluctuate. This leads to a wide distribution $P(T_K)$ of Kondo temperatures $T_K \sim \exp(-(1/N(E_F)J_{eff})$. An approximate power-law behavior $P(T_K) \sim T_K^{-\alpha_K}$ is found, with $\alpha_K \approx 0.9$, which may account for ΔC in the metallic state of Si:P (9).

The thermoelectric power S clearly gives evidence for scattering by localized magnetic moments (18) because it is particularly sensitive to the Kondo effect. Fig. 2 shows $S(T)$ for a Si:P sample with $N = 4.1 \cdot 10^{18}\mathrm{cm}^{-1}$.

The maximum of $S(T)$ is attributed to magnetic scattering since it is observed only for N slightly above N_c where an appreciable density of localized moments exists. Even more convincing is the suppression of the $S(T)$ maximum in large magnetic fields, also shown in Fig. 2. In $B = 6\,\mathrm{T}$ we recover the negative diffusion thermoelectric power $S \sim -T$ observed for $N \gg N_c$ in zero field (18). A very similar behavior of $S(T)$ has been found for compensated Si:(P,B) (19). Assuming a Kondo-derived thermoelectric power one can compare the $S(T)$ maximum to a corresponding single-ion expression derived by Maki (20), cf. Fig. 2. The deviations between data and best fit might be due to the neglect of a T_K distribution in the fit where a single-valued $T_K = 0.8\,\mathrm{K}$ is assumed.

Despite a number of important differences, transport processes of compensated vs. uncompensated Si:P on the metallic side not too close to the MIT are overall similar. For instance, we have just seen that the scattering by magnetic moments leads to a similar behavior of the thermoelectric power in both types of materials. It has been demonstrated with infrared reflection measurements that the MIT in Si:P occurs for electronic states in the impurity band which even for P concentrations $N \approx 2N_c$ is still energetically separated from the conduction band (21, 22). This was inferred from features at 10 and 45 meV in the reflection spectra that could be attributed to optical transitions within the valley-orbit split impurity multiplet, i.e. from the $1s(A_1)$ state to the (nearly degenerate) $1s(E)$ or $1s(T)$ states, and to the conduction band, respectively. That the impurity band is energetically separated from the conduction band at the MIT was actually suggested long time ago by Mott, based on some more indirect evidence (23). Again, overall similar behavior is found for the infrared reflection of compensated Si:(P,B) (21, 22).

3. Insulating Si:P: evidence for Hubbard splitting from transport properties

Marked differences in transport properties between Si:P and Si:(P,B) occur well on the insulating side (24). Fig. 3 shows the thermoelectric power plotted as $-S \cdot N$ vs. T on a log-log plot. (Here the T range is above 1 K so that the data should not be confused with the low-T Kondo anomaly in metallic samples, i.e. the positive $S(T)$ maximum superimposed on an overall negative diffusion thermopower with a concomitant sign change of $S(T)$.) For sufficiently high T ($\gtrsim 15\,\mathrm{K}$) S is always negative. While the $-S \cdot N$ curves for compensated Si:(P,B) fall on an almost universal curve for carrier concentration N between 1.5 and $2.95 \cdot 10^{18}\mathrm{cm}^{-3}$, such a scaling is seen for uncompensated Si:P only in the concentration range above $N_0 = 2.78 \cdot 10^{18}\mathrm{cm}^{-3}$. Below that concentration $S(T)$ exhibits a sign change

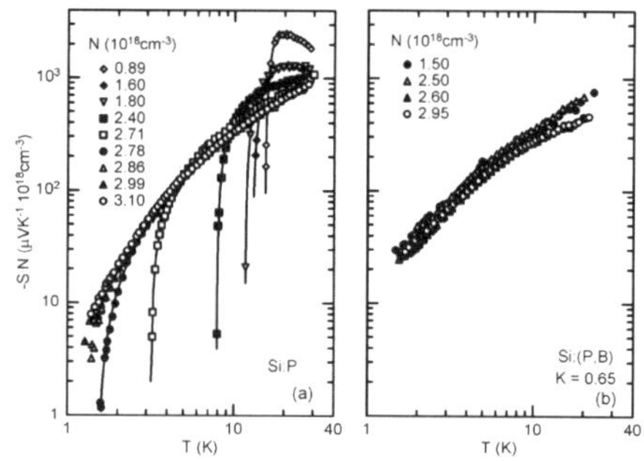

Figure 3. Negative thermoelectric power times carrier concentration $-SN$ vs. temperature T on a log-log plot for insulating samples of (a) Si:P and (b) Si:(P,B) (24)
.

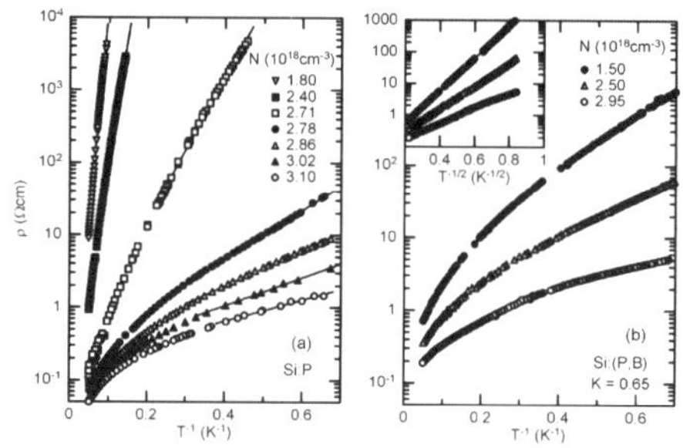

Figure 4. Electrical resistivity ρ plotted vs. inverse temperature for (a) Si:P and (b) Si:(P,B). Straight lines in (a) indicate fits to obtain the activation energy E_2. Inset in (b) shows the Si:(P,B) data plotted as log ρ vs $T^{-1/2}$. Straight lines indicate Efros-Shklovskii hopping (24).

at a temperature $T_{S=0}$ which rapidly shifts to higher values with decreasing N. This sign change from $S < 0$ to $S > 0$ with decreasing T is visible in Fig. 3 as a precipitous drop of $\log(-S \cdot N)$ with decreasing T.

A further strong difference between uncompensated and compensated samples is seen in the electrical resistivity $\rho(T)$ (Fig. 4). For $N < N_0$, ρ rises much faster with decreasing T for Si:P than for Si:(P,B), while the behavior is similar for both types of material $N > N_0$. The strong qualitative difference in $S(T)$ and $\rho(T)$ in Si:P upon crossing N_0 points to different dominant transport processes above and below N_0. Interpreting the steep $\rho(T)$ increase of Si:P below N_0 as an activated process, we can extract an activation energy E_2.

Fig. 5 shows a comparison of E_2 and $T_{S=0}$. Despite the order-of-magnitude difference in absolute values of E_2/k_B and $T_{S=0}$, the similarity of the con-

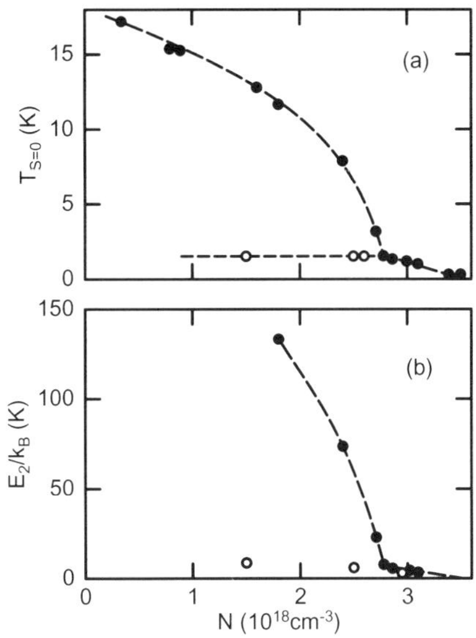

Figure 5. (a) Temperature $T_{S=0}$ of the thermoelectric power zero and (b) activation energy E_2 vs. carrier concentration N for uncompensated Si:P (closed circles) and Si:(P,B) (open circles) (24).

centration dependence is striking. Hence we interpret the sign change of $S(T)$ as the onset of an activated process. If one were forced to assume an E_2 process also for Si:(P,B) – although the data actually suggest Efros-Shklovskii variable range hopping (25) with an exponent $p = 1/2$, see inset of Fig. 4 – one would obtain the open circles in Fig. 5, i.e. no feature appears at N_0.

The sudden appearence of a hard gap only in Si:P at N_0 well below N_c suggests that we are observing the Hubbard gap due to the on-site Coulomb repulsion. It has been proposed already many years ago that the E_2 process is indeed due to the Hubbard gap (26). While the negatively charged isolated P donor is barely stable, i.e. U is of the order of the ionization energy, it is very likely that the on-site electron repulsion weakens progressively as the P concentration (and carrier concentration) increases, until at $N_0 = 2.78 \cdot 10^{18}\text{cm}^{-3} \approx 0.8\,N_c$ the two Hubbard bands start to overlap. Because of disorder, however, metallic behavior does not incur immediately since the tail states of the Hubbard bands are localized. It is only at the critical concentration N_c that the Hubbard bands are so close that the chemical potential is within the range of extended states. This scenario is schematically depicted in Fig. 6, yielding a physical picture of how disorder and electron-electron interactions drive the MIT in Si:P. Of

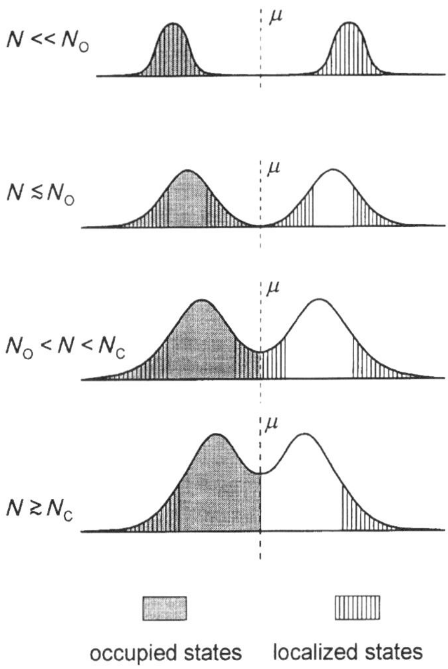

Figure 6. Qualitative sketch of the density of states of the impurity band of uncompensated Si:P for several P concentrations N, indicating the splitting into lower and upper Hubbard band around the chemical potential. See text for details.

course, the Hubbard features are absent in Si:(P,B) as they should because compensated semiconductors are away from half-filling. It is interesting to note that the density of localized magnetic moments in Si:P is not maximal at N_c but peaks rather precisely at N_0 where the two Hubbard bands in our scenario are suggested to merge, see Fig. 6.

4. Critical behavior of the conductivity at the metal-insulator transition

While early suggestions considered the MIT to be discontinuous (23), scaling approaches for noninteracting electrons suggested the existence of a continuous second-order phase transition for three-dimensional systems (27) where the dc conductivity at $T = 0$ is expected to vary as $\sigma(0) \sim | t - t_c |^{\mu}$. Theoretically, μ is usually inferred from the correlation-length critical exponent ν via Wegner scaling $\mu = \nu(d - 2)$ where d is the spatial dimension. In a self-consistent theory of Anderson localization (neglecting interactions), an exponent $\mu = 1$ for $d = 3$ is suggested (28). Field theoretical approaches are discussed in Ref. (29). Values of ν derived from numerical studies of noninteracting systems range between 1.3 and 1.6 (30, 31, 32).

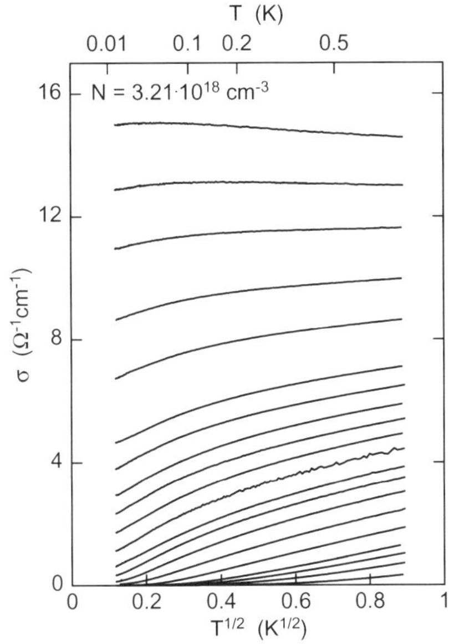

Figure 7. Conductivity σ of a Si:P sample with P concentration $N = 3.21 \cdot 10^{18}\mathrm{cm}^{-3}$ versus $\sqrt{T}$ for several values of uniaxial stress S applied along the [100] direction. From top to bottom: S = 3.05, 2.78, 2.57, 2.34, 2.17, 2.00, 1.94, 1.87, 1.82, 1.77, 1.72, 1.66, 1.61, 1.56, 1.50, 1.41, 1.33, 1.26, 1.18, 1.00 kbar. Solid lines are connecting the very finely spaced individual data points.

Experimentally, uncompensated semiconductors, were reported to show $\mu = 0.5$, in contrast to $\mu = 1$ generally found for compensated semiconductors and most amorphous metals (33). The exponent $\mu = 0.5$ was largely based on experiments (35, 34, 36) where uniaxial stress was used to drive an initially insulating Si:P sample metallic. We suggested a few years ago, on the basis of an observed crossover in concentration-tuned uncompensated Si:P from an exponent $\mu \approx 1.3$ close to N_c where $\mathrm{d}\sigma/\mathrm{d}T > 0$, to $\mu = 0.64$ above a crossover concentration $N_{cr} \approx 1.1\,N_c$ where $\mathrm{d}\sigma/\mathrm{d}T$ changes sign, to limit the critical region of the MIT on the metallic side to concentrations where σ decreases with decreasing T, i.e. to the range $N_c < N < N_{cr}$ (37).

Fig. 7 shows the electrical conductivity $\sigma(T)$ of a barely insulating sample ($N = 3.21 \cdot 10^{18}\mathrm{cm}^{-3}$) for uniaxial stress S between 1 and 3.05 kbar (38). The data are plotted vs. $\sqrt{T}$ which is the T dependence expected due to e-e interactions and indeed observed well above the MIT (36). The curves are in fact polygons connecting adjacent data points. For S between 1 and 2.57 kbar the $\sigma(T)$ curves evolve smoothly from insulating to metallic behavior with $m > 0$, and $\sigma(T)$ becomes nearly independent of T with a value $\sigma_{cr} \approx 12\,\Omega^{-1}\mathrm{cm}^{-1}$ at ≈ 2.7 kbar. Our data do not exhibit the

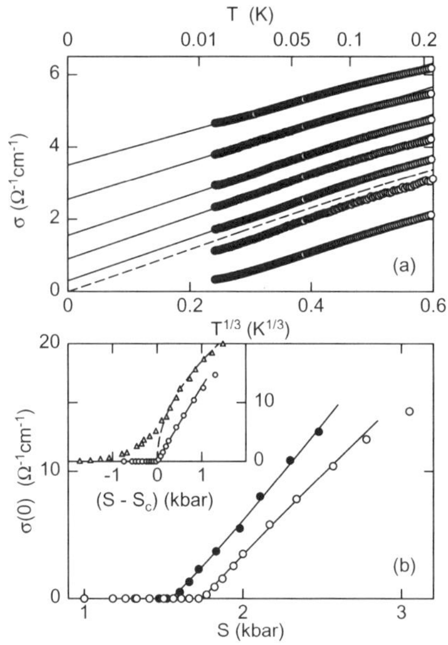

Figure 8. (a) Low-temperature data of σ of Fig. 1 for stress in the immediate vicinity of the metal-insulator transition plotted against $T^{1/3}$. Dashed line indicates the conductivity at the critical stress (see text). (b) Extrapolated conductivity $\sigma(0)$ for $T \to 0$ versus uniaxial stress S for two P concentrations $N = 3.21$ and $3.43 \cdot 10^{18}\text{cm}^{-3}$ (open and closed circles, respectively). The inset shows earlier $\sigma(0)$ versus $S - S_c$ data (triangles) together with the reported $\mu = 0.5$ (dashed line) from (35) in comparison to our data for sample 1 (circles) (38).

precipitous drop of $\sigma(T)$ below $\sim 40\,\text{mK}$ for S close to the MI transition, in distinction to the earlier stress-tuning data on Si:P extending to 3 mK (35, 34).

Closer inspection shows that the data near the MIT are actually better described by a $T^{1/3}$ dependence for low T, see Fig. 8a. $\sigma(0)$ obtained from the $T^{1/3}$ extrapolation to $T = 0$ is shown in Fig. 8b, together with data for a sample closer to the critical concentration, yielding $S_c = 1.75\,\text{kbar}$ and 1.54 kbar for 3.21 and $3.43 \cdot 10^{18}\text{cm}^{-3}$, respectively. Note that the critical stress S_c is quite well defined, as $\sigma(0)$ breaks away roughly linearly from zero within less than 0.1 kbar, i.e. $\mu \approx 1$. This behavior contrasts with the earlier stress-tuning data (35) reproduced in the inset of Fig. 8b, where appreciable rounding close to N_c is visible. However, those $\sigma(0)$ data between 4 and $16\,\Omega^{-1}\text{cm}^{-1}$ are compatible with a linear dependence on uniaxial stress.

In order to analyze the scaling behavior of σ at finite temperatures using the data of the sample with $N = 3.21 \cdot 10^{18}\text{cm}^{-3}$, we employ the scaling

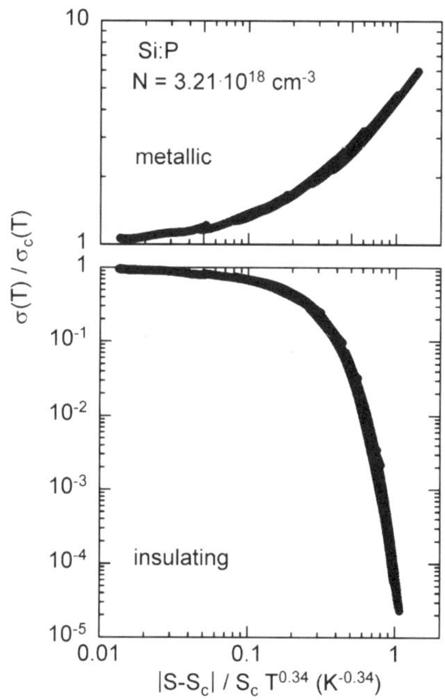

Figure 9. Scaling plot of σ/σ_c vs. $\mid S - S_c \mid /S_c T^y$ for Si:P with $N = 3.21 \cdot 10^{18}\text{cm}^{-3}$ at different uniaxial pressures S, with $S_c = 1.75\,\text{kbar}$ and $y = 0.34$ (38).

relation (39)

$$\sigma(t, T) = \sigma_c(T)\mathcal{F}'((t - t_c)/T^y) \tag{4}$$

where $\sigma_c(T) = \sigma(t_c, T)$ is the conductivity at the critical value t_c of the parameter t (in our case S) driving the MIT. If the leading term to $\sigma_c(T)$ is proportional to T^x, one obtains $x = \mu/\nu z$ and $y = 1/\nu z$ from a scaling plot. Fig. 7 shows that σ for S close to S_c does not exhibit a simple power-law T dependence over the whole T range investigated. We therefore describe $\sigma_c(T)$ by the function $\sigma_c(T) = aT^x(1 + dT^w)$ with $a = 6.01\,\Omega^{-1}\text{cm}^{-1}$, $x = 0.34$, $d =$ -0.202, $w = 0.863$, and T is expressed in K. Here the dT^w term presents a correction to the critical dynamics. All $\sigma(S, T)$ curves with $1.00\,\text{kbar} < S < 2.34\,\text{kbar}$ where $\text{d}\sigma/\text{d}T \geq 0$ are then used for the scaling analysis.

Fig. 9 shows the scaling plot of $\sigma(S, T)/\sigma_c(T)$ vs. $\mid S - S_c \mid /S_c T^y$. The data are seen to collapse on a single branch each for the metallic and insulating side, respectively. The best scaling, as shown, is achieved for $y = 1/z\nu = 0.34$. Together with $\mu = 1.0$ as obtained from Fig. 8b and assuming Wegner scaling $\nu = \mu$ for $d = 3$, we find $z = 2.94$, consistent with $\sigma_c \sim T^{1/z} \sim T^{1/3}$ for $T \rightarrow 0$ used to obtain $\sigma(0)$, see Fig. 8a. It is the consistent combination of both approaches, $\sigma(0)$ scaling and dynamic scaling, that lends confidence to the results.

From an analysis of $\sigma(T)$ of Si:P for different P concentrations N we had previously inferred $\mu = 1.3$ from $\sigma(0)$ vs. N (37) and $z = 2.4$ from a dynamic scaling analysis of metallic samples only (40), broadly consistent with the values obtained from the present stress-tuning study. In addition, the Mott temperature evaluated for insulating samples from the variable range hopping conductivity, vanishes at the same concentration N_c as determined for metallic samples (41). Recently, Bogdanovich et al. (39) demonstrated that $\sigma(S,T)$ for Si:B obeys very nicely dynamic scaling on both metallic and insulating sides, yielding $\mu = 1.6$ and $z = 2$, while concentration tuning of $\sigma(0)$ on the same system had suggested $\mu = 0.63$ (42). This large difference is not understood at present. For Si:P, the exponents μ and z are in broad agreement with the expectation for a noninteracting system. This is in line with our above schematic scenario for the MIT (cf. Fig. 6) where the transition proper is driven by disorder although interactions are vital in inducing the Hubbard splitting.

5. Acknowledgments

The work reviewed here grew out of a very fruitful collaboration with students, post-docs and colleagues. Their contributions can be identified from the references cited. In particular, I would like to acknowledge M. Lakner, X. Liu, C. Pfleiderer, H. G. Schlager, C. Sürgers, T. Trappmann and S. Waffenschmidt. I am grateful to P. Wölfle for numerous enlightening discussions on the theoretical aspects of the metal-insulator transition.

References

1. H. v. Löhneysen, Advances in Solid State Physics (Ed. by B. Kramer), Vieweg Braunschweig/Wiesbaden (2000), p. 143.
2. Y. Ootuka and N. Matsunaga, J. Phys. Soc. Jpn., **59**, 1801 (1990).
3. A. Roy and M. P. Sarachik, Phys. Rev. B, **37**, 5531 (1988).
4. M. A. Paalanen, S. Sachdev, A. E. Ruckenstein, Phys. Rev. Lett., **57**, 2061 (1986).
5. H. Alloul and P. Dellouve, Phys. Rev. Lett., **59**, 578 (1978).
6. J. R. Marko, J. P. Harison, J. D. Quirt, Phys. Rev. B, **10**, 2448 (1974).
7. N. Kobayashi, S. Ikehata, S. Kobayashi, W. Sasaki, Solid State Commun., **24**, 67 (1977).
8. M. Lakner and H. v. Löhneysen, Phys. Rev. Lett., **83**, 648 (1989).
9. M. Lakner, H. v. Löhneysen, A. Langenfeld, P. Wölfle, Phys. Rev. B **50**, 17064 (1994).
10. S. Wagner, M. Lakner, H. v. Löhneysen, Phys. Rev. B **55**, 4219 (1997).
11. R. N. Bhatt and P. A. Lee, Phys. Rev. Lett., **48**, 344 (1982).
12. T. Trappmann, C. Sürgers, H. v. Löhneysen, Europhys. Lett., **38**, 177 (1997).

13. H. G. Schlager and H. v. Löhneysen, Europhys. Lett., **40**, 661 (1997).
14. M. Milovanović, S. Sachdev, R. N. Bhatt, Phys. Rev. Lett., **63**, 82 (1989).
15. R. N. Bhatt and D. S. Fisher, Phys. Rev. Lett., **68**, 3072 (1992).
16. V. Dobrosavljević and G. Kotliar, Phys. Rev. Lett., **71**, 3218 (1993).
17. A. Langenfeld and P. Wölfle, Ann. Phys. (Leipzig), **4**, 43 (1995).
18. M. Lakner and H. v. Löhneysen, Phys. Rev. Lett., **70**, 3475 (1993).
19. P. Ziegler, M. Lakner, H. v. Löhneysen, Europhys. Lett., **33**, 285 (1996).
20. K. Maki, Prog. Theor. Phys., **41**, 586 (1969).
21. A. Gaymann, H. P. Geserich, H. v. Löhneysen, Phys. Rev. Lett., **71**, 3681 (1993).
22. A. Gaymann, H. P. Geserich, H. v. Löhneysen, Phys. Rev. B, **52**, 16486 (1995).
23. N. F. Mott, *Metal-insulator transitions*, London Taylor & Francis (1990).
24. X. Liu, A. Sidorenko, S. Wagner, P. Ziegler, H. v. Löhneysen, Phys. Rev. Lett., **77**, 3395 (1996).
25. A. L. Efros and B. I. Shklovskii, J. Phys. C, **8**, L49 (1975).
26. N. F. Mott and E. A. Davis, *Electronic processes in non-crystalline materials*, 2nd edn. Oxford: Clarendon (1979).
27. E. Abrahams, P. W. Anderson, D. C. Licciardello, T. V. Ramakrishnan, Phys. Rev. Lett., **42**, 673 (1979).
28. D. Vollhardt and P. Wölfle, in: *Electronic phase transitions*, ed. by W. Hanke and Yu. V. Kopaev, Elsevier B.V. (1992), 1.
29. D. A. Belitz and T. R. Kirkpatrick, Rev. Mod. Phys., **66**, 261 (1994).
30. B. Kramer, K. Broderix, A. MacKinnon, M. Schreiber, Physica A, **167**, 163 (1990).
31. B. Kramer and A. MacKinnon, Rep. Prog. Phys., **56**, 1469 (1993).
32. K. Slevin and T. Ohtsuki, Phys. Rev. Lett., **82**, 382 (1999).
33. G. A. Thomas, Phil. Mag. B, **52**, 479 (1985).
34. G. A. Thomas, M. A. Paalanen, T. F. Rosenbaum, Phys. Rev. B, **27**, 3897 (1983).
35. M. A. Paalanen, T. F. Rosenbaum, G. A. Thomas, R. N. Bhatt, Phys. Rev. Lett., **48**, 1284 (1982).
36. T. F. Rosenbaum, R. F. Milligan, M. A. Paalanen, G. A. Thomas, R. N. Bhatt, W. Lin, Phys. Rev. B, **27**, 7509 (1983).
37. H. Stupp, M. Hornung, M. Lakner, O. Madel, H. v. Löhneysen, Phys. Rev. Lett., **71**, 2634 (1993).
38. S. Waffenschmidt, C. Pfleiderer, H. v. Löhneysen, Phys. Rev. Lett., **83**, 3005 (1999).
39. S. Bogdanovich, M. P. Sarachik, R. N. Bhatt, Phys. Rev. Lett., **82**, 137 (1999).
40. H. Stupp, M. Hornung, M. Lakner, O. Madel, H. v. Löhneysen, Phys. Rev. Lett., **72**, 2122 (1994).
41. M. Hornung, M. Iqbal, S. Waffenschmidt, H. v. Löhneysen, phys. stat. sol. (b), **218**, 75 (2000).
42. P. Dai, Y. Zhang, M. P. Sarachik, Phys. Rev. Lett., **66**, 1914 (1991).

COMPETITION AND COEXISTENCE OF MAGNETIC AND QUADRUPOLAR ORDERING

Proceedings of the ARW NATO Workshop Hvar, Croatia, October 2002

P. FAZEKAS and A. KISS
Research Institute for Solid State Physics and Optics,
P.O.B. 49, H-1525 Budapest 114, Hungary

Abstract. The large number of low-lying states of d- and f-shells supports a variety of order parameters. The effective dimensionality of the local Hilbert space depends on the strength, and kind, of intersite interactions. This gives rise to complicated phase diagrams, and an enhanced role of frustration and fluctuation effects. The general principles are illustrated on the example of the effect of a magnetic field on quadrupolar phase transitions in the Pr-filled skutterudite $PrFe_4P_{12}$.

Key words: Quadrupolar Order, Multipolar Order, $PrFe_4P_{12}$, Skutterudites, Rare Earth Compounds, Pr Compounds, Mean Field Theory, Tricritical Point, Metamagnetic Transition, Quadrupolar Phase Transition in External Magnetic Field

1. Introduction

Transition metal and rare earth compounds show a rich variety of collective behavior: various kinds of ordered phases as well as strongly fluctuating states (spin and orbital liquids). The basic reason is that d- and f-shells have a relatively large number of low-energy states. Crystal field splitting usually reduces this number (the dimensionality $\mathcal{D}$ of the local Hilbert space) considerably below the free ion value, but complicated physics can arise even from $\mathcal{D} = 3$ or 4.

Let us briefly consider some examples. $\mathcal{D} = 3$ is, in one interpretation, the case of $S = 1$ spin models which turned out to have unanticipated phases like the spin nematics (1). A different realization is offered by interacting f-electron models based on the low-lying quasi-triplet of Pr ions in $PrBa_2Cu_3O_{7-\delta}$ where the nature of Pr ordering is still an open issue (2, 3). A literal realization of $\mathcal{D} = 4$ is offered by the Γ_8 ground state of Ce ions in CeB_6 which has a rich phase diagram (4). Alternatively, we

A.C. Hewson and V. Zlatić (eds.), Concepts in Electron Correlation, 169–177.

may think of the fourfold quasidegeneracy arising from the combination of twofold spin degeneracy with twofold orbital degeneracy, which is the simplest model of d-electrons which is capable of supporting either spin or orbital order, or a combination of both. Twofold orbital degeneracy may occur in cubic, tetragonal, and hexagonal environments, and there are quite different versions of the e_g-model. In the cubic e_g model, orbital momentum is completely quenched, but orbital order may still break time reversal invariance in the octupolar phase (5). In contrast, the trigonal e_g states sustain permanent orbital momentum along the threefold axis; such a model should be relevant for understanding the complex behavior of $BaVS_3$ (6, 7). The possibilities of $\mathcal{D} > 4$ models are largely unexplored, but we should mention the $\mathcal{D} = 6$ t_{2g}-models of $LaTiO_3$. There is an intricate relationship between orbital ordering and spin ferromagnetism which would be remarkably difficult to explain without due consideration of the orbital degrees of freedom (8).

As implied above, the definition of $\mathcal{D}$ is not straightforward: it is usually higher than the degeneracy of the ground state level (which would often be small because of small low-symmetry components of the crystal field), but it is not so high as the free-ion value. In CeB_6, the fourfold degenerate Γ_8 level is well separated from the higher-lying Γ_7 which still arises from the Hund's rule ground state; in $PrBa_2Cu_3O_{7-\delta}$, a low-lying doublet and a singlet can be lumped together to give a quasi-triplet which would have Γ_5 character if the symmetry were cubic; but it is, in fact, only tetragonal. In any case, the relevant dimensionality $\mathcal{D} = 3$ is much smaller than 9 which would be the Hund's rule value. It depends primarily on the strength of intersite interaction, which splittings should be considered small.

The highest symmetry of the the $\mathcal{D} = 3$ models would be SU(3), etc. Clearly, the exact realization of a high-symmetry model is more than improbable, and if it were really required, we should forget about it. However, there are indications that the domain of influence of such a seemingly artificially high symmetry point in parameter space extends over a substantial portion of the phase diagram (7). It stands to reason that SU($\mathcal{D}$) models (which have symmetries connecting spin and orbital axes in Hilbert space) are more quantum fluctuating than the pure SU(2) spin models: there are more transverse directions to fluctuate to. For instance, the SU(4) model on the triangular lattice has a plaquette resonating ground state (the SU(4) version of the resonating valence bond idea (9)), and the influence of this spin–orbital liquid state may extend to physically relevant regions of the parameter space (10).

Limitation of space forbids us to present more than one concrete example of spin–orbital models. We consider Pr-filled skutterudites, in particular $PrFe_4P_{12}$ which we model as a $\mathcal{D} = 4$ system.

2. The case of $PrFe_4P_{12}$

Pr-filled skutterudites show varied behavior: $PrRu_4P_{12}$ has a metal–insulator transition (13), $PrOs_4Sb_{12}$ is thought to be an exotic superconductor (14), while $PrFe_4P_{12}$ remains a normal metal in the entire temperature range studied so far. Our interest lies in $PrFe_4P_{12}$ which in a certain parameter range can be characterized as a heavy fermion system with exceptionally high electronic specific heat (11). $PrFe_4P_{12}$ has a phase transition at $T_{tr} \approx 6.5$K to an ordered phase which had first been thought to be antiferromagnetic, but mounting evidence indicates that it is, in fact, antiferroquadrupolar (AFQ) (11, 15, 12).

Our purpose is to model the AFQ transition by a crystal field model of the $4f$ electrons. We assume that Pr is trivalent ($4f^2$). This can not be literally true, since the driving force of the formation of a heavy band is presumably the admixture of other valence states. However, high-field studies show that the heavy fermion state competes with AFQ ordering (11), so a localized f-shell description should be acceptable within, or adjacent to, the AFQ phase. Besides, at low fields H and high temperatures T, thermal dehybridization acts to obviate the need to consider interband coherence effects.

2.1. THE CRYSTAL FIELD MODEL

The $J = 4$ manifold of Pr^{3+} is split by the approximately cubic crsytal field into the Γ_1 singlet, the Γ_3 doublet, and the Γ_4 and Γ_5 triplets[1]. Group theory does not tell us the sequence of the states, but fitting the measurements narrows the choice. Analyzing the anisotropy of the magnetization curves, it was concluded that the likely possibilities are: a Γ_1 ground state and a low-lying exited state Γ_4 (the Γ_1–Γ_4 scheme); or the Γ_1–Γ_5 scheme; or the Γ_3–Γ_4 scheme (11). Similar schemes were suggested for $PrOs_4Sb_{12}$ (14, 16).

The assumption that the low-T ordered state is AFQ, seems to speak in favour of the Γ_3–Γ_4 scheme, since then the ionic ground state Γ_3 posesses a (permanent) quadrupolar moment. It was also pointed out that the choice of the Γ_3 ground state is consistent with a symmetry analysis of the structural distortion accompanying the AFQ ordering (12). This latter argument relies only on the assumption of the Γ_3 ground state, and does not consider the effects of the low-lying excited state. Here we show that the assumption of

[1] The symmetry group is really not O_h, but the tetrahedral T_h. We nevertheless use the cubic classification, which is an approximation at zero field, but when $H \neq 0$, the symmetry will be in any case substantially lowered.

the Γ_1–Γ_4 scheme is also capable to account for most of the observed static properties of $PrFe_4P_{12}$.

We now discuss the consequences of assuming a Γ_1–Γ_4 level scheme. Since the singlet ground state

$$|\Gamma_1\rangle = \sqrt{\frac{5}{24}}\,(|4\rangle + |-4\rangle) + \sqrt{\frac{7}{12}}|0\rangle \tag{1}$$

does not carry any kind of moment, the ordered quadrupolar moment has to be induced by intersite interactions, assuming that the local Hilbert space contains also the triplet

$$|\Gamma_4^+\rangle = \frac{1}{4}\left\{|3\rangle + |-3\rangle + \sqrt{7}\,(|1\rangle + |-1\rangle)\right\} \tag{2}$$

$$|\Gamma_4^0\rangle = \frac{1}{\sqrt{2}}\,(|-4\rangle - |4\rangle) \tag{3}$$

$$|\Gamma_4^-\rangle = \frac{1}{4}\left\{|3\rangle - |-3\rangle + \sqrt{7}\,(|-1\rangle - |1\rangle)\right\}, \tag{4}$$

where we have chosen the basis of quadrupolar eigenstates. Choosing the energy of (1) as the zero, the states (2)–(4) lie at the level Δ.

The possible moments in the four-dimensional local Hilbert space spanned by (1)–(4) are given by the decomposition

$$(\Gamma_1 \oplus \Gamma_4) \otimes (\Gamma_1 \oplus \Gamma_4) = 2\Gamma_1 + \Gamma_3 + 3\Gamma_4 + \Gamma_5\,. \tag{5}$$

Evidently, the system could support either dipolar (Γ_4), or either of two kinds of quadrupolar (Γ_3 or Γ_5) order[2]. The quadrupolar order parameters are the same that appear in the decomposition of a purely Γ_4 system

$$\Gamma_4 \otimes \Gamma_4 = \Gamma_1 + \Gamma_3 + \Gamma_4 + \Gamma_5\,, \tag{6}$$

i.e., they are not sustained by inter-level matrix elements[3]. – It may be of some interest to mention that the Γ_1–Γ_4 scheme does not offer the possibility of octupolar order (but the Γ_3–Γ_4 scheme would).

---

[2] This is a classification of order parameters which can be defined purely locally. $Q \neq 0$ order needs further discussion.

[3] In contrast, matrix elements between Γ_1 and Γ_4 would bring extra possibilities of dipolar ordering. This does not seem to be relevant for $PrFe_4P_{12}$.

Our decomposition (5) allows to seek dipolar and/or quadrupolar ordering in the system. Experiments give the clue that we should, in fact, look for (antiferro)quadrupolar order. We may rather arbitrarily assume that it is of the Γ_3 kind[4], i.e., the possible order parameters are $\mathcal{O}_2^0 = 3J_z^2 - J(J+1)$ and $\mathcal{O}_2^2 = J_x^2 - J_y^2$. Furthermore, since the total energy expression for a pair of sites has only tetragonal (as opposed to cubic) symmetry, we need not assume that the $\mathcal{O}_2^0$ and $\mathcal{O}_2^2$ couplings would be equal, and we may seek, say, $\mathcal{O}_2^2$-type order. Using a mean field decoupling, our problem would be rather similar to a four-state Blume–Emery–Griffiths model.

It is a well-known feature of quadrupolar ordering that its phenomenology closely imitates that of antiferromagnetic transitions, though the underlying order parameter is non-magnetic. The phase diagram in the H–T plane (H: magnetic field) was mapped in (11). The salient features are the following: A sufficiently strong field applied in any direction will suppress AFQ ordering completely. On the phase boundary, a low-field regime of continuous transitions is separated by a tricritical point ($H^* \approx 2$Tesla, $T^* \approx 4$K) from the high-field regime of first-order transitions. This change in the character of the phase transition is shown in the field dependence of the specific heat. The nature of the magnetization curve changes drastically at T^*. For $T < T^*$, there is a steplike metamagnetic transition corresponding to the first-order transition from the low-T ordered phase to the disordered phase. For $T^* < T < T_{\text{tr}}(H = 0) \approx 6.5$K, there is a kind of a smooth metamagnetic transition, where the system crosses the second-order part of the phase boundary. For $T > T_{\text{tr}}(H = 0)$, the magnetization curve is completely smooth. We will show that a mean field treatment of the AFQ transition in the Γ_1–Γ_4 scheme accounts for these observations quite well.

In the absence of an external magnetic field, Γ_3 and Γ_4 type order parameters (i.e., quadrupolar moment and magnetization) are decoupled because the former is invariant under time reversal, while the latter changes sign. Switching on the magnetic field breaks time reversal invariance, allowing that quadrupolar moment and magnetization get mixed. We can also say that magnetic field, though it couples directly to the angular momentum $\mathbf{J}$, may also induce quadrupolar moment.

This is best illustrated by looking at the matrix which contains the matrix elements of the crystal field, the Zeeman energy for a field in the x-direction $-h_x J^x$, and also a term containing the quadrupolar moment

[4] Assuming $\mathcal{O}_{xy}$-type Γ_5 ordering would give similar results.

$\lambda\mathcal{O}_2^2$, within the basis (1)–(4)

$$\mathcal{M}(\lambda, h_x, \Delta) = \begin{pmatrix} 0 & -2\sqrt{5/3}h_x & 0 & 0 \\ -2\sqrt{5/3}h_x & \Delta + 7\lambda & 0 & 0 \\ 0 & 0 & \Delta & -h_x/2 \\ 0 & 0 & -h_x/2 & \Delta - 7\lambda \end{pmatrix} \qquad (7)$$

The field couples the singlet ground state $|\Gamma_1\rangle$ to the $\mathcal{O}_2^2$-moment bearing excited state $|\Gamma_4^+\rangle$. Therefore in the presence of a magnetic field, uniform quadrupolar moment is no longer "spontaneous". If at $H = 0$, we had to do with a transition to a ferroquadrupolar state, it would be smeared out in $H \neq 0$, and we no longer had a phase boundary to speak about[5]. However, for *anti*ferroquadrupolar coupling, the appearance of the staggered quadrupolar moment is still symmetry breaking, and therefore a sharp phase transition remains possible also in an external magnetic field. Therefore, if we had no other evidence than that $PrFe_4P_{12}$ has sharp phase transitions in external magnetic field, and we adopted the Γ_1–Γ_4 scheme, we would have to conclude that the ordered state could not be ferroquadrupolar, but only antiferroquadrupolar.

2.3. MEAN FIELD RESULTS

Doing the mean field theory (3) for AFQ ordering involves diagonalizing matrices like (7), and we do not give the details here.

Fig. 1 illustrates that an AFQ transition, though of non-magnetic nature, may give a susceptibility which looks very much like what you expect from an antiferromagnet. The cubic (001) and (100) directions are equivalent in the para phase but the appearance of $\mathcal{O}_2^2$-type AFQ order makes the x-field susceptibility appear as "transverse", while the z-field susceptibility looks "longitudinal". Of course, instead of $\mathcal{O}_2^2 = J_x^2 - J_y^2$ we might have chosen $J_y^2 - J_z^2$ or $J_z^2 - J_x^2$, so in a crystal one would expect an equal mixture of the corresponding AFQ domains, and the susceptibility suitably averaged. The experiments may be taken to correspond to this.

Fig. 2 (left) gives the phase diagram. The crystal field splitting Δ and the quadrupolar coupling λ were chosen so as to get at least a rough numerical agreement with the experimental phase diagram (11). There is still some freedom in the parameters, but we found that a rather low $\Delta \approx 4 - 6$K has to be chosen (with $z = 8$, $\lambda \approx 0.08 k_B$), if we want to get both $T_{tr}(H = 0)$ and the tricritical point right. These estimates are likely

[5] Purely in symmetry terms: applying a field in one of the cubic (100) directions, the symmetry would be lowered to C_{4h}, and the decomposition of Γ_4 of O_h in terms of the irreps of C_{4h} would contain the identity which also comes from Γ_1 ground state.

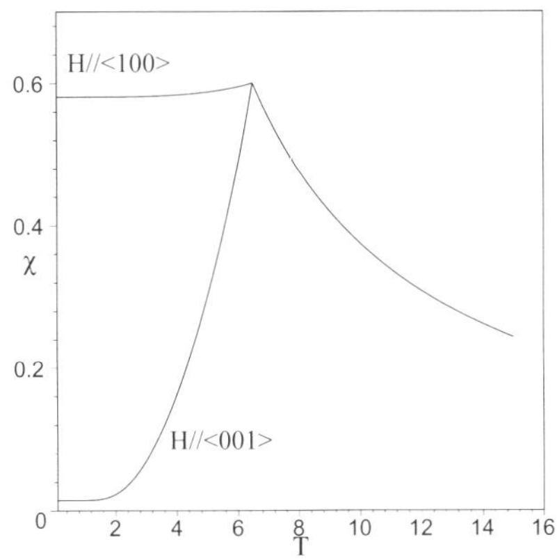

Figure 1. The temperature dependence of the magnetic susceptibility. The onset of $\mathcal{O}_2^2$-type quadrupolar order makes the two $H \parallel z$ and $H \parallel x$ behavior inequivalent.

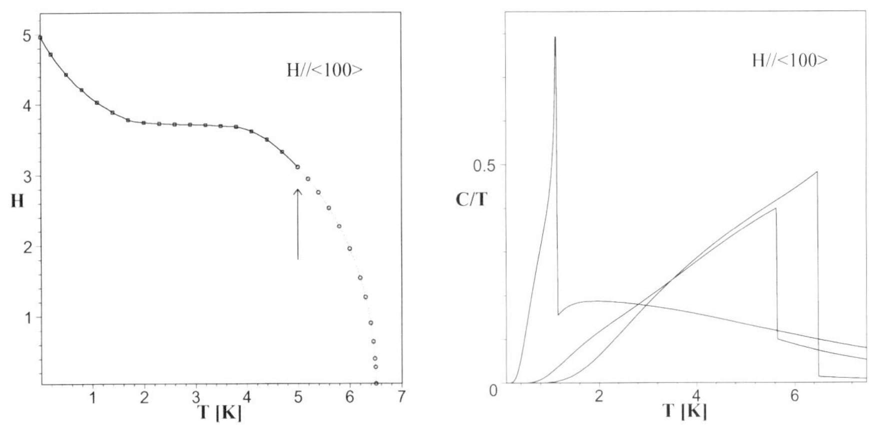

Figure 2. *Left*: The boundary of the antiferroquadrupolar phase in the H–T plane. The curve is drawn in black for first-order transitions, and in grey for continuous phase transitions. Arrow indicates the tricritical point. *Right*: The T-dependence of the temperature coefficient of the specific heat for $H = 0$, 2.5, and 4Tesla (in order of decreasing transition temperatures).

to be subject of some revision when further (especally dipolar) couplings are allowed for. – In spite of an overall similarity to the phase diagram based on experiments, we note that the low-T, high-H upcurving part of our present phase boundary represents a deviation, the reason for which remains to be clarified.

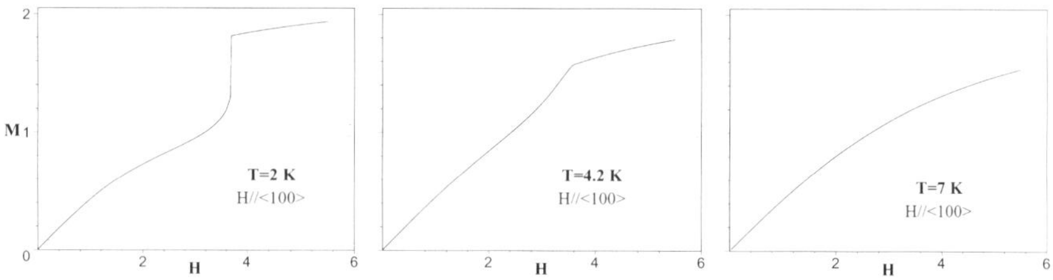

Figure 3. Antiferroquadrupolar order underlies the sharp metamagnetic transition at low T (left). With increasing T, we pass through second order transitions (middle) to smooth behavior (right).

The changeover to a regime of first-order transitions in higher fields is evident in the field dependence of the specific heat; the curves shown in Fig. 2 (right) bear a close resemblance to the measured ones. The same is true of the magnetization curves (Fig. 3) where we see a change from the regime of sharp metamagnetic transitions at low temperatures to continuous phase transitions at intermediate T's, and eventually smooth behavior in the para phase.

Acknowledgements

The authors profited much from cooperation with K. Penc, L. Forró, G. Mihály and I. Kézsmárki, and from discussions with H. Shiba. This work was supported by the grants OTKA T038162, and AKP 2000-123 2,2.

References

1. C.D. Batista, G. Ortiz, and J.E. Gubernatis, Phys. Rev. B**65**, 180402(R), (2002), and references therein.
2. A.T. Boothroyd, J. Alloys and Compounds 303-304, 489, (2000).
3. A. Kiss and P. Fazekas, cond-mat/0212345, and to be published.
4. R. Shiina, H. Shiba, and P. Thalmeier, J. Phys. Soc. Japan **66**, 1741, (1997).
5. A. Takahashi, H. Shiba, J. Phys. Soc. Japan **69**, 3328, (2000).
6. G. Mihály et al., Phys. Rev. B**61**, R7381, (2000); G. Mihály et al., at this conference.
7. K. Penc et al., to be published.
8. P. Fazekas, Found. Phys. **30**, 1999, (2000).
9. P.W. Anderson, Mat. Res. Bull. **8**, 173, (1973); P. Fazekas and P.W. Anderson, Phil. Mag. **30**, 423, (1974).
10. K. Penc, M. Mambrini, P. Fazekas, and F. Mila, cond-mat/0212211.
11. Y. Aoki et al., Phys. Rev. B **65** 064446, (2002); J. Phys. Chem. Sol. **63**, 1201, (2002).
12. S.H. Curnoe, K. Ueda, H. Harima, and K. Takegahara, J. Phys. Chem. Sol. **63**, 1207, (2002); S.H. Curnoe, H. Harima, K. Takegahara, and K. Ueda ,Physica B **312–313**, 837, (2002).

13. C. Sekine, T. Uchiumi, I. Shirotani, and T. Yagi, Phys. Rev. Lett. **79**, 3218, (1997).
14. E.D. Bauer et al., Phys. Rev. B **65**, 100506(R), (2002); K. Miyake et al., unpublished; D.E. MacLaughlin et al., Phys. Rev. Lett. **89**, 157001, (2002).
15. K. Iwasa et al., Physica B **312–313**, 834, (2002).
16. Y. Aoki et al., J. Phys. Soc. Japan **71**, 2098, (2002).

LOW DIMENSIONAL SPIN SYSTEMS IN HIGH MAGNETIC FIELDS: SPIN-PHONON INTERACTION

Proceedings of the ARW NATO Workshop Hvar, Croatia, October 2002

B. LÜTHI and B. WOLF
Physikalisches Institut, Universität Frankfurt, D-60054 Frankfurt
S. ZHERLITSYN
Forschungszentrum Rossendorf, PO Box 51 01 19, D-01314 Dresden

Abstract. We discuss ESR- and ultrasonic experiments for quasi- one and twodimensional spin-systems. We found several effects for $(VO)_2P_2O_7$ and $SrCu_2(BO_3)_2$. With ESR we determined the magnetic excitation spectrum and with ultrasonic measurements we were able to follow these excitations to lowest frequencies. The spin-phonon interaction is of the exchange-striction type and with ultrasonic experiments we are able to determine the strain dependence of the magnetic coupling constant.

Key words: Low Dimensional Spin Systems, Pulse Field Experiments, Spin Phonon Interaction, Ultrasonic Experiments, ESR Experiments, Magnetization Plateaus

1. Introduction

It is possible today to generate pulsed magnetic fields up to 60T in coils with 20 mm inner diameter and of special design. In our Frankfurt laboratory we specialized on developing measuring techniques for electron spin resonance (ESR) and ultrasonics in such high fields (1, 2). For the ultrasound we used a quadrature system capable of measuring simultaneously phase and amplitude in the frequency range of 10-200 MHz. From these data we calculate the change in sound velocity and the relative ultrasonic attenuation.

Such experiments (ESR and ultrasound) are very important for high magnetic fields, because they constitute the only spectroscopic tool for $B > 20T$, where neutron scattering experiments cannot be performed so far. In this article we concentrate on experiments on low dimensional spin

A.C. Hewson and V. Zlatić (eds.), Concepts in Electron Correlation, 179–187.

systems. There are several reviews dealing with this topic (2, 3, 4). Here we analyze in some detail the interaction between magnetic excitations and sound waves for the different spin systems investigated.

We discuss one- and two-dimensional dimerized spin-1/2 compounds, such as $(VO)_2P_2O_7$ and $SrCu_2(BO_3)_2$. Here V^{4+} and Cu^{2+} have spin-1/2 and form either chains or two dimensional dimer arrays. The lowest energy excitations are usually of the singlet-triplet type or modifications of it. These can be determined either by ESR excitations from the ground state or within the excited triplet. For $B = 0$ inelastic neutron scattering give the excitation spectrum for the whole k-space.

2. Spin-Phonon interaction in low dimensional spin systems

2.1. COUPLING MECHANISM

Usually the spin-phonon interaction for transition metal ions is of the single ion magnetoelastic type. The strain component couples to the electric quadrupole moment of the magnetic ion. However for spin 1/2 systems, involving e.g. Cu^{2+} or V^{4+} and for lower point symmetry than cubic (where the orbital moment is quenched) this magnetoelastic interaction is negligible. In this case two ion interactions of the form of exchange striction becomes important. A soundwave or more generally a phonon can modulate the exchange interaction thus creating a spin-phonon coupling as given by

$$E_{\text{exs}} = \Sigma_{ij}[J(\delta + \mathbf{u}_i - \mathbf{u}_j) - J(\delta)]\mathbf{S}_i \bullet \mathbf{S}_j \ . \tag{1}$$

Here $\delta = R_i - R_j$ measures the distance between two magnetic ions and $\mathbf{u}_i$ is the displacement vector for the ion $\mathbf{R}_i$. With a sound wave given by a plane wave $\mathbf{u} = \mathrm{U}\,\mathbf{e}\,e^{i(\mathbf{q} \bullet \mathbf{r} - \omega t)}$ with $\mathbf{e}$ the polarization vector, U amplitude, ω frequency of the sound wave the exchange striction term reads:

$$E_{\text{exs}} = \Sigma_i U(dJ/d\delta \bullet \mathbf{e})(\mathbf{q} \bullet \delta)(\mathbf{S}_i \bullet \mathbf{S}_{i+\delta})e^{i(\mathbf{q} \bullet \mathbf{r} - \omega t)} \tag{2}$$

In eq.(2) we have expanded the exponential to first order because $q\delta << 1$ for sound waves of typically 100 MHz. Eq.(2) tells us that shear waves propagating along symmetry directions do not couple to the spins with this mechanism whereas longitudinal waves do. This is seen in many experiments discussed below. It is worth pointing out that this coupling is also responsible for sound wave anomalies (attenuation and velocity) at magnetic phase transitions for magnetic ions with orbital moment $L = 0$ $(Mn^{2+}, Fe^{3+}, Gd^{3+})$(5).

2.2. TEMPERATURE AND MAGNETIC FIELD DEPENDENCE OF ELASTIC CONSTANTS IN DIMERIZED SPIN SYSTEM

One can describe the effective elastic constant in such systems by considering the strain susceptibility χ_d for an isolated dimer and introducing the dimer-dimer interaction K in RPA- approximation. One gets in this way for a symmetry elastic constant:

$$c_\Gamma = c_\Gamma^o - G^2 N\chi_d/(1 - K\chi_d) \tag{3}$$

Here N is the number of dimers per unit volume and G is a strain-dimer coupling constant. For a simple singlet-triplet dimer we have for the free energy $F = -kTlnZ$ with $Z = 1 + 3exp(-\Delta/kT)$ and Δ the singlet-triplet splitting. For χ_d we get with $G = d\Delta/d\varepsilon_{xx}$: $\chi_d = G^{-2}d^2F/d\varepsilon^2 = 3e^{-\Delta/kT}/[kTZ^2]$.

In the next section we will give some examples where such effects have been observed. Note that χ_d is the strain susceptibility analogue of the magnetic susceptibility. Analogously to the case of the temperature dependence we can calculate the field dependence of the symmetry elastic constant. Instead of the free energy used above we have to take the field splitting of the excited triplet state into account: $F = -kTlnZ$ with $Z = 1 + exp((-\Delta + g\mu_B B)/kT) + exp(-\Delta/kT) + exp((-\Delta - g\mu_B)/kT)$. Since in the systems investigated $\Delta \approx$ 30-60K and $T < 2K$, we have only to consider the lowest triplet level which cuts the singlet branch for $g\mu_B B_c = \Delta$. With this simplification we get for the single dimer strain susceptibility

$$\chi_d(B) = (1/kT)exp[(-\Delta + g\mu_B B)/kT]/[1 + exp((-\Delta + g\mu_B B)/kT)]^2 \tag{4}$$

and the elastic constant c_Γ is given again by eq.(3) with χ_d taken from eq.(4). The coupling constant G is given again by $G = d\Delta/d\varepsilon$.

There is another way to calculate the field dependent elastic constant using the Landau theory of phase transitions (6). For a longitudinal wave with strain ε_{xx} one can write for the spin-phonon interaction $F_{me} = G'\varepsilon_{xx}m_x^2$ with $G' \sim dJ/d\varepsilon_{xx}$. One gets with $c_{11} = d^2F/d\varepsilon_{xx}^2$ with $F = F_L + F_{me}$ the expression (here F_L is Landau-order parameter free energy)

$$c_{11} = c_{11}^0 - 4G'^2\chi(B)(m/m_0)^2 \tag{5}$$

with $\chi(B)$ the differential magnetic susceptibility and m/m_o the reduced magnetization. For $(VO)_2P_2O_7$ a comparison of the magnetic field dependence of the elastic constant we are using either the thermodynamic expression (eq.4) or the Landau formula (eq.6), the latter gives a much better fit.

2.3. THERMODYNAMIC TREATMENT VERSUS EQUATION OF MOTION

Above we have given a thermodynamic treatment of the temperature and field dependence of the symmetry elastic constant. This treatment should be applicable if one is in the hydrodynamic limit, i.e. $\omega\tau << 1$, with ω the sound wave frequency and τ a typical spin fluctuation relaxation time. In the opposite limit $\omega\tau \geq 1$ one has to take the coupled equations of motion for the magnetic excitations and for the sound waves and to calculate the sound velocity change from them. Close to the crossover of the sound wave frequency $\omega = vk$ and the magnetic excitation ω_m one expects a coupled oscillator formula

$$(\omega - \omega_m)(\omega - vk) = f(\omega) \tag{6}$$

From sound wave velocity and attenuation experiments one can estimate the value of $\omega\tau$ (7). For $\omega\tau < 1$ one obtains

$$\alpha = -\tau \cdot (\omega^2/2v_0) \cdot (\Delta c/c_0) \tag{7}$$

Here α denotes the attenuation and $\Delta c/c_0$ the relative elastic constant change. From eq.(7) one can determine the spin fluctuation relaxation time and therefore one can determine the regime (hydrodynamic($\omega\tau < 1$) or non-hydrodynamic ($\omega\tau > 1$)), in which the experiment is performed.

3. Examples for Spin-Phonon Effects:

In the following we discuss some compounds where ESR and ultrasonic measurements have been performed in high magnetic field. We will discuss also the temperature dependence of the elastic constants and some attenuation data.

3.1. DIMERIZED SPIN 1/2 CHAIN:

$(VO)_2P_2O_7$

This compound exists in two modifications, an ambient pressure phase and a high pressure phase (8). ESR and ultrasonics have been performed only in the former phase, therefore we concentrate on this one. A review on this compound has been given recently (9). It consists of strongly dimerized chains along the c-axis in alternate layers along the b-axis. These two chains have energy gaps at the Γ point of 35K and 60K respectively. The gaps have been determined with inelastic neutron scattering (10) and with ESR (11, 3). From the latter experiment we get by extrapolation a critical field value of $B_c = 26T$, where the lower triplet branch meets the singlet ground

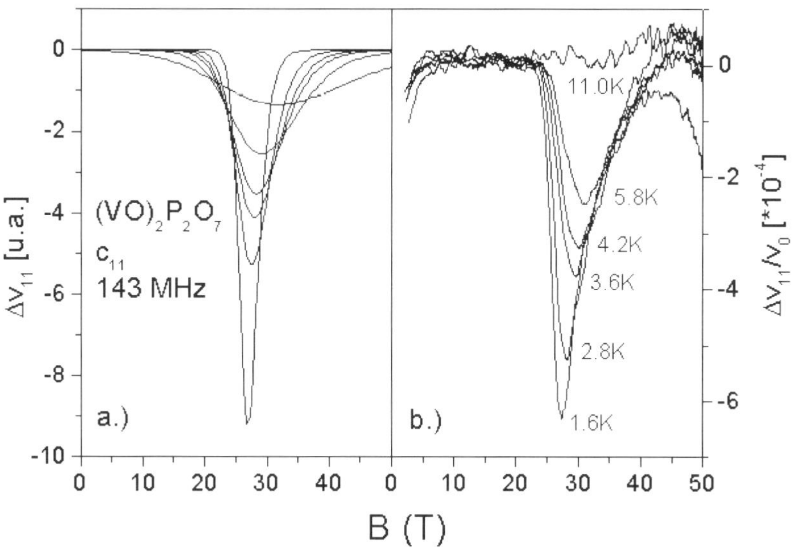

Figure 1. **a** : theoretical curves of the sound velocity for VOPO using eq.5 and the same temperatures as in fig.1b. **b** : field dependence of the longitudinal sound velocity in VOPO for temperatures from 1.6K to 11.0K measured in pulsed fields

state. Precisely around this field value we observe a sound velocity anomaly as shown in fig.1.

The velocity anomaly observed is of the order of $\Delta v/v_o \approx 5 \cdot 10^{-4}$. This rather small value for the c_{11} mode is due to the small exchange (and correspondingly small $dJ/d\varepsilon$) in the a-direction. For c_{33} along the c-axis we would expect a much larger effect, because in this case the propagation is along the dimerized chain. Nevertheless we have measured a distinct temperature dependence of this elastic anomaly. The effect being largest at low temperature of 1.6K and disappearing already at 11K. We notice that eq.(5) describes the experimental data of fig.1 much better than eq.(4). As seen from a comparison of fig.1a and 1b the shift of the sound velocity minima to higher fields for higher temperatures is well described by fig.1a and eq.(5). Preliminary sound attenuation data corroborate our thermodynamic data: Using eq.(7) we obtain $\omega\tau \approx 0.1$ so this approach is justified.

3.2. PLATEAU SUBSTANCE: $SrCu_2(BO_3)_2$

This substance forms two dimensional orthogonal $Cu-$ dimers as shown in fig.2a. High field magnetization data exhibit clear plateaus at low temperatures for $m/m_o = 1/8, 1/4, 1/3$ (12, 13) as shown in fig.2c. It was shown theoretically that this dimer arrangement corresponds to the Shastry - Sutherland model, with an exact singlet ground state wavefunction (14). The plateaus were calculated using two methods. One was the model of the square unit cells and the second was a hard core boson model (14).

It is interesting to notice that one can describe the temperature dependence of the magnetic susceptibility with an RPA-formula $\chi_m = \chi o/(1-j\chi_0)$ with χ_o being the magnetic susceptibility of a single dimer and j being the interdimer coupling (15). The fit is very good and $j = 273K$ gives exactly 4 times the dimer-dimer exchange J' indicated in fig.2a. In a similar way we can interpret the temperature dependence of the elastic constants (16). The polarization and propagation directions for the different modes (c_{11}, c_{44}, c_{66}) are indicated in fig.2a. Note that we obtain an exchange striction coupling also for the shear wave c_{66} because the propagation direction makes an angle of 45 degree with the dimer axis. We can interpret the $c_{ij}(T)$ using eq.(3). Using the lowest singlet-triplet splitting as $\Delta_1 = 30K$, the same value used for the magnetic susceptibility fit, we obtain a very good fit for c_{11} (15, 16). For the c_{66} mode (polarization and propagation in the dimer plane), which exhibits the strongest temperature dependence, we have to include a further coupling to the next higher triplet level $\Delta_2 = 58K$ (taken from inelastic neutron scattering) (17). Again we get a very good fit with coupling parameters as listed in ref. (16). The c_{44} mode exhibits a very small temperature dependence because the polarization vector points along the c-axis out of the dimer plane and the exchange striction coupling is therefore of higher order. Nevertheless we can fit also this mode satisfactorily (16). In ref. (16) we have given a table with the different coupling constants G_1, G_2, K for the three modes c_{11}, c_{44} and c_{66}. It is seen that K is ferrodistortive for c_{11} and antiferrodistortive for c_{66}.

We turn now to the magnetic field dependence of the elastic modes. ESR experiments determined the lowest singlet-triplet transition (18). We extended these measurements to lower frequencies and found still a linear field dependence down to 55GHz. In fig.2b,c we show the 3 elastic modes together with the magnetization taken from ref. (13). Similar to the temperature dependence the c_{66} mode exhibits the strongest field effect, a total softening of more than 25%, whereas the c_{44} mode shows no softening at all. We observe clear minima for c_{11} and especially for c_{66} exactly at the field values where new plateaus appear, indicated by dashed lines in fig.2b,c. These plateaus occur at 27T, 36T and 43T for $m/m_o = 1/8, 1/4$ and $1/3$. Note that the next plateau 1/2 should occur high above 60T according theoretical predictions.

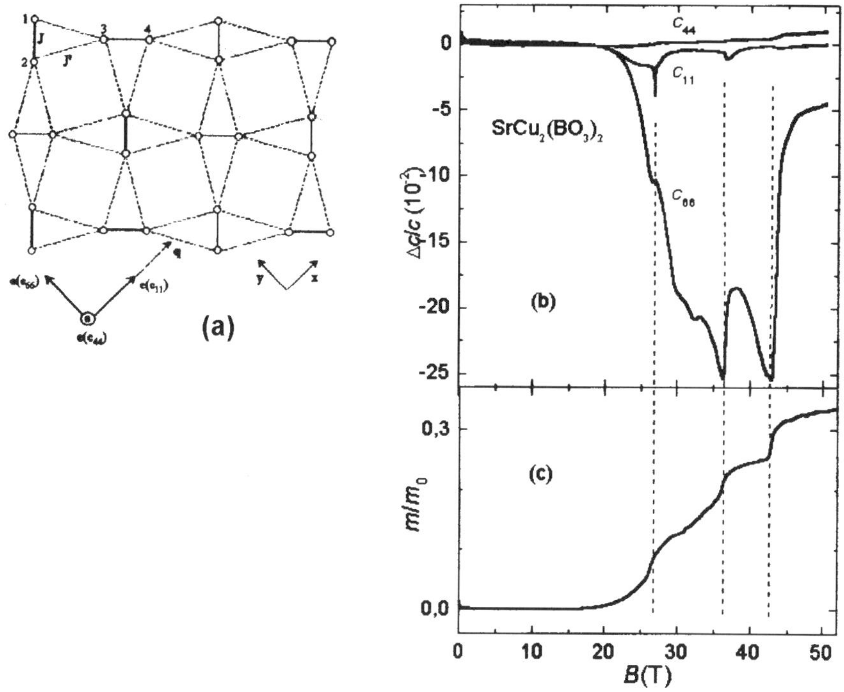

Figure 2. **a** : The polarization and propagation directions for the different modes (c_{11}, c_{44}, c_{66}) together with the two coupling constants J and J'.**b** : The field dependence of the elastic modes c_{11}, c_{44} and c_{66} at 1.5K. **c** : The magnetization taken from ref(13). The broken lines indicate the the positions of the different plateaus

How can one interpret these results of fig.2b,c? Attenuation measurements have been performed for the c_{11} mode. With eq.6 we get $\omega\tau \sim 1$ which means that one can no longer take thermodynamic expressions, eqs.3,4, but one has to calculate sound velocity changes from the coupled equations of motion, eq.6. In this case we have therefore a resonant interaction between the sound waves and the magnetic excitations, expressed by the coupled dispersion expression of eq.6. The important c_{66} mode has B_{2g}-symmetry for this tetragonal compound. Therefore the strong coupling to the magnetic excitation for the plateaus 1/4 and 1/3 implies that these magnetic modes have the same B_{2g}-symmetry. Consequently the condensed triplet states in the plateaus should have also the same order parameter symmetry. Therefore we can select from the calculations of this triplet condensation the one with the same symmetry. For the 1/4 plateau it is the calculated for the square unit cell (14). For the 1/8 plateau c_{11} shows also a considerable effect. We therefore conclude that a mixture of B_{1g} and $_{B2g}$ holds for this plateau.

4. Conclusion and outlook

This short review shows that the combination of ESR and ultrasonics gives an efficient spectroscopic tool for soft magnetic modes. This spectroscopy was applied in field regions where other experimental tools like elastic and especially inelastic neutron scattering no longer work ($> 15T$). ESR- and ultrasonic measurements were also performed on the low dimensional spin-plateau systems NH_4CuCl_3 and $CsCuCl_3$ (21, 23, 22). But our experiments have a much wider range of applications. We have performed ultrasonic experiments to determine the B-T-diagram for a mixed valence compound $YbIn_{1-x}Ag_xCu_4$(24) and the high field ultrasonic investigation of the heavy fermion compound URu_2Si_2 (25). Other pulsed field experiments with ESR and ultrasonics are possible, e.g. an investigation of semiconductors, especially heterostructure materials exhibiting quantum Hall effect.

Acknowledgements

The crystals for these investigations were given to us by A. V. Prokofiev and W. Assmus ($(VO)_2P_2O_7$) and H. Kageyama and Y. Ueda ($SrCu_2(BO_3)_2$). We thank H. Schwenk and S.Schmidt for the collaboration.

References

1. B. Wolf, B. Lüthi, S. Schmidt, H. Schwenk, M. Sieling, S. Zherlitsyn, I. Kouroudis, Physica B 294-295, 612 (2001).
2. B. Wolf, B. Lüthi, S. Zherlitsyn, in: *Very High Magnetic Fields and their Applications*, ed. by F.Herlach and N.Miura, World Scientific.
3. B. Lüthi, B. Wolf, S. Zherlitsyn, S. Schmidt, H. Schwenk, M. Sieling, Physica B 294-295, 20 (2001).
4. B. Wolf, S. Zherlitsyn, S. Schmidt, B. Lüthi, Phys. stat. sol. (a) 189, 389 (2002).
5. B. Lüthi in: *Dynamical properties of Solids*, Vol.3, North-Holland (1980).
6. B. Wolf, S. Schmidt, H. Schwenk, S. Zherlitsyn, B. Lüthi. J. Appl. Phys. 87, 7055 (2000).
7. B. Lüthi, W. Rehwald in: *Structural Phase Transitions I*, Topics in Current Physics (1981).
8. M. Azuma, T. Saito, Y. Fujishiro, Z. Hiroi, M. Takano, F. Izumi, T. Kamiyama, T. Ikeda, Y. Narumi, K. Kindo,Phys. Rev. B 60, 10145 (1999).
9. D.C. Johnston, T. Saito, M. Azuma, M. Takano, T. Yamauchi, Y. Ueda, Phys. Rev. B 64, 134403 (2001).
10. A.W. Garrett, S. E. Nagler, D.A. Tennant, B.C. Sales, T. Barnes, Phys. Rev. Lett. 79, 745 (1997).
11. A.V. Prokofiev, F. Büllesfeld, W. Assmus, H. Schwenk, D. Wichert, U. Löw , Eur. Phys. J. B 5, 313 (1998).

12. H. Kageyama, Yoshimura, R. Stern, N.V. Mushnikov, M. Kato, K. Kosuge, C.P. Slichter, T. Goto, Y. Ueda , Phys. Rev. Lett. 82, 3168 (1999).
13. K. Onizuka, H. Kageyama, Y. Narumi, K. Kindo, Y. Ueda, T. Goto, J. Phys. Soc. Jpn. 69, 1016 (2000).
14. S. Miyahara and K. Ueda, Phys. Rev. B 61, 3417 (2000).
15. S. Zherlitsyn, S. Schmidt, B. Wolf, H. Schwenk, B. Lüthi, K. Onizuka, Y. Ueda, K. Ueda, Phys. Rev. B 62, R6097 (2000).
16. B. Wolf, S. Zherlitsyn, S. Schmidt, B. Lüthi, H. Kageyama, Y. Ueda, Phys. Rev. Lett. 86, 4847 (2001).
17. H. Kageyama, M. Nishi, N. Aso, K. Onizuka, T. Yosihama, K. Nukui, K. Kodama, K. Kakurai, Y. Ueda, Phys. Rev. Lett. 84, 5876 (2000).
18. H. Nojiri, H. Kageyama, K. Onizaka, Y. Ueda, M. Motokawa, J. Phys. Soc. Jpn. 68,2906 (1999).
19. H. Tanaka, W. Shiramura, T. Takatsu, B. Kurniawan, M. Takahashi, K. Kamishima, K. Takizawa, H. Mitamura, T. Goto, Physica B246-247, 230 (1998).
20. B. Kurniawan, H. Tanaka, K. Takatsu, W. Shiramura, T. Fukuda, H. Nojiri, M. Motokawa, Phys. Rev. Lett. 82, 1281 (1999).
21. S. Schmidt, S. Zherlitsyn, B. Wolf, H. Schwenk, B. Lüthi, H. Tanaka, Europhys. Lett. 53, 591 (2001); addendum 54, 554 (2001).
22. B. Wolf, S. Zherlitsyn, S. Schmidt, B. Lüthi, Europhys. Lett. 48, 182 (1999).
23. S. Schmidt, B. Wolf, M. Sieling, S. Zvyagin, I. Kouroudis, B. Lüthi, Solid state Comm. 108, 509, (1998).
24. S. Zherlitsyn, B. Lüthi, B. Wolf, J.L. Sarrao, Z. Fisk, V. Zlatic, Phys.Rev.B 60, 3148 (1999).
25. B. Wolf, S. Zherlitsyn, H. Schwenk, S. Schmidt, B. Lüthi, J.Magn.Magn.Mat. 226-230, 107 (2001).

'TAXONOMY' OF ELECTRON GLASSES

Proceedings of the ARW NATO Workshop Hvar, Croatia, October 2002

N.P. ARMITAGE, E. HELGREN and G. GRÜNER
Department of Physics and Astronomy,
University of California, Los Angeles, CA 90095

Abstract. We report measurements of the real and imaginary parts of the AC conductivity in the quantum limit, $\hbar\omega > k_B T$ of insulating nominally uncompensated n-type silicon. The observed frequency dependence shows evidence for a crossover from interacting Coulomb glass-like behavior at lower energies to non-interacting Fermi glass-like behavior at higher energies across a broad doping range. The crossover is sharper than predicted and cannot be described by any existing theories. Despite this, the measured crossover energy can be compared to the theoretically predicted Coulomb interaction energy and reasonable estimates of the localization length obtained from it. Based on a comparison with the amorphous semiconductor NbSi, we obtain a general classification scheme for electrodynamics of electron glasses.

Key words: Long Range Interaction, Electron Glass, Silicon, Si:P, NbSi, Electrodynamics, Localization, Insulator, Disorder

1. Introduction

Strong electronic interactions are known to play a central role in disordered solids, of which Coulomb glasses are a canonical example. The lack of metallic screening on the insulating side of the metal-insulator transition (MIT) enables long-range Coulomb interactions (1). Efros and Shklovskii (ES), following the original considerations for the non-interacting Fermi glass case of Mott (2), derived a form for the T = 0 K photon assisted frequency dependent conductivity describing the crossover from interacting Coulomb glass-like behavior to Fermi glass-like behavior (3). These derivations were based on a theory of resonant absorption (4) and take into account the mean Coulomb interaction between two sites forming a resonant pair $U(r_\omega) = e^2/\varepsilon_1 r_\omega$, where $r_\omega = \xi[ln(2I_0/\hbar\omega)]$ is the most probable hop distance between pairs and ε_1 is the dielectric constant. The real part of the ES crossover form for the frequency dependent conductivity is:

A.C. Hewson and V. Zlatić (eds.), Concepts in Electron Correlation, 189–197.

$$\sigma_1 = \beta e^2 g_0^2 \xi^5 \omega [ln(2I_0/\hbar\omega)]^4 [\hbar\omega + U(r_w)] \tag{1}$$

where β is a constant of order one, g_0 is the non-interacting single particle density of states (DOS), I_0 is the pre-factor of the overlap integral and ξ is the localization length. The concentration dependent localization length is predicted to diverge as $(1 - x/x_c)^{-\nu}$ as the MIT is approached, where x is the dopant concentration, x_c is the critical dopant concentration of the MIT ($x_c = 3.52 \times 10^{18} cm^{-3}$ in Si:P (5)) and ν is the localization length exponent.

Neglecting logarithmic factors, Eq. (1) predicts a gradual crossover from linear to quadratic behavior as the incident photon energy exceeds the interaction energy of a typical charge excitation. For the case where the photon energy, $\hbar\omega > U(r_\omega)$, one recovers the quadratic frequency dependence, plus logarithmic corrections, that Mott originally derived for the non-interacting Fermi glass case (2). In the opposite limit, $\hbar\omega < U(r_\omega)$ the conductivity shows an approximately linear dependence on frequency, plus logarithmic corrections, and the material is called a Coulomb glass. We should note that Eq. (1) was derived for the case where $\hbar\omega > \Delta$, the Coulomb gap width. However a quasi-linear dependence (albeit with a different pre-factor) and an eventual crossover to Mott's non-interacting quadratic law is still expected even for the case where $\hbar\omega < \Delta$,

$$\sigma_1 \approx \frac{1}{10} \frac{\omega \varepsilon_1}{ln(2I_0/\hbar\omega)}. \tag{2}$$

There is a lack of experimental evidence to either corroborate or disprove Eq. (1) due to the difficulties associated with performing frequency dependent measurements in the so-called quantum limit, i.e. $\hbar\omega > k_B T$, but at small enough photon energies so as to not be exciting charge carriers to the conduction band. Moreover, in order to study the possible crossover from Mott to ES type behavior, one must measure across a broad enough bandwidth centered about the characteristic crossover energy scale for instance the Coulomb interaction energy U or the Coulomb gap width (6), Δ.

There have been some very recent experiments that have attempted to address these issues. M. Lee et al. found that for concentrations close to the MIT the expected linear to quadratic crossover occurs, but is much sharper than predicted (7). They proposed that this sharp crossover was controlled not by the average interaction strength U as in Eq.(1) (3), but instead by a sharp feature in the density of states, i.e. the Coulomb gap (8). They postulated that this Coulomb gap was not the single particle one measured in tunneling, but rather a smaller "dressed" or renormalized Coulomb gap that governs transport. There is some evidence from DC transport that such a feature exists, at least close to the MIT (9).

2. Experiment

Nominally uncompensated n-type silicon samples were obtained from Recticon Enterprises Inc. A Czochralski method grown boule with a phosphorous gradient along its length was cut into 1 mm thick discs. Room temperature resistivity was measured using an ADE 6035 gauge and the dopant concentration calibrated using the Thurber scale (10). The Si:P samples discussed here span a range from 39% to 69%, stated as a percentage of the sample's dopant concentration to the critical concentration at the MIT. A number of samples were measured before and after etching with a $4\%HF + 96\%HNO_3$ solution; this resulted in no difference in the results.

In the millimeter spectral range, 80 GHz to 1000 GHz, backward wave oscillators (BWO) were employed as coherent sources in a transmission configuration (11). The transmitted power through the Si:P samples as a function of frequency was recorded. For plane waves normally incident on a material, resonances occur whenever the thickness of the material is an integer number of half wavelengths. Both components of the complex conductivity can be uniquely determined for each resonance. The real part of the conductivity was evaluated at microwave frequencies from the measured loss of highly sensitive resonant cavities at 35 and 60 GHz via the perturbation method. This is a common technique and is described in the literature (12). The conductivity as determined from the resonant cavity data was normalized to the DC conductivity at higher temperatures, at above approximately 25 K. The resonant cavity data confirmed the linear dependence on frequency of the real part of the complex conductivity into the microwave regime for the samples closest to critical.

3. Results

In Fig. 1, we show the T→0 frequency dependent conductivity for two samples. This data, representative of all samples in our range, shows an approximately linear dependence at low frequencies and then a sharp crossover to an approximately quadratic behavior at higher frequencies. This is the qualitatively expected behavior from Eq. (1). However, as seen by the overlayed fits, Eq. (1) provides only a rough guide. The solid lines are linear and quadratic fits to the low frequency and high frequency data respectively. The dotted line is a fit to the form of the ES crossover function achieved by summing the separately determined linear and quadratic fits. As can be seen, the crossover between linear and quadratic portions is much more

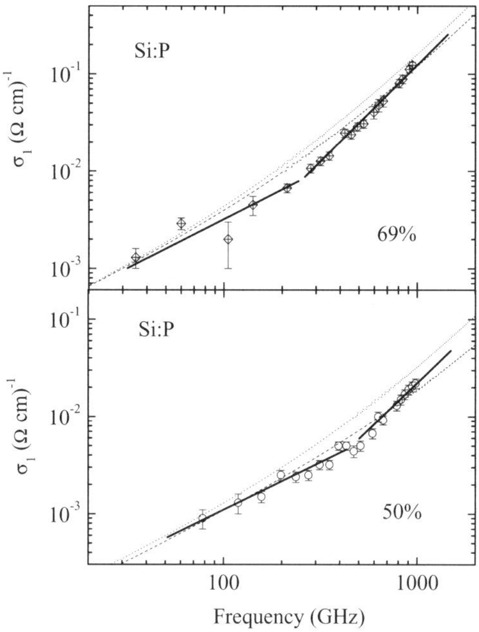

Figure 1. Low temperature frequency dependent conductivity data plotted as a function of frequency. The Si:P samples shown are at 50% and 69% dopant concentration relative to the critical concentration. The solid lines are linear and quadratic fits to the lower and upper portions of the data respectively. The dotted and dashed lines are fits following the form of Eq. (1) using two different methods described in the text.

abrupt than the ES function predicts. The dashed line is a fit using the same method as Ref. (7), namely forcing the linear portion to pass through the low frequency data, as well as the origin and leaving the pre-factor of the quadratic term as a free variable. The fit is not satisfactory in either case. A sharp crossover as such is observed over our entire doping range and has been observed previously in an analogous system, Si:B, for samples closer to the MIT (7). Note, that a linear dependence is seen in the imaginary part of the conductivity σ_2 over the whole measured frequency and doping range (13). This is consistent with theoretical predictions (14).

Because our data spans a large range of concentrations, the doping dependence of the crossover energy scale can be analyzed to see whether its dependence is consistent with other energy scales, e.g. the Coulomb interaction energy U or the Coulomb gap width Δ as *per* Ref. (7). Recall that the Coulomb interaction energy between two sites forming a resonant pair is $U(r_\omega) = e^2/\varepsilon_1 r_\omega$ which is dependent on concentration via the dielectric constant (measured, but not shown) and the localization length dependent most probable hop distance. By equating the crossover energy scale to the expected functional form for this Coulomb interaction energy we are able to determine the magnitude of the localization length and its scaling exponent. With an appropriate pre-factor in the overlap integral (15), $I_0 = 10^{13}s^{-1}$ for the expression for the most probable hop distance term, r_ω, we get a localization length dependence of $\xi \propto (1 - x/x_c)^{-0.83}$ with a magnitude of 21.2, 19.9, 20.1, 14.5, 14.3 and 13.0 nm for the 69%, 62%, 56%, 50%, 45% and 39% samples respectively. The localization length exponent is close to unity, the value originally predicted by McMillan in his scaling theory of the MIT (16), and the magnitude of the localization length is reasonable. Due to the fact that we obtain reasonable estimates for the relevant physical parameters over the whole doping range, we do not favor the previous speculation that it is in fact the Coulomb gap energy that creates the sharp crossover and hence sets its energy scale (7).

The approximately linear power law seen in the Coulomb glass regime at low ω in Fig. 1 of the conductivity can be expressed with the imaginary part as a simple Kramer-Kronig compatible form, $\sigma(\omega) = A(i\omega)^\alpha = A\omega^\alpha cos\left(\frac{\pi\alpha}{2}\right) + iA\omega^\alpha sin\left(\frac{\pi\alpha}{2}\right)$. In order to determine the power α one can take the ratio of $|\sigma_2|$ versus σ_1 (with the frequency as a variable). The power α is given by,

$$\alpha = \frac{2}{\pi} tan^{-1}\left(\frac{|\sigma_2|}{\sigma_1}\right). \tag{3}$$

Fig. 2a shows the ratio mentioned above of the imaginary to the real part of the dielectric constant for Si:P. Similar data from amorphous NbSi is included for comparison purposes (1). We note that this ratio for Si:P is large and essentially constant across the entire doping range. From Eq. (2), one expects $|\sigma_1|$ to be approximately equal to $|\sigma_2|$ to within a factor of 2-5 (with a reasonable estimate for I_0) because $|\sigma_2| \propto \varepsilon_1 \cdot \omega$. Applied to Si:P, the theory correctly predict a linear correspondence between σ_1 and $|\sigma_2|$, but incorrectly predicts the proportionality by at least a factor of thirty. The proportionality is near the predicted value for NbSi, but has a dependence on the doping concentration which is presumably due to entering the quantum critical (QC) regime as discussed below. Here we have used the susceptibility $4\pi\chi$ of the dopant electrons (i.e. with the background dielectric constant of silicon subtracted) in the expression for the magnitude of the imaginary component of the conductivity in Eq. 3.

The middle panel in Fig. 2 shows the power α as determined by Eq. (3). The values for Si:P are approximately equal to, but slightly less than one, consistent with Fig. 1. This indicates that the prefactor of the real and imaginary components of the complex conductivity have the same concentration dependence. The situation is different for NbSi. Near the

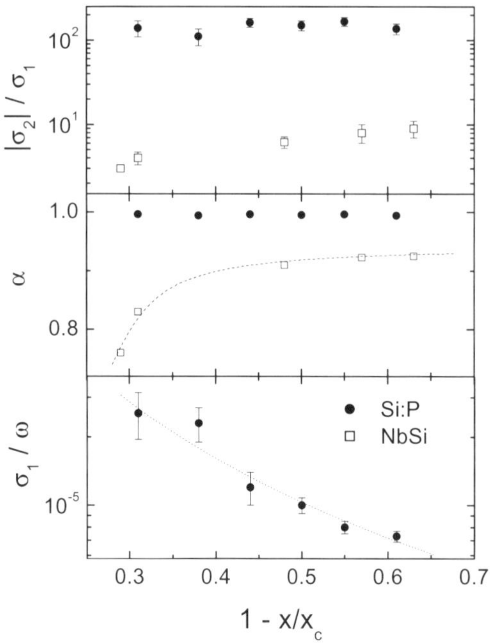

Figure 2. The upper panel shows the ratios of the imaginary to the real part of the complex conductivity for samples of Si:P and amorphous NbSi. The NbSi data is adapted from Ref 1. The middle panel shows the calculated powers of α as determined from Eq. 3. The dashed line through the NbSi data is a guide to the eye. The bottom panel shows the divergence of the prefactor of the real part of the conductivity, and the dotted line is a simple power law fit.

MIT, $\sigma(\omega)$ is expected to cross over to the QC dynamics (17, 18), i.e. $\sigma_1 \propto \omega^{1/2}$ when ξ, the localization length, is of the same scale as ℓ_ω, the dephasing length (the characteristic frequency dependent length scale) (19). This should be a smooth crossover and therefore looking at a fixed window of frequencies, a continuous change from $\omega \rightarrow \omega^{1/2}$ is expected, similar to that measured for NbSi shown in the middle panel of Fig 2.

Setting the relations for localization length and dephasing length equal (19), one finds the crossover condition for the frequency in terms of the normalized concentration, $\omega \propto \ell_0^z (1 - x/x_c)^{z\nu}$ where z is the dynamic exponent. As the prefactor can vary from system to system, the fact that we see an $\alpha \approx 1$ across our entire doping range in Si:P, but an α that approaches 1/2 in NbSi indicates that the critical regime in Si:P is much narrower and out of our experimental window. This is consistent with simple dimensional arguments (20) that show the crossover should be inversely proportional to the dopant density of states. The much smaller dopant density in Si:P vs.

NbSi (a factor of 10^3) is consistent with a narrower QC regime as compared to NbSi.

The bottom panel in Fig. 2 shows magnitude of the real part of the conductivity as the MIT is approached. This demonstrates that the prefactor A can be written as a function of the normalized concentration, i.e. $A \propto f(1 - x/x_c)$ for Si:P.

4. Discussion

In *typical* interacting systems, the effects of correlations become simpler as one goes to lower energies and/or lower temperatures. The canonical example of this is a Fermi liquid where at T=0 and ω=0 one recovers the non-interacting theory, but with parameters that are substantially modified (renormalized) from the free electron ones. The Coulomb glass is a fundamentally important example in solid state physics because it belongs to a class of systems where this does not occur and the non-interacting functional forms are not recovered at asymptotically low energies.

As predicted in the crossover function Eq. 1, we have observed in Si:P that the non-interacting functional form is recovered in the high-frequency limit and the low-frequency response shows interactions. Within the theory this is a result of the additional internal excitation structure of a resonant pair caused by interactions and the fact that any one pair can be thought of as a distinct entity, well separated in energy from other spatially nearby pairs because of the large disorder induced energy spread. This internal structure enables the excitation of pairs relatively deep within the Fermi sea and changes the factor in the initial state phase space from ω to $U(r) + \omega$. In contrast, in amorphous NbSi a smooth crossover is observed as the MIT is approached from a linear frequency dependence to a power law characterized by an exponent smaller than unity. This is consistent with the eventual frequency dependence $\sigma \propto \omega^{1/2}$ expected from quantum critical scaling arguments. No crossover to ω^2 was observed in NbSi.

Since it is the exponent α that distills the important physics (it indicates the phase space of initial states for $\alpha = 1$ or 2 and the presence of critical dynamics for $\alpha = 1/2$), we propose that one can classify the electrodynamics of electron glasses based on their α value. A schematic showing the parameter space for α is shown in Fig. 3. Here one has α's close to 1/2 near the MIT. There is a smooth crossover to Coulomb glass-like $\alpha = 1$ at lower doping levels and an intervening non-interacting Fermi glass regime at even higher energies and lower dopings. We expect that these general considerations are valid, despite the fact that some of the parameter space may not be accessible in certain systems. For instance, one may begin to excite structural modes at energies high enough to see ω^2 behavior in NbSi.

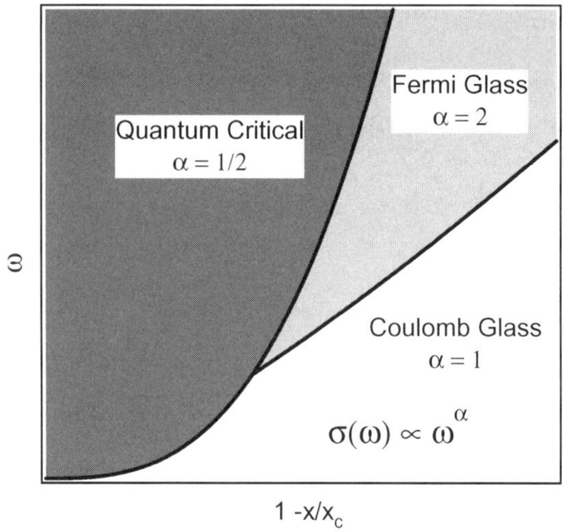

Figure 3. A schematic showing the parameter space for values taken by α for $\sigma \propto \omega^{\alpha}$. Note that the boundaries drawn on the plot are smooth crossovers and not sharp onsets. A classification based on α gives a taxonomy for the electrodynamics of electron glasses.

For dopings not close enough to critical in Si:P, excitation to the conduction band may be observed before critical dynamics are.

5. Conclusion

In summary, we have observed a crossover in the frequency dependence of the conductivity from Coulomb glass-like behavior to Fermi glass-like behavior across our entire range of doping concentrations in Si:P. The existence of a crossover is consistent with theoretical predictions, but it is sharper than predicted. The fact that we see the same functional form over the whole doping range (even deep into the insulating regime, where Eq. 1 is expected to be more valid) shows that the nature of the low energy charge excitations is qualitatively the same over the whole doping range; the inadequacy of Eq. 1 in describing the frequency dependent conductivity quantitatively is not limited to concentrations close to critical. In the amorphous semiconductor NbSi, we observe a gradual crossover from $\alpha = 1$ Coulomb glass-like behavior to quantum critical-like dynamics. This comparison allows us to obtain a general classification scheme for the electrodynamics of electron glasses based on the exponent of the frequency dependence α. We expect that this classification or 'taxonomy' will be valid even when certain regimes are not experimentally accessible.

Acknowledgements

We wish to thank Phu Tran for assisting with the cavity measurements and Barakat Alavi for assisting with the sample preparation. We would also like to thank Steve Kivelson and Boris Shklovskii for helpful conversations. This research was supported by the National Science Foundation grant DMR-0102405.

References

1. Erik Helgren, George Grüner, Martin R. Ciofalo, David V. Baxter, and John P. Carini, Phys. Rev. Lett. **87**, 116602 (2001).
2. N. F. Mott and E. A. Davis, *Electronic Processes in Non-Crystalline Materials, Second Edition,* (Oxford University Press, Oxford, 1979).
3. A. L. Efros and B. I. Shklovskii, in *Electron-electron Interactions in Disordered Systems*, edited by A. L. Efros and M. Pollak (Elsevier New York, 1985), p. 409-482.
4. S. Tanaka and H. Y. Fan, Phys. Rev. **132**, 1516 (1963).
5. H. Stupp, M. Hornung, M. Lakner, O. Madel, and H. v. Löhneysen, Phys. Rev. Lett. **71**, 2634 (1993).
6. A. L. Efros and B. I. Shklovskii, J. Physics C **8**, L49 (1975).
7. M. Lee and M. L. Stutzmann, Phys. Rev. Lett. **30**, 056402 (2001).
8. M. Lee and J. G. Massey, Phys Rev. B **60**, 1582(1999).
9. J. G. Massey and M. Lee, Phys. Rev. B **62**, R13270 (2000).
10. W. R. Thurber *et al.*, J. Electrochem. Soc. **127**,1807 (1980).
11. A. Schwartz *et al.*, Rev. Sci. Instrum. **66**, 2943 (1995).
12. G. Grüner, *Millimeter and Submillimeter Wave Spectroscopy of Solids* (Springer Verlag, Berlin, 1998).
13. E. Helgren, N.P. Armitage, and G. Grüner, Phys. Rev. Lett. **89**, 246601 (2002).
14. A. L. Efros, Sov. Phys. JETP **62**, 1057 (1985). An analagous three-dimensional form of the two-dimensional theory described in this paper can be derived.
15. B. I. Shklovskii. *private communication.* The I_0 used is the Bohr energy of phosphorous.
16. W. L. McMillan, Phys. Rev. B **24**, 2739 (1981).
17. H.-L. Lee, J. P. Carini, D. V. Baxter, and G. Grüner, Phys. Rev. Lett. **80**, 4261 (1998).
18. H.-L. Lee, John P. Carini, David V. Baxter, W. Henderson, and G. Grüner, Science **287**, 633 (2000).
19. S. L. Sondhi, S. M. Girvin, J. P. Carini, and D. Shahar, Rev. Mod. Phys. **69**, 315 (1997).
20. S. Kivelson, *private communication.*

RENORMALIZATION GROUP APPROACHES FOR SYSTEMS WITH ELECTRON-ELECTRON AND ELECTRON-PHONON INTERACTIONS

Proceedings of the ARW NATO Workshop Hvar, Croatia, October 2002

A.C. HEWSON and D. MEYER
Department of Mathematics, Imperial College,
London SW7 2AZ, UK.

Abstract. We present results of numerical renormalization group calculations for an impurity Anderson model with a coupling to a local phonon mode. The calculations are then extended, by incorporating the self-consistency condition of dynamic mean field theory, to consider the behaviour of the infinite dimensional Holstein model.

Key words: Holstein Model, Anderson-Holstein Model, Numerical Renormalization Group (NRG), Dynamical Mean-Field Theory Electron-Phonon Coupling Metal-Insulator Transition

1. Introduction

The application of the renormalization group approaches to strongly correlated systems has now a long history stretching back to the seminal work of Wilson on the Kondo problem (1), and before that to the work of Anderson (2). The Wilson numerical approach (NRG) has been particularly useful in providing a very reliable way of calculating the thermodynamic behaviour of magnetic impurity models, over the whole parameter range from weak to strong coupling (3). Its later extension to the calculation of dynamic response functions has considerably increased the range of physical properties that can now be calculated for comparison with experiment (4, 5). It has also enabled the method to be applied to lattice models, such as the Hubbard and periodic Anderson models, within the framework of dynamical mean field theory (DMFT), where the lattice problem is mapped into an effective impurity problem with an additional self-consistency constraint (for a comprehensive review of DMFT see reference (6)). Though this

A.C. Hewson and V. Zlatić (eds.), Concepts in Electron Correlation, 199–207.

mapping is only approximate, except in the limit of infinite dimensionality $d \to \infty$, there are indications that it provides a significant first step to understanding the behaviour of these lattice models in the difficult strong correlation regime. For an introduction to the NRG approach and a survey of the applications to both impurity and lattice models we refer the reader to recent review articles (7, 8).

In this paper we concentrate on using the NRG method to calculate the effects, within a generalized Anderson model, of including a coupling of the impurity with a local phonon mode. The generalization to the lattice case, using DMFT, is then used to calculate the behaviour of the Holstein model (9) in the strong and intermediate coupling regimes.

2. Anderson-Holstein impurity model

The Hamiltonian for the Anderson-Holstein model (10) is

$$H = \sum_{\sigma} \epsilon_f c^{\dagger}_{f\sigma} c_{f\sigma} + U n_{f\uparrow} n_{f\downarrow} + g(b^{\dagger} + b)(\sum_{\sigma} n_{f_\sigma} - 1)$$

$$+ \sum_{\mathbf{k},\sigma} V_{\mathbf{k}}(c^{\dagger}_{f\sigma} c_{\mathbf{k}\sigma} + c^{\dagger}_{\mathbf{k}\sigma} c_{f\sigma}) + \sum_{\mathbf{k}\sigma} \epsilon_{\mathbf{k}} c^{\dagger}_{\mathbf{k}\sigma} c_{\mathbf{k}\sigma} + \omega_0 b^{\dagger} b, \tag{1}$$

which corresponds to an Anderson model with an additional coupling of the impurity occupation number to a local optical phonon mode of frequency ω_0, as in the Holstein model.

We can get some insight into the behaviour of this model by considering the large frequency limit $\omega_0 \to \infty$. In this limit the model can be mapped into a Anderson model without the phonon term but with a renormalized level $\tilde{\epsilon}_f = \epsilon_f + v$ and an interaction $U_{\text{eff}} = U - 2v$, where $v = g^2/\omega_0$ and the limit $g \to \infty$ is taken such that v remains finite. In the particle-hole symmetric case with $U = 0$, and $\epsilon_f = 0$, the model then becomes a negative-U Anderson model. The behaviour of the negative U model is essentially that of the positive U model with spin and charge interchanged. This means that if we do move away from particle-hole symmetry for the negative U model by switching on a finite value for ϵ_f, then this is equivalent to adding a magnetic field to the positive U model.

The NRG technique as used to calculate the response functions for the standard Anderson model can be generalized to this model also (10), as the phonon term interacts with the electrons only at the impurity site. The number of phonon excitations has to be truncated and this has negligible effect provided that it exceeds a value of the order $4g^2/\omega_0^2$. In practice it is easy to take many more than this and about 200 phonon excitations are retained.

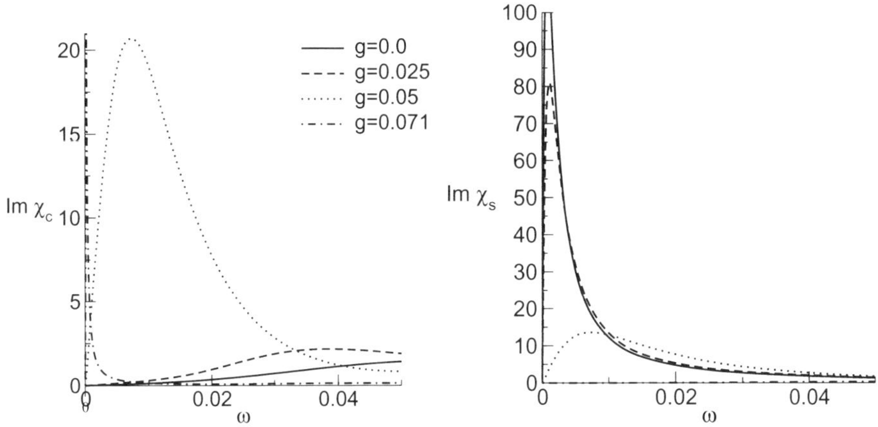

Figure 1. Imaginary part of the dynamic charge (χ_c) and spin susceptibilities (χ_s) for the symmetric parameters, $U = 0.1$, $\epsilon_f = -0.05$, $\Delta = 0.04$ and values of $U_{\text{eff}} = 0.1, 0.075, 0.0, -0.1$.

The change in the nature of the charge and spin excitations can be seen in the spectral densities of the corresponding charge and spin susceptibilities which are shown in figure 1. For a particle-hole symmetric model with $U = 0.1$, $\epsilon_f = -0.05$, and $g = 0$, the spin susceptibility has a low energy peak, while the peak for the charge susceptibility is suppressed and pushed to higher energies due to the local repulsive interaction U. As the coupling strength g to the phonons is increased these features are gradually reversed. For the largest value of the phonon coupling shown, $g = 0.071$, the effective local interaction is negative $U_{\text{eff}} = U - 2v = -0.1$, and now the spin excitations are suppressed and the charge excitations develop a low energy peak. The susceptibilities shown in figure 1 are for $\omega_0 = 0.05$, and the low energy peak in the charge susceptibility at $U_{\text{eff}} = -0.1$ ($g = 0.071$) can be seen to be much narrower than that of the corresponding spin susceptibility at $U_{\text{eff}} = 0.1$ ($g = 0$). As $\omega_0 \to \infty$ the peaks are identical as one would expect from our earlier discussion.

The narrowing of the low energy peak in the charge susceptibility with decrease in ω_0 is reflected in the spectral densities of the local electron Green's function for the particle-hole symmetric model, which are plotted in figure 2 for three values of $\omega_0 = 0.05, 0.1, 0.5$ with $U = 0$ and a fixed value of U_{eff}. The first plot has $U_{\text{eff}} = -0.05$ and so is in the weak to intermediate coupling regime. The narrowing of the resonance as a function of ω_0 is quite pronounced and distinct shoulders on either side of the resonance can be seen in the plot with the smallest value of ω_0. The second plot is for $U_{\text{eff}} = -0.1$ which is in the strong coupling regime. Again the central resonance can be seen to narrow as ω_0 is decreased. There are also distinct

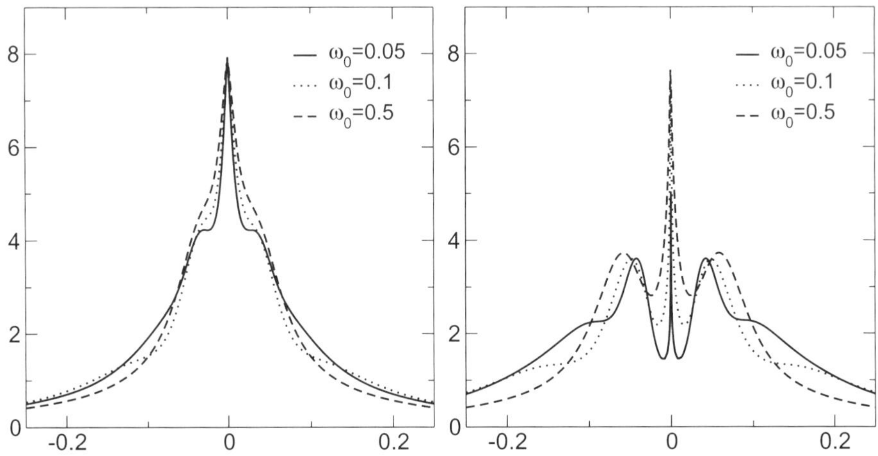

Figure 2. In these two figures the spectral density of the local Green's function for the particle-hole symmetric model with $\Delta = 0.04$, and $U = 0$, are plotted for three values of ω_0. The value of U_{eff} is maintained as constant in each plot with $U_{\text{eff}} = -0.05$ for the first figure and $U_{\text{eff}} = -0.1$ for the second.

upper and lower peaks in all three plots. The lower energy peak is associated with the adding or removing of an electron from a double or zero occupied impurity state, which are degenerate in the particle-hole symmetric case. The corresponding peak above the Fermi-level is associated with the adding or removing an electron from a singly occupied impurity level. These peaks are the broadened counterparts of those that exist in the 'atomic limit' $V = 0$. Further shoulders on these peaks can be seen in the case of the smallest value of ω_0 taken, which correspond to additional excitations in which phonons are emitted or absorbed. In the opposite limit $\omega_0 \to \infty$ such excitations are pushed out to very high energies, and in the strong coupling regime the central peak can be interpreted as a Kondo resonance, with a Kondo temperature T_{K} given by $T_{\text{K}} = (|U_{\text{eff}}|\Delta/2)^{1/2}\exp(-\pi|U_{\text{eff}}|/8\Delta + \pi\Delta/|U_{\text{eff}}|)$. The Kondo physics in this case is associated with charge or pseudo-spin degeneracy and not spin degeneracy.

The effects of the narrow peak in the charge susceptibility for the particle-hole symmetric model with $U = 0$ and $U_{\text{eff}} < 0$ can be seen the spectrum for the phonon Green's function $D(\omega) = \langle\langle b + b^\dagger : b + b^\dagger \rangle\rangle$. A relation between the two can be derived from the equations of motion (10),

$$D(\omega) = D^0(\omega) + g^2 D^0(\omega)\chi_c(\omega)D^0(\omega). \tag{2}$$

The low energy peak in $\chi_c(\omega)$ induces a low energy peak in the phonon spectrum, which survives even if we take the limit $\omega_0 \to \infty$.

It is also possible to interpret the results of these calculations in terms of the development of a double potential well. At large coupling strengths, or in the atomic limit, there are two equivalent potential energy minima (for $U = 0$) as a function of $x = (b + b^\dagger)/\sqrt{2\omega_0}$, which is treated as a classical variable, corresponding to double or zero occupation of the impurity site. The tunnelling between these two minima is responsible for the central resonance. In this strong coupling limit the interaction can be described in terms of a pseudo-spin s-d type of model (12). In the limit $\omega_0 \to \infty$ this s-d model is isotropic and leads to the usual Kondo physics in terms of a pseudo-spin. However, for finite values of ω_0 the transverse coupling terms in the pseudo-spin model are reduced by a factor of e^{-4g^2/ω_0^2} over the longitudinal terms, as they are associated with tunnelling between the two minima, and the model then corresponds to an anisotropic s-d model. It would be of interest to compare the results with those of this anisotropic s-d model in detail in the appropriate parameter regime.

These results for the spectra of the one-electron Green's function for the particle-hole symmetric model throw some light on the corresponding results for the infinite dimensional lattice Holstein model which we consider in the next section, but before considering that case we will discuss a few more results for the Anderson-Holstein model in other parameter regimes.

Away from particle-hole symmetry the results for the Anderson-Holstein model are quite straightforward to interpret. For finite U and $U_{\text{eff}} > 0$ the Kondo resonance first broadens as the coupling strength g increases from zero, because this reduces U_{eff}, and then rapidly disappears. In the strong coupling regime ($U_{\text{eff}} < 0$) the Kondo resonance is not present as there is no longer any degeneracy of the doubly and zero occupied states. If the double and zero occupation states are regarded as the up and down states of a pseudo-spin then in this regime there is an effective field acting on the pseudo-spin and the Kondo resonance is suppressed.

3. The infinite dimensional Holstein model

The NRG calculations for the impurity models can be extended to the calculation of the response functions for lattice models in the framework of dynamical mean field theory (DMFT). The DMFT has been applied widely and is extensively reviewed in reference (6). The lattice problem is mapped onto an effective impurity problem, which is then subject to a self-consistency constraint. There are many ways of carrying out calculations for the effective impurity model, and the Monte Carlo approach is one that has been used extensively (13). However, the Monte Carlo calculations cannot be performed for $T = 0$, or for very low temperatures, so the NRG approach has a distinct advantage in this regime. The NRG method is also capable of resolving very low energy scales. Most applications so far of the NRG-DMFT technique have been to the Hubbard and periodic Anderson models. Using the approach developed in the previous section, we are in a position to carry out calculations which also include a coupling to phonons, and we have focussed our attention so far on the Holstein model (9). In this model the electron occupation number at each site is linearly coupled to a local boson mode of frequency ω_0. The corresponding Hamiltonian has the form,

$$H = \sum_{\mathbf{k}\sigma} \epsilon_{\mathbf{k}} c^{\dagger}_{\mathbf{k}\sigma} c_{\mathbf{k}\sigma} + g \sum_{i} (b^{\dagger}_{i} + b_{i})(\sum_{\sigma} n_{i_{\sigma}} - 1) + \omega_0 \sum_{i} b^{\dagger}_{i} b_{i}, \qquad (3)$$

where i is a site index. In the limit $\omega_0 \to \infty$ with $U_{\text{eff}} = -2g^2/\omega_0$ finite the model maps into an effective 'negative-U' Hubbard model. The dynamic mean field theory approach is strictly speaking only valid for the infinite dimensional model but should constitute a good starting point for understanding the physics of this model, particularly in the intermediate and strong coupling regime where most methods break down. The effective impurity model in this case is the $U = 0$ Anderson-Holstein model considered in the previous section. The effective density of states of the host, however, has to be calculated self-consistently from the DMFT constraint. Results of applying the DMFT-NRG approach to this model are shown in figure 3. The first plot gives the spectral density of the local one-electron Green's function for the particle-hole symmetric model. As the coupling strength g is increased from zero, a narrow peak develops at the Fermi-level, which is quite similar to that seen in the results for the impurity model in figure 2 in the same parameter regime. The results are also quite similar to those obtained by application of the Migdal-Eliashberg approach (14) in which all perturbational theory diagrams are summed subject to the neglect of all vertex corrections.

As the coupling is further increased a pseudo-gap with a very narrow central peak begins to develop. At a critical coupling strength g_c the central peak disappears and a gap in the excitation spectrum develops continously. The quasi-particle spectral weight z decreases monotonically with increasing coupling strength and goes to zero at $g = g_c$. Due to the attractive effective interaction the electrons pair to form bipolarons as the interaction becomes stronger, and the lower peak is associated with the bipolaron formation. The Migdal-Eliashberg (ME) approach breaks down before a pseudogap develops and the results are limited to the weak coupling regime. If the ME approach is applied to the Anderson-Holstein model its limitations can be clearly seen as it does not develop the higher and lower atomic-like peaks.

It is interesting to contrast these results with those for the positive U-Hubbard model where a metal-insulator gap develops at a critical coupling

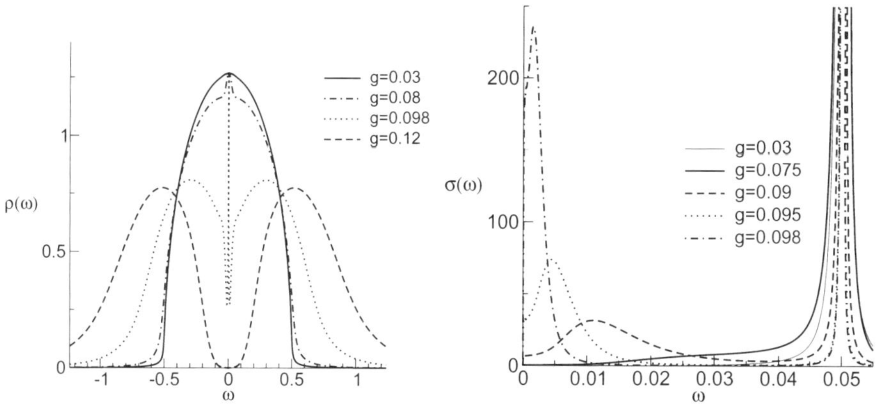

Figure 3. The spectral density for the local one-electron Green's function $\rho(\omega)$ and that for the phonon Green's function $\sigma(\omega)$ for the infinite dimensional Holstein model plotted for various values of the coupling strength g.

strength U_c (6, 15). In the Hubbard case a central quasi-particle peak occurs at the Fermi-level but on the approach to U_c this central peak is largely isolated from the upper and lower Hubbard bands, and at the metal insulator transition when the central peak disappears there is a finite gap.

We can also calculate the corresponding spectral density for the phonon spectrum, either directly or from an equation of motion similar to that given in equation 2 in terms of the effective impurity charge susceptibility. As the coupling strength increases a lower excitation mode develops in the spectrum in addition to the broadened original mode at ω_0. This additional low energy peak softens as the critical value of g is approached from below, and diverges in the limit $g \rightarrow g_c$. A similar low energy peak is seen in the calculations for the Anderson-Holstein model but this mode never completely softens, as no gap develops and there is always a central resonance. In the impurity case the induced lower phonon peak can be seen clearly to reflect the low energy peak in the local charge susceptibility. The picture of the change in the phonon spectrum in this case is quite different from that found in the ME calculations, where there is only a single peak which softens monotonically with increasing g (14). In the ME case there is no complete softening, as the method breaks down before that can occur.

4. Conclusions

The combination of the numerical renormalization group and dynamical mean field theory gives an effective way of tackling a number of difficult problems relating to strongly correlated systems. Here we have shown that it can be used to provide insights into models with coupling to phonons in the intermediate and strong coupling regime, which have so far proved to be intractable. These calculations are only a start, and there are many other interesting questions that one should be able to address using this approach. The Holstein model neglects the Coulomb interaction and so over emphasizes the tendency for bipolaron formation. Introducing a short range repulsive interaction, as in the Hubbard model, should enable strong polaron physics to be studied. Polaronic physics described by the spinless Holstein model (which has no induced attractive electron-electron interaction) with one or two electrons present has been extensively investigated (see for example (16)). There has however been very little work on systems with finite electron density, which is the most relevant for applications (for example to the fullerides (17)). The extension of the DMFT-NRG to the Holstein-Hubbard model should be a good way of investigating the physics in the strong coupling, finite electron density regime. It should also be possible to extend the technique to include a coupling to a local spin as a model to investigate the rich physics of manganite systems (18). Both these possibilities are currently being investigated.

Acknowledgements

We thank the EPSRC for the support of a research grant (GR/J85349), Ralf Bulla for cooperation on aspects of this project, and David Edwards for helpful and stimulating conversations.

References

1. K.G. Wilson, Rev. Mod. Phys. **47**, 773 (1975).
2. P.W. Anderson, J. Phys. C **3**, 2349 (1970).
3. H.R. Krishna-murthy, J.W. Wilkins, and K.G. Wilson, Phys. Rev. B **21**, 1003 and 1044 (1980).
4. O. Sakai, Y. Shimizu and T. Kasuya, J. Phys. Soc. Jap. **58**, 3666 (1989).
5. T. A. Costi and A. C. Hewson, Phil. Mag. B **65**, 1165, (1992) : J. Phys. Cond. Mat. **30**, L361 (1993): T.A. Costi, A.C. Hewson and V. Zlatić, J. Phys. Cond Mat **6**, 2519 (1994).
6. A. Georges, G. Kotliar, W. Krauth and M.J. Rozenberg, Rev. Mod. Phys. **68**, 13 (1996).
7. R. Bulla, Adv. in Solid State Physics, **40**, 169 (2000).
8. A.C. Hewson, S.C. Bradley, R. Bulla, and Y. Ono, Int. J. Mod. Phys. B, **15**, 2549 (2001).
9. T. Holstein, Ann. Phys. **8**, 325 (1959).
10. A.C. Hewson and D. Meyer, J. Phys.:Cond.Mat., **14**, 427, 2002.
11. D. Meyer, A.C. Hewson and R. Bulla, Phys. Rev. Lett. **89**, 196401 (2002).
12. J.K. Freericks, Phys. Rev. B **48**, 3881 (1993).

13. J.K. Freericks, M. Jarrell and D.J. Scalapino, Phys. Rev. B **48**, 6302 (1993).
14. J.P. Hague and N. d'Ambrumenil, cond-mat/0106355 (2001).
15. R. Bulla, Phys. Rev. Lett. **83**, 136 (1999).
16. S.A. Trugman, J. Bonca, and L.-C. Ku, Int. J. Mod. Phys. B **15**, 2707 (2001).
17. O. Gunnarsson, Rev. Mod. Phys. **69**, 575 (1997).
18. A.J. Millis, P.B. Littlewood and B.I. Shraiman, Phys. Rev. Lett. **74**, 5144 (1995).

QUANTUM PHASE TRANSITIONS IN MODELS OF MAGNETIC IMPURITIES

Proceedings of the ARW NATO Workshop Hvar, Croatia, October 2002

R. BULLA and M. VOJTA
Theoretische Physik III, Elektronische Korrelationen und Magnetismus, Universität Augsburg, 86135 Augsburg, Germany

Abstract. Zero temperature phase transitions not only occur in the bulk of quantum systems, but also at boundaries or impurities. We review recent work on quantum phase transitions in impurity models that are generalizations of the standard Kondo model describing the interaction of a localized magnetic moment with a metallic fermionic host. Whereas in the standard case the moment is screened for any antiferromagnetic Kondo coupling as $T \to 0$, the common feature of all systems considered here is that Kondo screening is suppressed due to the competition with other processes. This competition can generate unstable fixed points associated with phase transitions, where the impurity properties undergo qualitative changes. In particular, we discuss the coupling to both non-trivial fermionic and bosonic baths as well as two-impurity models, and make connections to recent experiments.

Key words: Quantum Phase Transition, Kondo Effect, Local Criticality, Numerical Renormalization Group, Two-Impurity Kondo Model, Two-Channel Kondo Model

1. Introduction

Quantum mechanical systems can undergo zero-temperature phase transitions upon variation of a non-thermal control parameter (1), where order is destroyed solely by quantum fluctuations. Quantum phase transitions occur as a result of competing ground state phases, and can be classified into first-order and continuous transitions. The transition point of a continuous quantum phase transition, the so-called quantum-critical point, is typically characterized by a critical continuum of excitations, and can lead to unconventional behavior – such as non-trivial power laws or non-Fermi liquid physics - over a wide range of the phase diagram (see Fig. 1a).

A.C. Hewson and V. Zlatić (eds.), Concepts in Electron Correlation, 209–217.

An interesting class of quantum phase transitions are so-called *boundary* transitions where only the degrees of freedom of a subsystem become critical. In this paper we consider impurity transitions – the impurity can be understood as a zero-dimensional boundary – where the *impurity* contribution to the free energy becomes singular at the quantum critical point. Such impurity quantum phase transitions require the thermodynamic limit in the bath system, but are completely independent of possible *bulk* phase transitions in the bath.

Our model systems are built from magnetic moments which show the Kondo effect (2). Originally, this effect describes the behavior of localized magnetic impurities in metals. The relevant microscopic models are the Kondo model and the single-impurity Anderson model. In the standard case (i.e. a single magnetic impurity with spin $\frac{1}{2}$ coupling to a single conduction band with a finite density of states (DOS) near the Fermi level) screening of the magnetic moment occurs below a temperature scale $T_{\rm K}$. The screening is associated with the flow to strong coupling of the effective interaction between impurity and host fermions. The Kondo temperature $T_{\rm K}$ depends exponentially on the system parameters.

In this paper we want to give a brief summary of impurity models where the flow to strong coupling is prevented by the competition with other processes, and, depending on the system parameters, weak and/or intermediate-coupling fixed points can be reached. We are particularly interested in intermediate-coupling fixed points, which typically show a finite ground-state entropy and are unstable, which means that to reach them requires either fine tuning of the couplings (typically associated with an impurity phase transition) or the presence of special symmetries (e.g., channel symmetry in multi-channel models). In the following sections we discuss single-impurity models with a coupling to a non-trivial fermionic bath (Sec. 2), to multiple fermionic baths (Sec. 3), to both fermionic and bosonic baths (Sec. 4), and finally systems of coupled magnetic impurities (Sec. 5). In quoting results, we refer to perturbative renormalization group (RG) treatments, to numerical RG (NRG) calculations, and to exact results if available. We restrict ourselves to models with spin-$\frac{1}{2}$ impurities.

2. Pseudogap Kondo and Anderson models

A straightforward possibility to suppress Kondo screening is to reduce the electron bath DOS at the Fermi level to zero – in a superconductor this can also be interpreted as competition between Kondo effect and superconducting pairing. Two cases have to be distinguished: so-called hard-gap and pseudogap (soft-gap) systems.

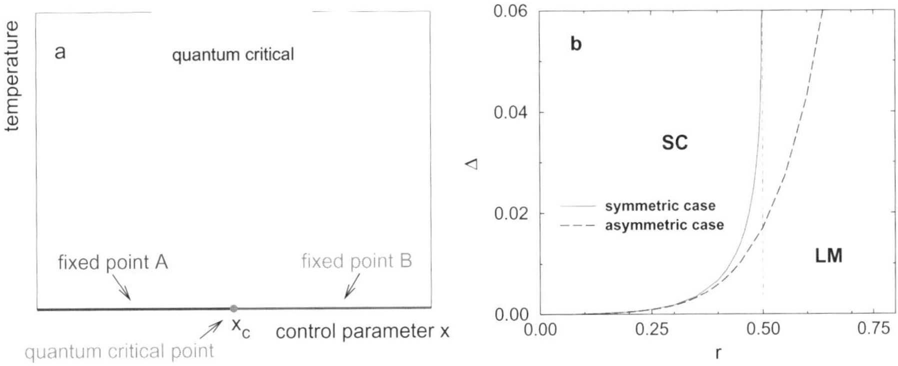

Figure 1. (a) Schematic phase diagram near a quantum phase transition between fixed points A and B upon variation of a control parameter x; the $T = 0$ critical point controls the dynamics in the $T > 0$ quantum critical region whose crossover boundaries are given by $T \sim |x - x_c|^{\nu z}$ where ν and z are the correlation length and dynamical critical exponents, respectively. (b) $T = 0$ phase diagram for the pseudogap Anderson model in the p-h symmetric case (solid line, $U = 10^{-3}$, $\varepsilon_f = -0.5 \cdot 10^{-3}$, conduction band cutoff at -1 and 1) and the p-h asymmetric case (dashed line, $\varepsilon_f = -0.4 \cdot 10^{-3}$); Δ measures the hybridization strength, $\widetilde{\Delta}(\epsilon) \equiv \pi V^2 \rho(\epsilon) = \Delta|\epsilon|^r$.

In the hard-gap case, the DOS $\rho(\varepsilon)$ is zero in a finite energy interval around the Fermi level. Then, the absence of low-energy states prevents screening for gap values exceeding the energy gain due to Kondo screening. The resulting transition between a local-moment (LM) phase without Kondo screening, realized at small Kondo coupling J, and a screened strong-coupling (SC) phase, reached for large J, is of first order, and it occurs only in the presence of particle-hole (p-h) asymmetry. In the p-h symmetric case, the local-moment state persists for arbitrary values of the coupling (3).

The pseudogap case, first considered by Withoff and Fradkin (4), corresponds to a bath with $\rho(\varepsilon) \propto |\varepsilon|^r$ ($r > 0$), i.e., the DOS is zero only *at* the Fermi level. The corresponding Kondo and single-impurity Anderson models interpolate between the metallic case ($r = 0$) and the hard-gap case ($r \to \infty$). The pseudogap case $0 < r < \infty$ leads to a very rich behaviour, in particular to a continuous transition between a local-moment and a strong-coupling phase. Figure 1b shows a typical phase diagram for the pseudogap Anderson model. In the p-h symmetric case (solid) the critical coupling Δ, measuring the hybridization between band electrons and local moment, diverges at $r = \frac{1}{2}$, and no screening occurs for $r > \frac{1}{2}$ (5, 6). No divergence occurs for p-h asymmetry (dashed) (5).

We now briefly describe the properties of the fixed points in the pseudogap Kondo problem (5). Due to the power-law conduction band DOS, already the stable LM and SC fixed points show non-trivial behavior (5, 6). The LM phase has the properties of a free spin $\frac{1}{2}$ with residual entropy $S_{\rm imp} = k_B \ln 2$

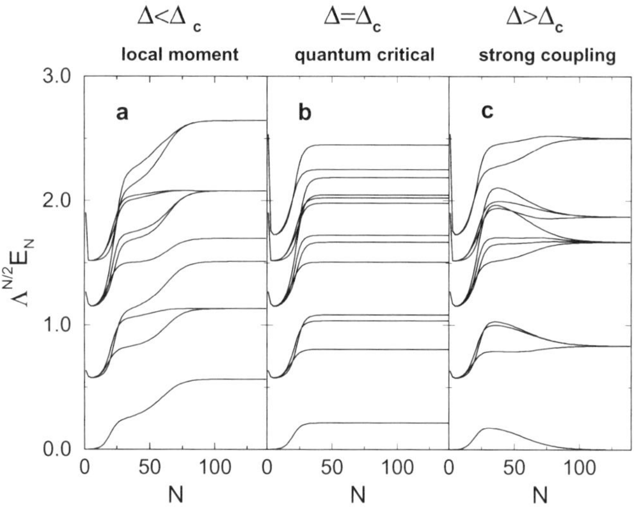

Figure 2. Flow diagrams for the low-energy many-body excitations obtained from the numerical renormalization group for the three different fixed points of the soft-gap Anderson model. N is the number of iterations of the NRG procedure, Λ the NRG discretization parameter.

and low-temperature impurity susceptibility $\chi_{\rm imp} = 1/(4k_BT)$, but the leading corrections show r-dependent power laws. The p-h symmetric SC fixed point has very unusual properties, namely $S_{\rm imp} = 2rk_B \ln 2$, $\chi_{\rm imp} = r/(8k_BT)$ for $0 < r < \frac{1}{2}$. In contrast, the p-h asymmetric SC fixed point simply displays a completely screened moment, $S_{\rm imp} = T\chi_{\rm imp} = 0$. The impurity spectral function follows a ω^r power law at both the LM and the asymmetric SC fixed point, whereas it diverges as ω^{-r} at the symmetric SC fixed point – this "peak" can be viewed as a generalization of the Kondo resonance in the standard case ($r = 0$), and scaling of this peak is observed upon approaching the SC-LM phase boundary (6, 7). At the critical point non-trivial behavior corresponding to a fractional moment can be observed: $S_{\rm imp} = k_B\mathcal{C}_S(r)$, $\chi_{\rm imp} = \mathcal{C}_\chi(r)/(k_BT)$ with $\mathcal{C}_S$, $\mathcal{C}_\chi$ being universal functions of r (5, 8). The spectral function displays a ω^{-r} power law (for $r < 1$) with a remarkable "pinning" of the critical exponent.

We note that the critical point at small r is perturbatively accessible in a double expansion in r and J. However, the NRG results suggest that the physics is different for $r > \frac{1}{2}$; in particular, $r = 1$ appears to play the role of an upper critical "dimension" (5, 8, 9). The universal critical theory of the transitions in the pseudogap Kondo model is not yet known. A piece of information, provided by the numerical renormalization group method, is shown in Fig. 2. Here we plot the energies of the many-body states as a function of the iteration number N of the NRG procedure. Increasing values of N correspond to decreasing temperature, $N \propto -\ln T$, so we can easily recognize the three different fixed points for $\Delta < \Delta_c$ (Fig. 2a), $\Delta = \Delta_c$ (Fig. 2b), and $\Delta > \Delta_c$ (Fig. 2c). The structure of the LM and SC fixed points can be easily understood as that of a free conduction electron chain (10, 11). The combination of the single-particle states of the free chain leads to the degeneracies seen in the many-particle states of the LM and SC fixed points (Fig. 2a and Fig. 2c). In contrast, the structure of the quantum critical point is unclear. Degeneracies due to the combination of single-particle levels are missing, probably because the quantum critical point is *not* build up of non-interacting single-particle states.

The pseudogap Kondo model has been proposed (8) to describe impurity moments in d-wave superconductors ($r = 1$), where signatures of Kondo physics have been found in NMR experiments (12). Furthermore, a large peak seen in STM tunneling near Zn impurities (13) can be related to the impurity spectrum in the asymmetric pseudogap Kondo model.

3. Multi-channel Kondo model

Kondo screening is strongly modified if two or more fermionic screening channels compete. Nozières and Blandin (14) proposed a two-channel generalization of the Kondo model, which shows overscreening associated with an intermediate-coupling fixed point and non-Fermi liquid behaviour in various thermodynamic and transport properties. In general, such behavior occurs for any number of channels $K > 1$ coupled to a spin $\frac{1}{2}$, and does not require fine-tuning of the Kondo coupling, however, it is unstable w.r.t. a channel asymmetry.

The anomalous static properties at the two-channel non-Fermi liquid fixed point are (15) a residual entropy $S_{\mathrm{imp}} = \frac{k_B}{2} \ln 2$ (indicating that 'half'-fermionic excitations play a crucial role for the structure of the fixed point), a logarithmic divergence of the susceptibility $\chi_{\mathrm{imp}} \propto \ln T$ and of the specific heat ratio $\gamma = C_{\mathrm{imp}}/T \propto \ln T$, and an anomalous Wilson ratio $R = \chi/\gamma = 8/3$, in contrast to the result for the standard Kondo model $R = 8/4 = 2$.

Analogous to the discussion in the previous section the many-particle excitation spectrum of this intermediate-coupling fixed point cannot be decomposed in terms of usual free fermions (16, 17), however, a description in terms of non-interacting *Majorana* fermions is possible (18). Many of the low-energy properties of the two-channel and related models have been studied using conformal field theory techniques (19, 20). Interestingly, the multi-channel Kondo fixed point is perturbatively accessible in the limit of large channel number ($K \gg 1$) (14, 20). Experimental realizations have been discussed in the context of rare-earth compounds (15); furthermore, proposals based on quantum-dot devices have been put forward.

4. Bose-Fermi Kondo model

Novel phenomena occur for magnetic impurities coupled to *both* a fermionic and a bosonic bath, where the bosons represent collective host spin fluctuations. In the resulting Bose-Fermi Kondo model, the two interactions compete in a non-trivial manner (21, 22). Particularly interesting is the case of a bosonic bath with zero or small gap, corresponding to the vicinity to a magnetic quantum critical point in the host. In $(3-\epsilon)$ dimensions, the bosonic spectral density then follows a power law $\propto \omega^{1-\epsilon}$. For $\epsilon > 0$, the interaction between the impurity and the bosonic bath can completely suppress fermionic Kondo screening. The resulting phase corresponds to an intermediate-coupling fixed point w.r.t. the impurity–boson interaction. Here, the impurity moment shows universal fluctuations, with local spin correlations characterized by a power law, $\langle S(\tau)S\rangle \propto \tau^{-\epsilon}$, and a Curie contribution to the impurity susceptibility equivalent to a fractional spin, $\chi_{\rm imp} = \widetilde{\mathcal{C}}_\chi(\epsilon)/(k_B T)$ (21, 22, 23). On the other hand, large fermionic Kondo coupling leads to a strong-coupling phase with conventional Kondo screening.

The resulting phase diagram for the Bose-Fermi Kondo model thus shows a Kondo-screened phase, a bosonic fluctuating phase, and a continuous quantum phase transition in between. The boundary quantum critical point has magnetic properties similar to the bosonic fluctuating fixed point (22). Both intermediate-coupling fixed points are perturbatively accessible for small ϵ; it is likely that the structure of the phase diagram also applies to $\epsilon = 1$, however, no accurate numerical calculations are available to date. Both the critical and the bosonic fluctuating fixed points are unstable w.r.t. breaking of SU(2) symmetry, but the structure of the phase diagram is similar for both XY and Ising symmetries (22).

The Bose-Fermi Kondo model has recently received a lot of interest in the context of the extended dynamical mean-field theory (24) where a lattice model is mapped onto a self-consistent impurity model with both fermionic and bosonic baths. In particular, based on neutron scattering experiments on the heavy-fermion compound $CeCu_{5-x}Au_x$ (25), a self-consistent version of the Bose-Fermi Kondo model has been proposed to describe a "local" critical point in Kondo lattice models.

We further note that the Bose Kondo model with Ising symmetry and an additional external field is related to the spin-boson model (26), which also shows phase transitions upon variation of the coupling between spin and dissipative bath. Spin-boson models have been studied extensively in the context of dissipative two-level systems, and have applications in many fields like glass physics, quantum computation etc.

5. Two-impurity Kondo models

Models of two impurities offer a new ingredient, namely the exchange interaction, I, between the two impurity spins which competes with Kondo screening of the individual impurities. This inter-impurity interaction, which can lead to a magnetic ordering transition in lattice models, arises both from direct exchange and from the Ruderman-Kittel-Kasuya-Yosida (RKKY) interaction mediated by the conduction electrons.

In the simplest model of two $S = \frac{1}{2}$ impurities, a ground state singlet ($S_{\text{tot}} = 0$) can be realized either by individual Kondo screening (if $I < T_K$) or by formation of an inter-impurity singlet (if $I > T_K$). It has been shown that these two parameter regimes are continuously connected (without a $T = 0$ phase transition) as I is varied in the generic situation without particle-hole symmetry. Notably, in the particle-hole symmetric case one finds a transition associated with an unstable non-Fermi liquid fixed point (27, 28, 29).

Quantum phase transitions generically occur in impurity models showing phases with *different* ground state spin. For two impurities, this can be realized by coupling to a *single* conduction band channel only (30). In this case, a Kosterlitz-Thouless-type transition between a singlet and a doublet state occurs, associated with a second exponentially small energy scale in the Kondo regime (30). The physics becomes even richer if multi-channel physics is combined with multi-impurity physics – here, a variety of fixed points including such with local non-Fermi liquid behavior can be realized.

Experimentally, quantum dots provide an ideal laboratory to study systems of two (or more) "impurities" – note that the local "impurity" states can arise either from charge or from spin degrees of freedom on each quantum dot. In particular, a number of experiments have been performed on coupled quantum dot systems which can be directly mapped onto models of two Kondo or Anderson impurities (31). In addition, experimental realizations of two-impurity models using magnetic adatoms on metallic surfaces appear possible.

6. Summary

We have reviewed a variety of interesting zero-temperature critical points, which exist in impurity models where conventional Kondo screening is suppressed by competing physics. Significant progress has been made in recent years, both in analytical and numerical work; however, in a number of cases our understanding concerning, e.g., the nature of the quantum critical points is rather limited. Clearly, a further development of theoretical techniques, both perturbative and non-perturbative, is essential.

Acknowledgements

It is a pleasure to acknowledge fruitful discussions and collaborations with A. Hewson, W. Hofstetter, D. Logan, Th. Pruschke, and S. Sachdev, as well as financial support by the DFG through SFB 484.

References

1. S. Sachdev, *Quantum Phase Transitions*, Cambridge University Press, Cambridge (1999).
2. A. C. Hewson, *The Kondo Problem to Heavy Fermions*, Cambridge University Press, Cambridge (1993).
3. K. Chen and C. Jayaprakash, Phys. Rev. B **57**, 5225 (1998).
4. D. Withoff and E. Fradkin, Phys. Rev. Lett. **64**, 1835 (1990).
5. C. Gonzalez-Buxton and K. Ingersent, Phys. Rev. B **57**, 14254 (1998).
6. R. Bulla, T. Pruschke and A. C. Hewson, J. Phys.: Condens. Matter **9**, 10463 (1997); R. Bulla, M. T. Glossop, D. E. Logan and T. Pruschke, *ibid* **12**, 4899 (2000).
7. D. E. Logan and M. T. Glossop, J. Phys.: Condens. Matter **12**, 985 (2000).
8. M. Vojta and R. Bulla, Phys. Rev. B **65**, 014511 (2002).
9. K. Ingersent and Q. Si, Phys. Rev. Lett. **89**, 076403 (2002).
10. K. G. Wilson, Rev. Mod. Phys. **47**, 773 (1975).
11. H. R. Krishna-murthy, J. W. Wilkins and K. G. Wilson, Phys. Rev. B **21**, 1003 (1980).
12. J. Bobroff, W. A. MacFarlane, H. Alloul, P. Mendels, N. Blanchard, G. Collin and J.-F. Marucco, Phys. Rev. Lett. **83**, 4381 (1999).
13. E. W. Hudson, S. H. Pan, A. K. Gupta, K. W. Ng and J. C. Davis, Science **285**, 88 (1999).
14. P. Nozières and A. Blandin, J. Physique **41**, 193 (1980).
15. D. L. Cox and A. Zawadowski, Adv. Phys. **47**, 599 (1998).
16. D. M. Cragg, P. Lloyd and P. Nozières, J. Phys. C, **13**, 803 (1980).
17. H.-B. Pang and D. L. Cox, Phys. Rev. B **44**, 9454 (1991).
18. R. Bulla, A. C. Hewson and G.-M. Zhang, Phys. Rev. B **56**, 11721 (1997).
19. I. Affleck and A. W. W. Ludwig, Nucl. Phys. B **352**, 849 (1991) and **360**, 641 (1991), Phys. Rev. B **48**, 7297 (1993).
20. O. Parcollet, A. Georges, G. Kotliar and A. Sengupta, Phys. Rev. B **58**, 3794 (1998).
21. J. L. Smith and Q. Si, cond-mat/9705140, Europhys. Lett. **45**, 228 (1999); A. M. Sengupta, Phys. Rev. B **61**, 4041 (2000).
22. L. Zhu and Q. Si, Phys. Rev. B **66**, 024426 (2002); G. Zarand and E. Demler, Phys. Rev. B **66**, 024427 (2002).
23. S. Sachdev, C. Buragohain and M. Vojta, Science **286**, 2479 (1999); M. Vojta, C. Buragohain and S. Sachdev, Phys. Rev. B **61**, 15152 (2000).
24. Q. Si, S. Rabello, K. Ingersent and J. L. Smith, Nature **413**, 804 (2001).
25. A. Schröder, G. Aeppli, R. Coldea, M. Adams, O. Stockert, H. v. Löhneysen, E. Bucher, R. Ramazashvili and P. Coleman, Nature **407**, 351 (2000).
26. A. J. Leggett, S. Chakravarty, A. T. Dorsey, M. P. A. Fisher, A. Garg, and W. Zwerger, Rev. Mod. Phys. **59**, 1 (1987).
27. B. A. Jones and C. M. Varma, Phys. Rev. Lett. **58**, 843 (1987); B. A. Jones, C. M. Varma and J. W. Wilkins, *ibid.* **61**, 125 (1988).

28. O. Sakai, Y. Shimizu and T. Kasuya, Solid State Comm. **75**, 81 (1990); O. Sakai and Y. Shimizu, J. Phys. Soc. Jpn **61**, 2333 (1992), *ibid*, **61**, 2348 (1992).
29. I. Affleck, A. W. W. Ludwig and B. A. Jones, Phys. Rev. B **52**, 9528 (1995).
30. M. Vojta, R. Bulla and W. Hofstetter, Phys. Rev. B **65**, 140405(R) (2002).
31. W. G. van der Wiel, S. De Franceschi, J. M. Elzerman, S. Tarucha, L. P. Kouwenhoven, J. Motohisa, F. Nakajima and T. Fukui, Phys. Rev. Lett. **88**, 126803 (2002).

INSTABILITY OF THE FERMI-LIQUID FIXED POINT IN AN EXTENDED KONDO MODEL

Proceedings of the ARW NATO Workshop Hvar, Croatia, October 2002

M. LAVAGNA*
Commissariat à l'Energie Atomique, DRFMC/SPSMS, 17, rue des Martyrs, 38054 Grenoble Cedex 9, France

A. JEREZ
European Synchrotron Radiation Facility, 6, rue Jules Horowitz, 38043 Grenoble Cedex 9, France

D. BENSIMON†
Department of Appled Physics, Hongo 7-3-1, University of Tokyo, Tokyo 113-8656, Japan

Abstract. We study an extended SU(N) single-impurity Kondo model in which the impurity spin is described by a combination of Abrikosov fermions and Schwinger bosons. Our aim is to describe both the quasiparticle-like excitations and the locally critical modes observed in various physical situations, including non-Fermi liquid (NFL) behavior in heavy fermions in the vicinity of a quantum critical point and anomalous transport properties in quantum wires. In contrast with models with either pure bosonic or pure fermionic impurities, the strong coupling fixed point is unstable against the conduction electron kinetic term under certain conditions. The stability region of the strong coupling fixed point coincides with the region where the partially screened, effective impurity repels the electrons on adjacent sites. In the instability region, the impurity tends to attract $(N-1)$ electrons to the neighboring sites, giving rise to a double-stage Kondo effect with additional screening of the impurity.

Key words: Non-Fermi Liquid, Strong versus Intermediate Coupling Fixed Point, Extended SU(N) Kondo Model

* Also at the Centre National de la Recherche Scientifique (CNRS)
† Also at the Commissariat à l'Energie Atomique, DRFMC/SPSMS, 17, rue des Martyrs, 38054 Grenoble Cedex 9, France

A.C. Hewson and V. Zlatić (eds.), Concepts in Electron Correlation, 219–227.

1. Introduction

One of the most striking properties discovered in Heavy-Fermion systems in recent years is the existence of a non-Fermi-liquid behavior (11, 21, 12) in the disordered phase close to the magnetic quantum critical point with a temperature dependence of the physical quantities which differs from that of a standard Fermi-liquid. Recent results obtained in $CeCu_{5.9}Au_{0.1}$ by inelastic neutron scattering experiments (19, 20) have shown the existence of anomalous ω/T scaling law for the dynamical spin susceptibility at the antiferromagnetic wavevector which persists over the entire Brillouin zone. This indicates that the spin dynamics are critical not only at large length scales as in the itinerant magnetism picture (7, 13) but also at atomic length scales. It strongly suggests the presence of locally critical modes beyond the standard spin-fluctuation theory. Alternative theories have been proposed to describe such a local quantum critical point. In this direction, we will mention recent calculations (18) based on dynamical mean field theory (DMFT) which seem to lead to encouraging results such as the prediction of a scaling law for the dynamical susceptibility. The other approach which has been put forward to describe the local QCP is based on supersymmetric theory (6, 17, 3, 4) in which the spin is described in a mixed fermionic-bosonic representation. The interest of the supersymmetric approach is to describe the quasiparticles and the local moments on an equal footing through the fermionic and the bosonic part of the spin respectively. It appears to be specially well-indicated in the case of the locally critical scenario in which the magnetic temperature scale T_N, and the Fermi scale T_K (the Kondo temperature) below which the quasiparticles die, vanish at the same point δ_C.

Let us now emphasize another aspect in the discussion of the breakdown of the Fermi-liquid theory. It has to do with the question of the instability of the strong coupling (SC) fixed point. A stable SC fixed point is usually associated with a local Fermi-liquid behavior. On the contrary, an instability of the SC fixed point is an indication for the existence of an intermediate coupling fixed point with non-Fermi-liquid behavior. The traditional source of instability of the SC fixed point in the Kondo model is the presence of several channels for the conduction electrons with the existence of two regimes, the underscreened and the overscreened ones with very different associated behavior. In the one-channel antiferromagnetic single-impurity Kondo model, it is well known from renormalization group arguments, that the system flows to a stable SC fixed point (2) with a behavior of the system identified with that of a local Fermi liquid (14). The situation is rather different in the case of the multichannel Kondo model with a number K of channels for the conduction electrons ($K > 1$) (15). In the underscreened regime when $K < 2S$ (where S is the value of the spin in $SU(2)$), the SC fixed point is stable. It becomes unstable in the other regime $K > 2S$, known as the overscreened regime. The underscreened regime corresponds to the one-stage Kondo effect with the formation of an effective spin $(S - 1/2)$ resulting from the screening of the impurity spin by the conduction electrons located on the same site. The system described by the SC fixed point behaves as a local Fermi liquid. The instability of the SC fixed point in the overscreened regime is associated with a multi-stage Kondo effect in which successively the impurity spin is screened by conduction electrons on the same site, and then the dressed impurity is screened by conduction electrons on the neighboring site and so forth. The overscreened regime leads to the existence of an intermediate coupling fixed point with non-Fermi-liquid excitation spectrum and an anomalous residual entropy at zero temperature. It has been recently put forward (3, 4) that other sources of instability of the SC fixed point may exist other than the multiplicity of the conduction electron channels. Recent works have shown that the presence of a more general Kondo impurity where the spin symmetry is extended from $SU(2)$ to $SU(N)$, and the impurity has mixed symmetry, may also lead to an unstability of the SC fixed point already in the one-channel case. In the large N limit, Coleman et al. (3) have found that the SC fixed point becomes unstable as soon as the impurity parameter q (defined below) is larger than $N/2$ whatever the value of $2S$ (defined below) is, giving rise to a two-stage Kondo effect. This result opens the route for the existence of an intermediate coupling fixed point with eventually non-Fermi-liquid behavior.

It is worth noting that the supersymmetry theory, or more specifically the taking into account of general Kondo impurities appears to offer valuable insights into the two issues raised by the breakdown of the Fermi liquid theory that we have summarized above, i.e. both the existence of locally critical modes and the question of the instability of the SC fixed point. Somehow it seems that the consideration of general Kondo impurities captures the physics present in real systems with the coexistence of the screening of the spin by conduction electrons responsible for the formation of quasiparticles, and the formation of localized magnetic moment that persists and eventually leads to a phase transition as the coupling to other impurities becomes dominant.

The aim of this work is to study the $SU(N)$, extended single-impurity Kondo model in the one-channel case (10). We would like to investigate how the system behaves when not only the values of the parameters $(2S, q)$ of the representation vary, but also the number n_d of conduction electrons on the neighboring site does. We want to discuss the effect of n_d on the stability of the SC fixed point and to further understand the nature of the screening with the possibility of achieving either a one-stage, two-stage or multi-stage Kondo effect depending on the regime considered. Implications for the behavior of physical quantities will be given.

2. Extended $SU(N)$ single-impurity Kondo model

We consider a generalized, single-impurity, Kondo model with one channel of conduction electrons and a spin symmetry group extended from $SU(2)$ to $SU(N)$. An impurity spin, $\mathbf{S}$, is placed at the origin (site 0). We deal with impurities that can be realized by a combination of $2S$ bosonic ($b^\dagger_\alpha$) and $q-1$ fermionic ($f^\dagger_\alpha$) operators. The hamiltonian describing the model is written as

$$H = \sum_{\mathbf{k},\alpha} \varepsilon_{\mathbf{k}} c^\dagger_{\mathbf{k},\alpha} c_{\mathbf{k},\alpha} + J \sum_A \mathbf{S}^A \sum_{\alpha,\beta} c^\dagger_\alpha(0) \tau^A_{\alpha\beta} c_\beta(0) \ , \tag{1}$$

where $c^\dagger_{\mathbf{k},\alpha}$ is the creation operator of a conduction electron with momentum $\mathbf{k}$, and $SU(N)$ spin index $\alpha = a, b, ..., r_N$, $c^\dagger_\alpha(0) = \frac{1}{\sqrt{N_S}} \sum_{\mathbf{k}} c^\dagger_{\mathbf{k},\alpha}$, where N_S is the number of sites, and $\tau^A_{\alpha\beta}$ ($A = 1, \ldots, N^2-1$) are the generators of the $SU(N)$ group in the fundamental representation, with $Tr[\tau^A \tau^B] = \delta_{AB}/2$. The conduction electrons interact with the impurity spin $\mathbf{S}^A$ ($A = 1, \ldots, N^2-1$), placed at the origin, via Kondo coupling, $J > 0$. When the impurity is in the fundamental representation, we recover the Coqblin-Schrieffer model (5, 8) describing conduction electrons in interaction with an impurity spin of angular momentum j, ($N = 2j+1$, $a = j$, $b = j-1, \ldots, r_N = -j$), resulting from the combined spin and orbit exchange scattering.

2.1. STRONG-COUPLING FIXED POINT

We consider the case where $J \to \infty$ and we can neglect the kinetic energy in Eq. 1. In this limit the model can be solved exactly in terms of the invariants associated to the spin of the electrons and of the impurities (9). The eigenvalues are of the form

$$E_{SC} = \frac{J}{2} \left[\mathcal{C}_2(R_{SC}) - \mathcal{C}_2(I) - \mathcal{C}_2(Y)\right] \tag{2}$$

where I denotes the impurity spin, Y the spin of the n_c conduction electrons coupled to the impurity at the origin, and R_{SC} the spin of the resulting SC state at the impurity site. The quantities $\mathcal{C}_2$ are the $SU(N)$ generalization of the $SU(2)$ eigenvalues, $S(S+1)$, and can be readily evaluated (for details, see Ref. (10)).

The ground state corresponds to having $n_c = (N - q)$ electrons at the origin, partially screening the impurity. It can be written explicitly as the action of $(2S - 1)$ bosonic operators on a singlet state. For instance the *highest spin* state can be written as

$$|GS\rangle^{[2S-1]}_{\{a\}aa} = \frac{1}{\sqrt{(2S-1)!}}(b^\dagger_a)^{2S-1}\left[\frac{1}{\gamma}\mathcal{A}(b^\dagger_{i_1}(\prod_{\alpha=i_2}^{i_q} f^\dagger_\alpha)(\prod_{\beta=i_{q+1}}^{i_N} c^\dagger_\beta))\right]|0\rangle \qquad (3)$$

with $\gamma \equiv \sqrt{(2S+N-1)C^{q-1}_{N-1}}$. Here, $c^\dagger_\alpha \equiv c^\dagger_\alpha(0)$. We denote the ground state energy by E_0.

The effect of the kinetic term in Eq. 1 is, to lowest order in perturbation theory, to mix the ground state with excited states where the number of electrons changes by one. There are three such states, which we denote by $|GS+1\rangle^S$, $|GS+1\rangle^A$ and $|GS-1\rangle$. The labels S(A) indicate that the additional electron is coupled symmetrically(antisymmetrically) to the ground state. The states are readily obtained by deriving the relevant SU(N) Clebsch-Gordan coefficients (10).

2.2. STABILITY OF THE STRONG COUPLING FIXED POINT

In order to understand better the low-energy physics of the system, we should consider the finite Kondo coupling, allowing virtual hopping from and to the impurity site. These processes generate interactions between the composite at site 0 and the conduction electrons on neighboring sites, that can be treated as perturbations of the SC fixed point. The energy shifts due to the perturbation can be reproduced by introducing an interaction between the spin at the impurity site and the spin of the electrons on the neighboring site, with effective coupling J_{eff}. Applying an analysis similar to that of Nozières and Blandin (15) to the nature of the excitations, we can argue whether or not the SC fixed point remains stable once virtual hopping is allowed.

Thus, if the coupling between the effective spin at the impurity site and that of the electrons on site 1 is ferromagnetic we know, from the scaling analysis at weak coupling, that the perturbation is irrelevant, and the low energy physics is described by a SC fixed point. That is, an underscreened, effective impurity weakly coupled to a gas of free electrons with a phase shift indicating that there are already $(N - q)$ electrons screening the original impurity.

If, on the contrary, the effective coupling is antiferromagnetic, the perturbation is relevant, the SC fixed point is unstable and the low-energy physics of the model corresponds to some intermediate coupling fixed point, to be identified.

We explicitly calculate the effects of hopping on the SC fixed point to the lowest order in perturbation theory, that is, second order in t. We consider the case with an arbitrary number n_d of conduction electrons on site 1 generalizing the case $n_d = 1$ considered in ref. (4).

Adding n_d electrons on site 1 leads to two different states, that we will call symmetric (S) and antisymmetric (A). These states, degenerate in the SC limit, acquire energy shifts, ΔE_0^A and ΔE_0^S due to the perturbation given, in the large-N limit, by

$$\Delta E_0^A = -\left(\frac{2t^2}{J}\right)\left[\left(\frac{N-q}{q}\right) - \frac{n_d}{N}\left(\frac{N(N-2q)}{q(N-q)}\right)\right] , \tag{4}$$

$$\begin{aligned} \Delta E_0^S - \Delta E_0^A &= -\left(\frac{2S+n_d-1}{2S+N-q}\right)\left(\frac{2t^2}{JN}\right)\left(\frac{N(N-2q)}{q(N-q)}\right) , \\ &= \frac{J_{eff}}{2}(2S+n_d-1) . \end{aligned} \tag{5}$$

Notice that the behavior of both Eq. 4 and Eq. 5 are controlled by the same factor. This result has the immediate following physical consequence. The change of sign of J_{eff} (Fig. 1, Right) -and hence of the stability of the SC fixed point- is directly connected to the change in the behavior of $\Delta E_0^A \sim \Delta E_0^S$ with n_d. In particular, when $\Delta E_0^S = \Delta E_0^A$, $\Delta E_0^A = -(2t^2/J)$ independently [1] of n_d, q or $2S$. In the regime where the SC fixed point is stable, $q/N < 1/2$, $J_{eff} < 0$, the lowest energy corresponds to $n_d = 1$, whereas for $q/N > 1/2$, $J_{eff} > 0$, the energy expressed in Eq. (4) is minimized for $n_d = (N-1)$ (Fig. 1, Left). This is precisely the mechanism behind the two-stage quenching. The accumulation of electrons on site 1 is not related to J_{eff} which is independent of n_d, but results from the dependence of $\Delta E_0^A \sim \Delta E_0^S$ with n_d.

We finish by making some remarks on the physical properties of the model in the different regimes. As is common to all models with an antiferromagnetic Kondo coupling, there will be a crossover from weak coupling above a given Kondo scale, T_K, to a low-energy regime. When the SC fixed point is stable, we should expect for $T \ll T_K$ a weak coupling of the effective impurity at site-0 with the rest of the electrons. The physical properties at low temperature are controlled by the degeneracy of the effective impurity, $d([2S-1]) = C_{N+2S-2}^{N-1}$. Thus, we should expect a residual entropy $\mathcal{S}^i \sim \ln C_{N+2S-2}^{N-1}$ and a Curie susceptibility, $\chi^i \sim C_{N+2S-2}^{N-1}/T$, with logarithmic corrections (9, 16). This is the result that we would expect for a purely symmetric impurity. Contrary to the purely bosonic case, only

[1] This is true for arbitrary values of N, and $2S$.

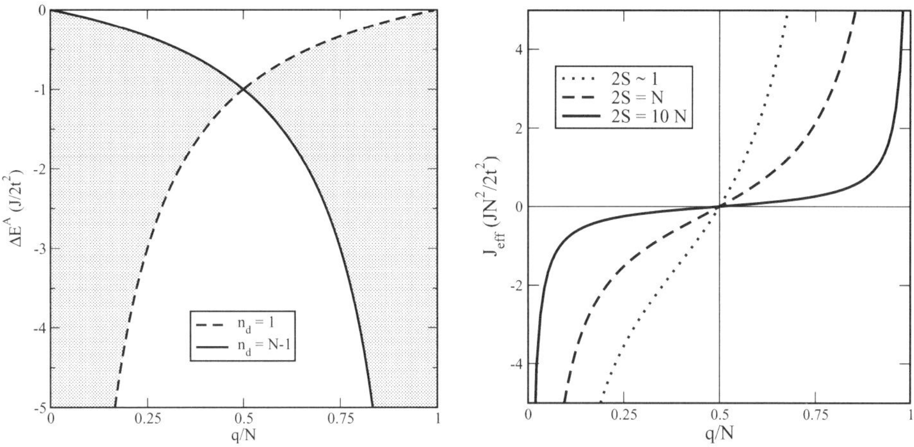

Figure 1. (Left) Leading term in the energy shift, $\Delta E_0^A \sim \Delta E_0^S$, as a function of q/N, for $1 < n_d < N-1$ (shaded region), and in the limiting cases $n_d/N \ll 1$ (dashed line), and $n_d/N \approx 1$ (solid line). Notice that the value at $q/N = 1/2$ is equal to $-2t^2/J$, for any n_d. (Right) Energy difference, $(\Delta E_0^S - \Delta E_0^A)$, as a function of q/N, for different values of $2S$ in the large-N limit.

$(N - q)$ electrons are allowed at the origin in the SC limit, instead of $(N-1)$. Thus, we would expect to find different results for quantities that involve the scattering phase shift of electrons off the effective impurity.

In the $q > N/2$, we do not have access to the intermediate coupling fixed point that determines the low-energy behavior. Nevertheless, it is reasonable to think that there would be a magnetic contribution to the entropy, and a Curie-like contribution to the susceptibility, since the impurity remains unscreened. This behavior is different from that of the multichannel Kondo model, which is also characterized by an intermediate coupling fixed point, but where the impurity magnetic degrees of freedom are completely quenched. The degeneracy of the true ground state is an open question, but we can assume that the entropy will be smaller than that of the SC fixed point (1). It is in the scattering properties that we might be able to see the anomalous features of this new fixed point more clearly.

3. Conclusions

We have studied a Kondo model where the spin of the impurity has mixed symmetry, as a way of incorporating the phenomena of local moment screening and magnetic correlation that is observed in some heavy fermion compounds. Such model is naturally realized by extending the spin symmetry to $SU(N)$, and it displays a rich phase diagram. We find that as long as the bosonic component of spin is of order N, there is a transition around the point where the fermionic component of the impurity is $q = N/2$. At this particular point, the energy shift is, to lowest order in perturbation theory around the SC fixed point, equal to $-2t^2/J$, independent of the impurity parameters, q, S and N. When $q < N/2$, the low-energy physics corresponds to the SC fixed point. For $q > N/2$ the SC fixed point is unstable and anomalous behavior is expected, in particular, the two-stage quenching effect. This phase diagram is not accidental, but is due to the relation of the effective impurity site in the SC regime to the conduction electrons in neighboring sites, as our study of the dependence of the energy shifts on n_d reveals. If $q < N/2$, the energy is minimized when the dressed impurity repels the electrons on the next site. That is, when $n_d = 1$. At $q = N/2$, the energy shift is also independent of n_d. Finally, the lowest energy shift for $q > N/q$ corresponds to a maximal n_d, indicating the accumulation of electrons leading to the two-stage Kondo quenching.

Acknowledgements

We would like to thank V. Zlatic and A. Hewson for organizing this very interesting meeting. We are most grateful to N. Andrei for continuous encouragement and discussions. We also thank S. Burdin, P. Coleman, Ph. Nozières, and C. Pépin for helpful discussions.

References

1. I. Affleck and A. W. W. Ludwig, Phys. Rev. B **48**, 7297 (1993).
2. P. Anderson, Phys. Rev. B **164**, 352 (1967).
3. P. Coleman, C. Pépin, and A. M. Tsvelik, Phys. Rev. B **62**, 3852 (2000).
4. P. Coleman, C. Pépin, and A. M. Tsvelik, Nucl. Phys. B **586**, 641 (2000).
5. B. Coqblin and J. R. Schrieffer, Phys. Rev **185**, 847 (1969).
6. J. Gan, P. Coleman, and N. Andrei, Phys. Rev. Lett. **68**, 3476 (1992).
7. J. Hertz, Phys. Rev. B **14**, 1165 (1976).
8. A.C. Hewson, *The Kondo problem to heavy fermions.* (Cambridge University Press, 1993).
9. A. Jerez, N. Andrei, and G. Zaránd, Phys. Rev. B **58**, 3814 (1998).
10. A. Jerez, M. Lavagna, and D. Bensimon. Preprint.
11. H. v. Löhneysen, J.Phys.: Cond. Mat. **8**, 9689 (1996).
12. N. Mathur, F. Grosche, S. Julian, I. Walker, D. Freye, R. Haselwimmer, and G. Lonzarich, Nature **394**, 39 (1998).
13. A. Millis, Phys. Rev. B **48**, 7183 (1993).
14. P. Nozières, J. Phys. (Paris) **37**, C1–271 (1976).
15. P. Nozières and A. Blandin, J. Phys. (Paris) **41**, 193 (1980).
16. O. Parcollet, A. Georges, G. Kotliar, and A. Sengupta, Phys. Rev. B **58**, 3794 (1998).
17. C. Pépin and M. Lavagna, Phys. Rev. B **19**, 12180 (1999).

18. Q. Si, S. Rabello, K. Ingersent and J.L. Smith, Nature **413**, 804 (2001).
19. A. Schröder, G. Aeppli, E. Bucher, R. Ramazashvili, and P. Coleman, Phys. Rev. Lett. **80**, 5623 (1998).
20. A. Schröder, G. Aeppli, R. Coldea, M. Adams, O. Stockert, H. v. Löhneysen, E. Bucher, R. Ramazashvili, and P. Coleman, Nature **407**, 351 (2000).
21. Steglich, F., B. Buschinger, P. Gegenwart, M. Lohmann, R. Helfrich, C. Langhammer, P. Hellmann, L. Donnevert, S. Thomas, A. Link, C. Geiber, M. Lang, G. Sparn, and W. Assmus, J. Phys.: Cond. Mat. **8**, 9909 (1996).

PROJECTION OF THE KONDO EFFECT BY RESONANT EIGENSTATES INSIDE A CIRCULAR QUANTUM CORRAL

Proceedings of the ARW NATO Workshop Hvar, Croatia, October 2002

A. LOBOS and A. A. ALIGIA
Centro Atómico Bariloche and Instituto Balseiro, Comisión Nacional de Energía Atómica, 8400 Bariloche, Argentina

Abstract. We calculate the resonant eigenstates inside a circular quantum corral, modeling the confining potential by a continuous radial δ-function. In order to describe the corral with a magnetic impurity inside, we use the impurity Anderson model and second order perturbation theory in the Coulomb repulsion U. The effect of the impurity in the conduction electron density of states reveals a mirage of the Kondo resonance at a position different from that of the impurity.

Key words: Quantum Corrals, Mirage, Surface States, Magnetic Impurity, Kondo Effect, Anderson (Impurity) Model, Perturbation Theory, Differential Conductance, Nanoscale Structures, Scanning Tunneling Microscopy

1. Introduction

Nanoscale structures called "quantum corrals" have been constructed using recent developments in scanning tunneling microscopy (STM) techniques. These corrals are assembled by depositing a close line of atoms on Cu or noble metal (111) surfaces (1, 2). The Fermi surface in these metals is nearly spherical, with eight necks at the (111) and equivalent directions. For small wave vectors parallel to the (111) surface, these metals have a nearly parabolic band of two dimensional (2D) surface states, which are uncoupled to bulk states and cross the Fermi energy ϵ_F (3).

A circular corral of radius 71.3 Å was constructed depositing 48 Fe atoms on a Cu(111) surface (1). Measured differential conductance dI/dV can be fitted by a combination of the density of eigenstates close to ϵ_F, of a 2D electron gas inside a hard wall circular corral (1). In another experiment, a Co atom acting as a magnetic impurity was placed at one focus of an elliptical quantum corral, and the corresponding Kondo feature in dI/dV

A.C. Hewson and V. Zlatić (eds.), Concepts in Electron Correlation, 229–237.

was observed not only at that position, but also with reduced intensity at the other focus (2). This "mirage" is the projection of the Kondo effect, transmitted from the impurity to distant points by the corral eigenstates. Similar experiments were done in circular corrals (4). The corral confinement is needed in order to separate the conduction electronic energy levels (otherwise forming a continuum) and reduce the destructive interference produced by incoherent superposition of standing waves (5). Among these eigenstates, the density of the one closest to ϵ_F is clearly displayed in the change of differential conductance $\Delta dI/dV$ after adding the impurity (2).

Some theories of the Kondo mirage have attempted to fit or explain the experimental data *assuming* the existence of the Kondo resonance, without actually calculating it (5, 6, 7). The line shape of $\Delta dI/dV$, in the many-body approaches which can explain it, is very sensitive to the width of the resonance levels, and this width was taken as a parameter (8, 9, 10). If this width is reduced to zero, as in ordinary Lanczos calculations (11), the Kondo resonance splits into two delta functions out of ϵ_F and the experimental line shape cannot be reproduced. This problem persists if hopping between impurity and bulk is considered, in spite of the broadening of the delta functions (12). On the other hand, if the width of the resonance is very large, the mirage disappears (8).

In this work we calculate the widths of the conduction states δ_i for a circular corral and solve the impurity Anderson model which describes the mirage experiment using perturbation theory in the Coulomb repulsion U.

2. Resonances in a circular corral

Here we consider first the corral without impurity. We model the boundary of an empty circular corral by a continuous potential $W\delta(r-r_0)$, where r is the distance to the center of the circle and r_0 its radius. The approximation of a continuous boundary (instead of discrete adatoms forming the corral) simplifies considerably the mathematics and is justified by the fact that the Fermi wave length for surface electrons $2\pi/k_F \simeq 30$ Å is larger than the average distance between adatoms ~ 10 Å.

The angular momentum projection perpendicular to the surface $l_z = m$ is a good quantum number. For each m and energy $E = (\hbar k)^2/2m_e^*$, where m_e^* is the effective mass, the eigenstates can be written in the form $\varphi_{km}(r,\theta) = \psi_{km}(kr)\ e^{im\theta}$ in polar coordinates. The solutions of the 2D radial Schrödinger equation with this potential can be written in the form $\psi_{km} = \psi_{km}^{<}\theta(r_0-r) + \psi_{km}^{>}\theta(r-r_0)$, where $\theta(r)$ is the step function and

$$\psi_{km}^{<} = fJ_m(kr), \quad \psi_{km}^{>} = f(A_m(k)J_m(kr) + B_m(k)Y_m(kr)), \tag{1}$$

where J_m (Y_m) is the Bessel function of the first (second) kind. We normalize the wave functions $\varphi_{km}(r, \theta)$ inside a hard wall box of large radius R and take the limit $R \to \infty$. In this limit, the eigenstates of the problem form a continuum in the k variable. However, inside the corral, resonances are formed for finite W (13, 14). The resonant eigenstates of the problem can be obtained by scattering theory methods, specifically, by finding the complex poles of the scattering matrix $S(k)$ (13, 14).

The essential information on the spectral properties observed by STM is contained in the surface conduction electrons Green's function. In the absence of the impurity, it can be written in the form

$$G_s^0(z; r, \theta, r', \theta') = \sum_{km} \frac{\varphi_{km}(r, \theta)\overline{\varphi}_{km}(r', \theta')}{z - (k\hbar)^2/2m_e^*}, \tag{2}$$

where m_e^* is the effective electron mass. The sum can be converted to an integral by standard methods using the density of allowed k. Inside the corral we obtain:

$$G_s^0(z; r, \theta, r', \theta') = \sum_m \int \frac{dk\ k\ J_m(kr)J_m(kr')\ e^{im(\theta-\theta')}}{2\pi\left(A_m^2(k) + B_m^2(k)\right)\left(z - \frac{(\hbar k)^2}{2m_e^*}\right)}. \tag{3}$$

This expression for G_s^0 can be written as a sum of contributions from the resonances (12, 13):

$$G_s^0(z; r, \theta, r', \theta') = \sum_{n,m} \frac{C_n^m\ J_m(k_n^m r)J_m(k_n^m r')\ e^{im(\theta-\theta')}}{z - \epsilon_n^m + i\delta_n^m}, \tag{4}$$

where k_n^m are the poles in complex k-space of the scattering matrix (m and n are angular and radial quantum numbers respectively) (15), with $(\hbar k_n^m)^2/2m_e^* r_o^2 = \epsilon_n^m - i\delta_n^m$, and $\delta_n^m > 0$. Here, ϵ_n^m is the energy of the resonant eigenstate, $\delta_n^m > 0$ is its width, and C_n^m are the residues at the poles of the integral in Eq. (3) (12). We have taken an effective mass $m_e^* = 0.38\ m_e$(1, 16). $J_m(kr)$ has no singularities in the complex k-plane. The poles appearing in Eq. (4) are poles of the normalization factor f (when continued into the complex k-plane). From the properties of $\psi_{km}^<$ and $\psi_{km}^>$, and its derivatives at the boundary r_0, one obtains for f:

$$f^2 = \frac{k}{2R(A_m^2(k, W) + B_m^2(k, W))}, \tag{5}$$

where the coefficients $A_m(k, W)$ and $B_m(k, W)$ depend on $J_m(kr_0)$ and $Y_m(kr_0)$ and the dependence on W is now explicit. In Fig 1(a) we show the adimensional product $f^2 R r_0$, proportional to the density of states, as a function of k (real) for two different values of W and for states with

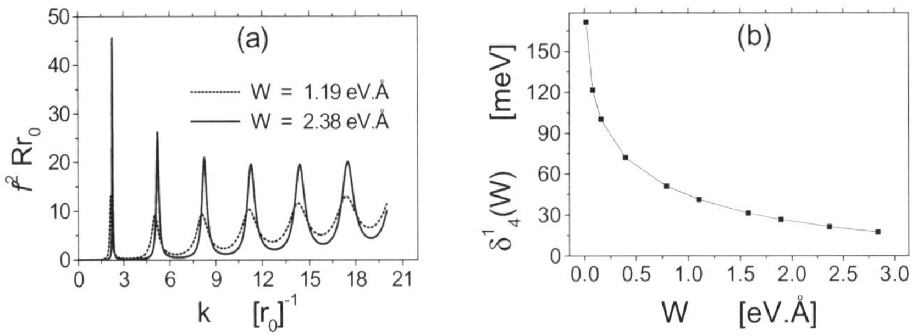

Figure 1. (a)$f^2 Rr_0$ as a function of radial momentum k, for m=0 states . (b)Width of the resonant state $n = 4$, $m = 1$, δ_4^1 as a function of the confinement parameter W.

quantum number m=0. As k approaches a pole in the complex plane k_n^m, $f^2 Rr_0$ tends to diverge, and the weight of the resonant eigenfunction outside the corral becomes very small in comparison to that inside (f depends essentially on $\psi_{km}^>$, since the area of the corral compared to that of the hard wall box is of order $(r_0/R)^2$) . The confinement of the resonant eigenstate within the corral, the width δ_i and the energy eigenvalue ϵ_i depend on the magnitude of the parameter W. In the limit $W \to \infty$, $f^2 Rr_0$ becomes a sum of delta functions at zeros of $J_m(\alpha)$ and the hard wall eigenstates are recovered (1). This can be seen in Fig 1(a). When W is increased, the poles move towards the real k axis in the complex plane, while the weight of the states with $k \neq \text{Re } k_n^m$ decreases, leading to the formation of sharp peaks in the conduction electron density of states for $k = \text{Re } k_n^m$.

As a first approximation, the resulting δ_n^m/ϵ_n^m is a constant, independent of m and n. As W decreases, the confinement properties of the corral decrease and the width δ_i must increase. This is shown in Fig 1(b) where we have studied the dependence of δ_4^1. Another effect of confinement is that ϵ_n^m decreases slightly with increasing W. This effect can be interpreted as a reduction of the effective mass, plus smaller corrections.

3. The many-body problem

In this section we introduce the magnetic impurity in the form of one non-degenerate localized d orbital. We consider surface states, described inside the corral by the Green's function G_s^0, hybridized with the d orbital, for which there is an important intra-orbital repulsion U (5, 8, 9, 10, 11). We also include a hybridization of the d orbital with bulk states. The

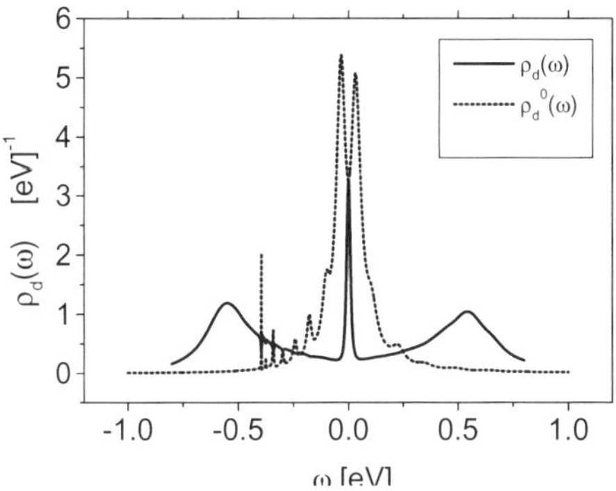

Figure 2. Impurity density of states for U=0 (dotted line) and U=1 eV (solid line). Parameters are δ_F = 20 meV, δ_b = 35 meV and V_s = 0.45 eV, and $E_d^{eff} = \epsilon_F$.

Hamiltonian can be written as:

$$\begin{aligned} H &= \sum_{j\sigma} \varepsilon_j s_{j\sigma}^\dagger s_{j\sigma} + \sum_{j\sigma} \varepsilon_j^b b_{j\sigma}^\dagger b_{j\sigma} + E_d \sum_{i\sigma} d_{i\sigma}^\dagger d_{i\sigma} + U d_\uparrow^\dagger d_\uparrow d_\downarrow^\dagger d_\downarrow \\ &+ \sum_{j\sigma} V_s \lambda [\varphi_j(R_i) d_\sigma^\dagger s_{j\sigma} + H.c.] + \sum_{j\sigma} V_b^j [d_\sigma^\dagger b_{j\sigma} + H.c.]. \end{aligned} \tag{6}$$

Neglecting small distortions very near the impurity, dI/dV is proportional to $\rho_s(\omega, r, \theta)$ = -Im G_s/π, where G_s is the Green's function for conduction electrons, in presence of the impurity (9). Using equations of motion it can be shown that:

$$\Delta G_s = G_s - G_s^0 = V_s^2 \lambda^4 G_s^0(R_t, R_i, \omega) G_s^0(R_i, R_t, \omega) G_d, \tag{7}$$

where G_d is the impurity Green's function, R_t and R_i are the position of the STM tip and the impurity respectively, and λ^2 is the area per Cu atom in the surface. To calculate G_d we used perturbation theory up to second order in U (17). We assume independence of spin projection (no magnetic field applied) and a situation in which the effective one-particle d level $E_d^{eff} = E_d + U\langle d_\sigma^\dagger d_\sigma \rangle$ is very near the Fermi level (18). We took U = 1 eV. Then:

$$[G_d(\omega)]^{-1} = [G_d^0(\omega)]^{-1} - \Sigma(\omega), \tag{8}$$

where G_d^0 is the impurity Green's function for $U = 0$ and is also obtained from the equations of motion (12):

$$G_d^0(\omega) = \frac{1}{\omega - E_d^{eff} + i\delta_b - (V\lambda)^2 G_s^0(R_i, R_i, \omega)}. \tag{9}$$

The term δ_b represents here the hybridization between bulk states and the impurity d level, and is a parameter in our calculations. The function $\Sigma(\omega)$

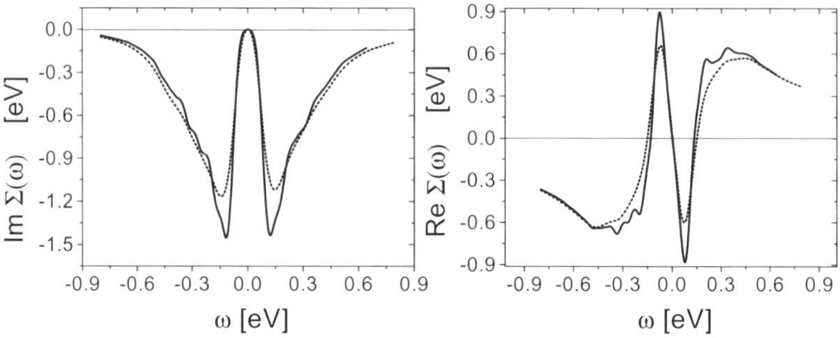

Figure 3. Real and imaginary parts of the self-energy for $\delta_4^1 = 20$ meV (solid line) and $\delta_4^1 = 40$ meV (dotted line).

is the self-energy of the problem up to second order in U (17):

$$\Sigma(i\omega_l, T) = U^2 T \sum_m G_d^0(i\omega_l - i\nu_m)\chi(i\nu_m), \tag{10}$$

$$\chi(i\nu_m) = -T \sum_n G_d^0(i\omega_n) G_d^0(i\omega_n + i\nu_m), \tag{11}$$

where T is the temperature, ω_n and ν_m are fermionic and bosonic Matsubara frequencies respectively. A more detailed explanation of this calculation is in Ref. (12).

In Fig. 2 we show the impurity spectral densities of states, for two different values of U, and $\delta_4^1 = 20$ meV. In Fig 3 we display a self-energy calculation for these parameters and another calculation for a different set of parameters. The origin of energies was set at $\epsilon_F = 0$. We see that, as a consequence of the structure of the real part of $\Sigma(\omega)$ near ϵ_F, a narrow peak in ρ_d develops there (see Fig. 2). This is the Kondo resonance. The imaginary part of $\Sigma(\omega)$ is proportional to $-\omega^2$, as it should be for a Fermi liquid. For $U = 1$ eV, also the usual broad peaks at energies near E_d and $E_d + U$ are clearly seen. For the results presented here, as in recent experiments (4), we have taken $r_0 = 63.5$ Å, so that the energy level of the state with $m = 1$, $n = 4$ falls at ϵ_F. We took $W = 7\hbar^2/2m_e^* r_0 \simeq 1.19$ eV Å in order that the width of this state would be $\delta_4^1 \simeq 40$ meV (8). This leads to a good agreement with experiment in the case of the elliptical corral (8, 9). The point of observation $r = 0.14\ r_0$ was taken at the maximum of $J_1(k_4^1 r)$. As δ_F decreases (see Fig. 3), the structure of the resonances in $G_s^0(\omega)$ are transmitted to $G_d(\omega)$ and $\Sigma(\omega)$, and several peaks appear. For $\delta_F \to 0$ the Kondo resonance is destroyed (8, 9, 10).

In Fig. 4 we show the space dependence of $\Delta\rho_s(\epsilon_F)$ and the density of the corresponding hard wall eigenstate with quantum numbers $m = 1$ and $n = 4$. We used $V_s = 0.48$ eV, $\delta_b = 0.032$ eV and the broadening of the

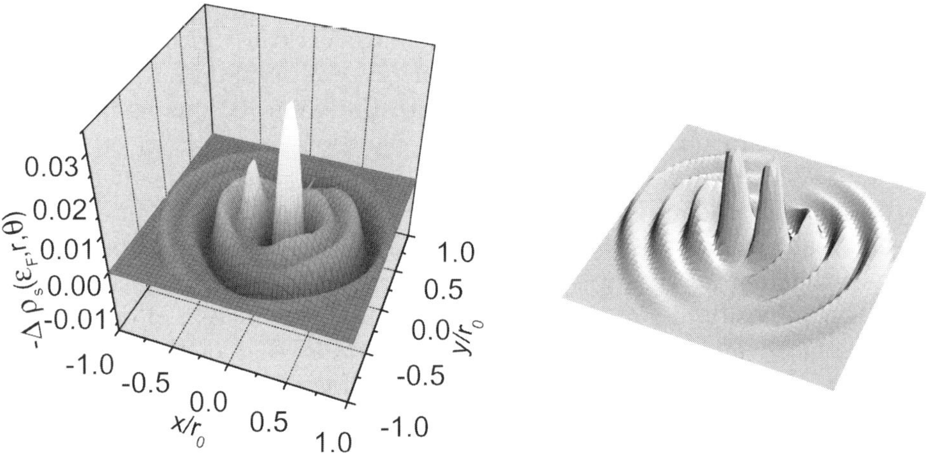

Figure 4. (a) Space dependence of $-\Delta\rho_s(\epsilon_F)$ and (b) density of the hard wall eigenstate with $m = 1$ and $n = 4$.

eigenstate in ϵ_F is $\delta_F = \delta_4^1 = 40$ meV. These parameters lead to a good description of the reported line shape of $\Delta dI/dV$ for the elliptical corral (2).The largest peak occurs at the position of the impurity and the other, weaker in intensity, is the so called mirage. The correspondence between both figures is clearly seen. However, the mirage is somewhat distorted because of destructive interference of other resonant states which energy eigenvalues lies near ϵ_F. An analysis of the voltage dependence of $\Delta\rho_s(\epsilon_F)$ is in Refs.(9, 12).

4. Summary and discussion

We have studied the resonances produced inside a circular quantum corral, assuming a continuous $W\delta(r - r_0)$ confining potential in a 2D electron gas. These resonant eigenstates are very similar to the eigenstates of a hard wall potential and they can be obtained from the poles of the normalization factor f. This fact can be easily interpreted, since a pole in f indicates that the state has little weight outside the corral, leading to a "quasi-confined" state. The introduction of a width δ_i for each resonance is the main difference with a calculation assuming a hard wall corral. Our model leads to a resonance δ_i, proportional to its energy ϵ_i, in a first approximation.

We used the Anderson model to describe the physics when a magnetic impurity is introduced inside the corral. At $T < T_K$, the Kondo effect is observed in the impurity spectral density of states as a sharp peak at ϵ_F. The resonant conduction state which lies at ϵ_F can be observed because of two factors: the low energy scale of the Kondo effect (the width of the Kondo peak is $2k_BT_K \simeq 10$ meV), and the energy level separation of resonant states which hybridizes significatively with the impurity, due to the corral confinement. In our calculations, this separation ranges from 50 to 100 meV. This avoids an incoherent superposition of waves. Nevertheless, the broadening introduced by finite W increases the destructive interference because at ϵ_F the density of states has non-negligible components of other states, different to that we want to project. A reduction in δ_i leads to a stronger intensity at the mirage point (because of reduction of destructive interference), and the space dependence tends to that of the conduction state at ϵ_F. However, for very small δ_i the voltage dependence, is strongly distorted and two peaks of positive weight of $\Delta dI/dV$ appears at moderate non-zero voltages (9).

So far, the theories of the quantum mirage have neglected the effect of s and p states brought by the impurity atom. The introduction of these states should affect the intensity of dI/dV near the impurity. Therefore, a study of this effect is desirable.

Acknowledgements

One of us (AA) is partially supported by CONICET, Argentina. This work benefited from PICT 03-06343 of ANPCyT.

References

1. M.F. Crommie, C.P. Lutz, and D.M. Eigler, Science **262**, 218 (1993).
2. H.C. Manoharan, C.P. Lutz, and D.M. Eigler, Nature (London) **403**, 512 (2000).
3. S.L. Hulbert, P.D. Johnson, N.G. Stoffel, W.A. Royer, and N.V. Smith, Phys. Rev. B **31**, 6815 (1985).
4. H.C. Manoharan, PASI Conference, *Physics and Technology at the Nanometer Scale* (Costa Rica, June 24 - July 3, 2001).
5. D. Porras, J. Fernández-Rossier, and C. Tejedor, Phys. Rev. B **63**, 155406 (2001).
6. O. Agam and A. Schiller, Phys. Rev. Lett. **86**, 484 (2001).
7. G.A. Fiete, J. S. Hersch, E. J. Heller, H.C. Manoharan, C.P. Lutz, and D.M. Eigler, Phys. Rev. Lett. **86**, 2392 (2001)..
8. A.A. Aligia, Phys. Rev. B **64**, 121102(R) (2001).
9. A.A. Aligia, Phys. Status Solidi (b) **230**, 415 (2002).
10. G. Chiappe and A.A. Aligia, Phys. Rev. B **66**, 075421 (2002).
11. K. Hallberg, A.A. Correa, and C.A. Balseiro, Phys. Rev. Lett. **88**, 066802 (2002).

12. A.Lobos and A.A.Aligia, cond-mat/0208533
13. G. García Calderón, Nucl. Phys. A **261**, 130 (1976).
14. See for example J.R. Taylor, *Scattering Theory: the quantum theory of non-relativistic collisions* (Wiley, New York, 1972).
15. The Green's function and the scattering matrix have the same distribution of poles in the complex k-plane (13).
16. A. Euceda, D.M. Bylander, and L. Kleinman, Phys. Rev. B **28**, 528 (1983).
17. B. Horvatić, D. Šokčević, and V. Zlatić, Phys. Rev. B **36**, 675 (1987); references therein.
18. M. Weissmann and A.M. Llois, Phys. Rev. B **63**, 113402 (2001).

NONEQUILIBRIUM ELECTRON TRANSPORT THROUGH NANOSTRUCTURES: CORRELATION EFFECTS

Proceedings of the ARW NATO Workshop Hvar, Croatia, October 2002

P. WÖLFLE, A. ROSCH, J. PAASKE and J. KROHA
Institut für Theorie der Kondensierten Materie, Universität Karlsruhe, 76128 Karlsruhe, Germany

Abstract. We consider electron transport through a quantum dot in the Coulomb blockade regime, when the dot has a total spin $\frac{1}{2}$. The low energy physics of this system is modeled by a Kondo-type Hamiltonian describing spin-dependent tunneling and exchange interaction of the electron spins in the leads and the local spin. We show that in the regime of large transport voltage V and magnetic field B with $\max(V, B) \gg T_K$, where T_K is the Kondo temperature, renormalized perturbation theory is valid. Physical quantities such as the differential conductance G and the local magnetization M may be calculated in a controlled way by summing the leading logarithmic terms in each order of perturbation theory. We develop a poor man's scaling renormalization group treatment for frequency dependent coupling functions, which allows one to calculate G and M to leading order in $1/\ell n[(V, B)/T_K]$.

Key words: Nonequilibrium Transport, Quantum Dot, Kondo Effect, Conductance, Poor Man's Scaling, Renormalization Group Method

1. Introduction

The transport of electrons through a quantum dot in the limit of weak coupling to the leads is dominated by Coulomb interaction effects, forcing integral electron charge on the dot (1) (Coulomb blockade). In the case that the total spin of the dot is nonzero, however, the antiferromagnetic exchange interaction of this local spin with the conduction electron spins in the leads gives rise to a Kondo resonance in the local density of states at the Fermi level. Electron transport may then take place via resonance tunneling, thus removing the Coulomb blockade (2, 3). This has been seen in a number of experiments (4, 5, 6, 7, 8).

A.C. Hewson and V. Zlatić (eds.), Concepts in Electron Correlation, 239–246.

Powerful methods have been developed to describe this strong coupling problem, among them methods based on the renormalization group idea, in a simplified form (9) ("poor man's scaling") and in a full numerical implementation by Wilson (10), exact methods for calculating the thermodynamic quantities using the Bethe ansatz (11), conformal field theory (12) and resummations of perturbation theory using auxiliary particle representations (13) to access dynamical quantities.

Necessary prerequisites for the existence of the Kondo effect are (i) the spin degeneracy of the ground state of the quantum dot, and (ii) quantum coherence over sufficiently long time periods. The characteristic energy scale against perturbations have to be measured is the Kondo temperature $T_K \simeq D\sqrt{JN_0}\exp(-1/2N_0J)$, where J is the exchange coupling constant, N_0 is the conduction electron density of states per spin at the Fermi level and D is the conduction band width.

Nonlinear transport through a Kondo system has been considered by several authors. Early on, Appelbaum (15) calculated the differential conductance $G(V, B)$ for large bias voltage V in a finite magnetic field B, assuming erroneously, however, that the local spin remains in thermal equilibrium. König et al. (16) calculated $G(V, B)$ for an Anderson model, employing a certain resummation of perturbation theory. Brief accounts of the work to be presented below have been published elsewhere (14, 17).

2. Magnetization and Conductance in Perturbation Theory

We study the following Kondo-type model Hamiltonian of a quantum dot:

$$\begin{aligned} H = & \sum_{\mathbf{k},\sigma,\alpha=L,R} (\varepsilon_{\mathbf{k}} - \mu_\alpha) c^\dagger_{\alpha\mathbf{k}\sigma} c_{\alpha\mathbf{k}\sigma} - BS_z \\ & + \frac{1}{2} \sum_{\mathbf{k},\mathbf{k}',\sigma,\sigma',\alpha,\alpha'=L,R} J_{\alpha'\alpha}\, \mathbf{S} \cdot (c^\dagger_{\alpha'\mathbf{k}'\sigma'} \tau_{\sigma'\sigma} c_{\alpha\mathbf{k}\sigma}), \end{aligned} \tag{1}$$

where $\mathbf{S}$ is the local spin operator on the dot (assumed to be $S = \frac{1}{2}$) and τ is the vector of Pauli matrices. The $c^+_{\alpha\mathbf{k}\sigma}$ create conduction electrons with momentum $\mathbf{k}$ and spin σ in lead α. The chemical potential shifts induced by the bias voltage are given by $\mu_{L,R} = \pm V/2$ for the left (L) and right (R) lead. The exchange energies $J_{\alpha\alpha'}$, $\alpha = L, R$, will be assumed to have the symmetry $J_{RR} = J_{LL}$ and $J_{LR} = J_{RL}$. We shall use the dimensionless coupling constants $g_d = N_0 J_{RR}$ and $g_{LR} = N_0 J_{LR}$, with N_0 the local density of states at the Fermi energy, assumed to be flat in the accessible regime $\mid \omega \mid \leq V, B$. A magnetic field is included, inducing a Zeeman splitting of the magnetic levels of the local spin $\omega_\gamma = -\gamma B/2$, $\gamma = \pm 1$. The effect of the magnetic field on the conduction electrons gives only rise to particle-hole asymmetry terms, which are usually small and may be neglected.

In lowest order perturbation theory the current through the dot is given by the Golden Rule expression

$$I = \frac{\pi e}{4\hbar} \int d\omega \sum_{\gamma=\pm 1} n_\gamma \Big[g_1^2 f_{\omega-\mu_L}(1 - f_{\omega-\mu_R}) + 2g_2^2 f_{\omega-\mu_L}(1 - f_{\omega-\mu_R-\gamma B}) \Big] - (L \leftrightarrow R) \tag{2}$$

where g_1 and g_2 are the bare dimensionless coupling constants for spin-nonflip and spin-flip interaction, $g_1 = g_2 = N_0 J_{LR}$, $f_{\omega-\mu_\alpha}$ is the Fermi function in the lead $\alpha = L, R$, and γB is the Zeeman energy transfer taking place in a spin flip process. At low temperatures, $T \ll V, B$, the product of Fermi functions limits the energy integration to the window $\mu_R < \omega < \mu_L$ and $\mu_R + \gamma B < \omega < \mu_L$, respectively. The current is seen to depend on the occupation $n_\gamma = \frac{1}{2}(1 + \gamma M)$ of the local spin states, where M is the magnetization.

For sufficiently small voltage, $V \ll T$, the local spin system is in thermal equilibrium. In the opposite limit, $V \gg T$, the stationary current through the dot drives the system out of equilibrium. The occupation numbers n_γ are then determined by the rate equation $\frac{\partial}{\partial t} n_\gamma = C_\gamma\{n_{\gamma'}\} = 0$, where C is the collision integral. This leads to the condition (assuming $J_{LR} = J_{LL}$ here)

$$n_\uparrow \sum_{\alpha,\alpha'=L,R} \int d\omega g_2^2 f_{\omega-\mu_\alpha}\Big(1 - f_{\omega-\mu_{\alpha'}-B}\Big) = = n_\downarrow \sum_{\alpha,\alpha'=L,R} \int d\omega g_2^2 f_{\omega-\mu_\alpha}\Big(1 - f_{\omega-\mu_{\alpha'}+B}\Big) . \tag{3}$$

The spin-flip coupling constant g_2 cancels out of this equation, yielding the nonequilibrium magnetization as given by (5) in the limit $J \to 0$.

In the limit $V \to 0$, the equilibrium result $M = \tanh \frac{B}{2T}$ is recovered, whereas in the limit $V \gg B, T$, $M = \frac{2B}{V}$, independent of temperature. This result has been obtained independently in (18, 20).

In order to calculate higher order contributions in perturbation theory in J we switch now to a more formal description in terms of nonequilibrium Green's functions. We find it convenient to represent the local spin operator in pseudofermion (PF) language (19), $\mathbf{S} = \frac{1}{2}\sum_{\gamma\gamma'} f_\gamma^+ \tau_{\gamma\gamma'} f_{\gamma'}$, where f_γ^+ creates a pseudofermion of spin $\gamma = \uparrow, \downarrow$ at the dot. The projection onto the physical sector of Hilbert space, with pseudofermion occupation number $Q = \sum_\gamma f_\gamma^+ f_\gamma = 1$, is done by adding a term λQ to the Hamiltonian and

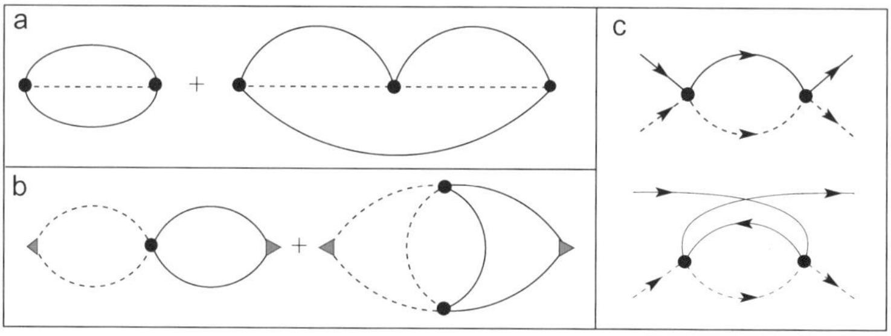

Figure 1. Feynman diagrams for (a) PF self-energies, (b) current and (c) vertices entering the 1-loop RG equation. PF (electron-) propagators are displayed as dashed (full) lines.

taking the limit $\lambda \to \infty$. This means that the PF system is taken in the low density limit. The Feynman diagrams of perturbation theory in the exchange interaction will therefore have one PF loop at most.

The local magnetization can be calculated in terms of the PF Green's function $G^<_\gamma(\omega)$ as $M = \sum_{\gamma=\pm1} \gamma \int \frac{d\omega}{2\pi i} G^<_\gamma(\omega)$, where $G^<$ is found by solving a quantum kinetic equation. In steady state one finds

$$G^<_\gamma(\omega)\Gamma_\gamma(\omega) = \Sigma^<_\gamma(\omega)A_\gamma(\omega) \tag{4}$$

where $A_\gamma(\omega)$ is the PF spectral function, $\Sigma^<_\gamma(\omega)$ is the lesser component of the self-energy and $\Gamma_\gamma(\omega)$ its imaginary part. In general Eq. (4) is an integral equation for $G^<_\gamma(\omega) = in_\gamma(\omega)A_\gamma(\omega)$, which, however, can be approximated in a controlled way by making use of the sharply peaked form of $A_\gamma(\omega)$ to give an algebraic equation for $n_\gamma(0)$. In this form Eq. (4) corresponds to the rate equation discussed above, with higher order processes included in the transition amplitudes. The diagrams of the self-energy up to third order in J are depicted in Fig. 1a. Note that the lines represent Keldysh matrix Green's functions dressed with arrows in all topologically different ways. The result including leading logarithmic corrections is found as

$$M = \frac{Z}{\coth\frac{B}{2T}\left[\frac{Z}{2} + g_-^2 B(1+\mathcal{L}(B))\right] + g_{LR}^2(c(B)+c(-B))} \tag{5}$$

where

$$Z = 2g_+^2 B(1+\mathcal{L}(B)) + 2g_{LR}^2\Big[2B + (V+B)\mathcal{L}(V+B) - (V-B)\mathcal{L}(V-B)\Big] \tag{6}$$

and we use the abbreviations $g_\pm^2 = g_d^2 \pm g_{LR}^2$, $\mathcal{L}(B) = 2g_d \ell n \frac{D}{|B|}$, and $c(B) = \coth\frac{V+B}{2T}[(V+B)(1+\mathcal{L}(V+B)) + V\mathcal{L}(V) + B\mathcal{L}(B)]$. We note that logarithmic corrections of order $g^3 \ell n D$ to the self-energies lead to corrections of order $g\ell nD$ to M. These corrections are larger than the leading logarithmic corrections in equilibrium ($V \to 0$), which are of order $g^2\ell nD$. A corresponding result of the current I is given in (17).

3. Resummation of Perturbation Theory in Nonequilibrium: a Poor Man's Scaling Approach

Even for small coupling constants g and for sufficiently large V and B, such that $V, B \gg T_K$, but still in the scaling regime $V, B \ll D$, such that $g\ell n(D/V) \ll 1$, bare perturbation theory converges slowly. It is necessary to sum the leading logarithmic contributions in all orders of PT. In the equilibrium state a powerful method is available to perform this resummation in a controlled way: the perturbative renormalization group method (9). It makes use of the fundamental idea that a change of the cut-off D can be fully absorbed into a redefinition of the coupling constant g. As long as the running coupling constant $g(D)$ is small, the change of g under an infinitesimal change of D, $\partial g/\partial \ell nD$, may be calculated in perturbation theory. It is well known that in the equilibrium Kondo problem the coupling constant is found to grow to infinity, thus leaving the perturbative regime beyond $g \sim 1$. A nonperturbative treatment is then necessary, as shown by Wilson (10) in his pioneering work on the numerical RG. In the nonequilibrium situation, the RG flow is cut-off by inelastic processes already within the weak coupling regime, such that perturbation theory is valid (14). This is true at least for the case $J_{LR} = J_{LL}, J_{RR}$, considered here. While the discussion of the RG flow in (14) was a qualitative one, motivated by results obtained in the so-called "Non-crossing approximation" for the Anderson model, and applied only to the limit $B = 0$, here we consider the RG formulation on a more fundamental level. A different and considerably more involved real-time RG scheme has been developed by Schoeller and König (21).

An analysis of the PT result Eq. (5) suggests that the frequency dependence of the couplings becomes important. In order to understand how these frequency dependencies are generated on the lowest level, i.e. in one-loop order, it is sufficient to analyze the behavior of the vertex corrections shown in Fig. 1c under a change of cut-off. Logarithmic corrections in perturbation theory are generated by PF - conduction electron bubbles containing the product of a real part of a PF Green's function, $1/(\omega \pm B/2)$, and the Keldysh component of the local conduction electron line, $-2\pi i N_0 \tanh[(\omega - \mu_\alpha)/2T]$. If the energy of the PF - conduction electron intermediate state is within the interval $[-D, D]$ the process will contribute to the renormalization of the coupling function $g(\omega)$, otherwise it will not. In general the coupling functions depend on three frequencies (taking energy conservation into account). Using the fact that the spectral function of the PFs is sharply peaked at $\omega = \mp B/2$, we set two of the three frequencies to $\omega = \mp B/2$, keeping only one frequency variable. In addition, the frequency dependence may be neglected within the running band-width, $|\,\omega\,| < D$. The spin structure of the general coupling functions is given by two invariant amplitudes, $\tilde{g}_2$ for spin flip and $\tilde{g}_1$ for spin non-flip processes

$$g^{\alpha\sigma,\omega;\alpha',\sigma',\omega-(\gamma-\gamma')B}_{\gamma,-\gamma B/2;\gamma',-\gamma' B/2} = \tau^z_{\sigma\sigma'}\tau^z_{\gamma\gamma'}\tilde{g}_1(\omega)+(\tau^x_{\sigma\sigma'}\tau^x_{\gamma\gamma'}+\tau^y_{\sigma\sigma'}\tau^y_{\gamma\gamma'})\tilde{g}_2(\omega-\gamma B/2) \quad (7)$$

where $g^{a;a'}_{b;b'}$ denotes the coupling function for conduction electrons of spin σ and energy ω in lead $\alpha(a = (\alpha, \sigma, \omega))$ interacting with a pseudofermion in state $b = (\gamma, \omega_f)$ and going into states a', b'. The two frequency-dependent running coupling functions $\tilde{g}_1(\omega)$ and $\tilde{g}_2(\omega)$ obey the following flow equations,

$$\begin{aligned}
\frac{\partial \tilde{g}_1(\omega)}{\partial \ln D} &= -\frac{1}{2}\sum_{\alpha=\pm 1}\tilde{g}_2^2\Big(\frac{B+\alpha V}{2}\Big)\Big(\Theta_{\omega+B+\alpha\frac{V}{2}} + \Theta_{\omega-B-\alpha\frac{V}{2}}\Big) \\
\frac{\partial \tilde{g}_2(\omega)}{\partial \ell n D} &= -\frac{1}{2}\tilde{g}_1\Big(\frac{V}{2}\Big)\sum_{\alpha=\pm 1}\tilde{g}_2\Big(\frac{B+\alpha V}{2}\Big)\Big(\Theta_{\omega+\frac{B+\alpha V}{2}} + \Theta_{\omega-\frac{B+\alpha V}{2}}\Big) \qquad (8)
\end{aligned}$$

with initial condition $\tilde{g}_1(\omega) = \tilde{g}_2(\omega) = JN_0$ and $\Theta_\omega = \Theta(D-\mid\omega\mid)$. In the limit $V, B \to 0$, Eq. (8) reduces to the well-known scaling equations (6).

There is, however, still one additional effect missing: the finite relaxation of the spins even in the limit $T \to 0$ leading to inelastic broadening of the local spin levels, which we identify with the transverse spin relaxation rate $\Gamma = 1/T_2$. Within the PF representation this effect shows up as a finite imaginary part of the PF self-energy.

Assuming the RG flow to be stopped at the scale Γ, we replace the step functions Θ_ω in Eq. (8) by $\Theta(D - \sqrt{\omega^2 + \Gamma^2})$. The decay rate Γ has to be calculated self-consistently with the solution of the flow equations (8). Starting from the golden rule expression for the transverse relaxation rate,

$$\begin{aligned}
\Gamma = \frac{\pi}{4\hbar}\sum_{\alpha,\alpha'=L,R,\gamma=\uparrow,\downarrow}\int d\omega\Big[\tilde{g}_1(\omega)^2 f_{\omega-\mu_\alpha}(1 - f_{\omega-\mu_{\alpha'}}) \\
+\tilde{g}_2(\omega-\gamma B/2)^2 f_{\omega-\mu_\alpha}(1 - f_{\omega-\mu_{\alpha'}-\gamma B})\Big] \qquad (9)
\end{aligned}$$

the renormalized Γ is obtained by replacing $g_{1,2}$ by $\tilde{g}_{1,2}(\omega)$ as determined from the solution of (2).

We are now ready to calculate further physical quantities. The renormalized value of the magnetization is obtained by substituting $\tilde{g}_{1,2}(\omega)$ in

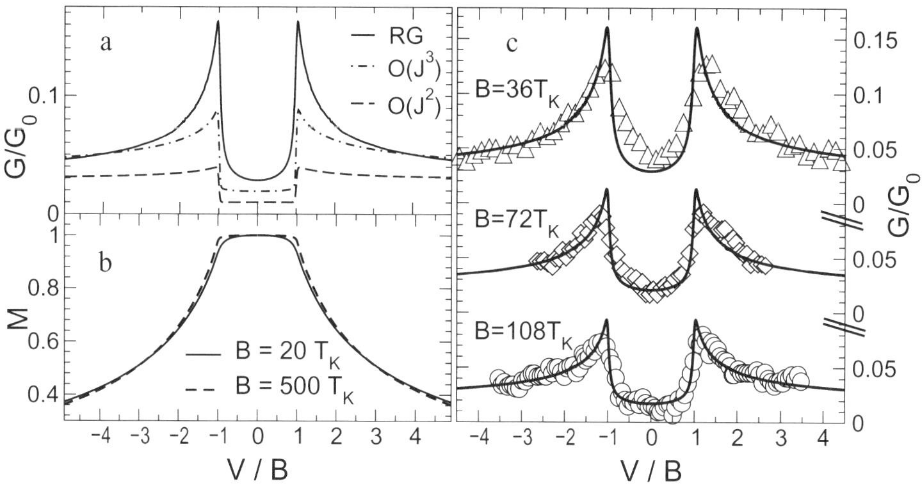

Figure 2. a) Conductance $G(V/B)$ calculated in leading (dashed line) and next-to--leading (dot-dashed line) order PT compared to the result of perturbative RG (solid line) for $B = 100T_K$, $D = 10^4 T_K$, $T_K = D\sqrt{g}e^{-1/2g}$). b) Local magnetization $M(V/B, B/T_K)$ of a symmetric dot for fixed magnetic field $B = 20, 500\,T_K$. c) Experiments by Ralph and Buhrman (22) (symbols) on transport through metallic point contacts in magnetic fields $0.85, 1.7, 2.55\,\mathrm{T}$ ($B = 36, 72, 104\,T_K$ where $T_K \approx 30\mathrm{mK}$ (22)).

place of $g_{1,2}$ in the Golden Rule expression (3). In Fig. 2b we show the fully renormalized result for M as a function of V/B for different values of B.

The charge current I is calculated from Eq. (2) inserting the renormalized coupling functions. Fig. 2a shows a comparison of the differential conductance $G(V) = dI/dV$ obtained in this way, with the bare result (2) and the PT result including leading logarithmic corrections.

The peak structures in $G(V, B)$ appearing at $V = \pm B$ have been detected in experiment. In order to reach large values of B/T_K, it is necessary to have relatively low T_K. This happened to be the case in transport through metallic point contacts, containing a magnetic impurity (22). In Fig. 2c the result of our theory using the value of $T_K \simeq 30mK$ given in (22) is compared to the experimental data, after subtracting a background contribution (see (17) for details). The agreement is seen to be excellent.

4. Conclusion

We considered the transport of electrons through a Kondo dot in the regime of large transport voltage V and in the presence of a magnetic field B, such that $\max(V, B) \gg T_K$ is satisfied. A finite difference of the chemical potentials in the two leads, $\mu_L - \mu_R = V$, changes the RG flow substantially.

Furthermore, it opens an energy window for inelastic processes even at $T = 0$, leading to a broadening Γ of the spin sublevels of the dot and providing a cut-off for the scaling towards the Kondo fixed point. A finite magnetic field induces a local magnetization M at the dot, which for $V \gg T_K, T$ is independent of temperature, being solely controlled by the voltage V. We derived a set of RG equations within a poor man's scaling approximation, which for once reproduce the bare PT result, but then may be integrated to give the fully renormalized result. The integration is done including the self-consistently determined cut-off Γ. The results obtained in this way are valid up to corrections of order $1/\ell n[(V, B)/T_K]$.

5. Acknowledgments

We would like to thank S. De Franceschi, J. König, O. Parcollet, H. Schoeller and A. Shnirman for helpful discussions and especially L. Glazman, who suggested investigating the case of finite B. Part of this work was supported by the Center for Functional Nanostructures and the Emmy-Noether program (A.R.) of the DFG.

References

1. I. Aleiner, P. Brouwer, and L. Glazman, Phys. Rep. **358**, 309 (2002).
2. L. Glazman and M. Raikh, JETP Letters **47**, 452 (1988).
3. T. Ng and P.A. Lee, Phys. Rev. Lett. **61**, 1768 (1988).
4. D. Goldhaber-Gordon, H. Shtrikman, D. Mahalu, D. Abusch-Magder, U. Meirav and M. Kastner, Nature **391**, 156 (1998).
5. S. Cronenwett, T. Oosterkamp and L. Kouwenhoven, Science **281**, 540 (1998).
6. J. Schmid, J. Weis, K. Eberl, and K. v. Klitzing, Physica B **258**, 182 (1998).
7. J. Nygard, D. Cobden and P. Lindelof, Nature **408**, 342 (2000).
8. S. De Franceschi *et al.*, cond-mat/0203146.
9. P. W. Anderson, J. Phys. C **3**, 2436 (1966).
10. K.G. Wilson, Rev. Mod. Phys. **47**, 773 (1975).
11. N. Andrei, Phys. Rev. Lett. **45**, 379 (1980); N. Andrei, K. Furuya and J.H. Lowenstein, Rev. Mod. Phys. **55**, 331 (1983).
12. I. Affleck and A. W. W. Ludwig, Nucl. Phys. B **360**, 641 (1991).
13. J. Kroha and P. Wölfle, in press (Springer), cond-mat/0105491.
14. A. Rosch, J. Kroha and P. Wölfle, Phys. Rev. Lett. **87**, 156802 (2001).
15. J. Appelbaum, Phys. Rev. Lett, **17**, 91 (1966); Phys. Rev. **154**, 633 (1967).
16. J. König, J. Schmid, H. Schoeller, and G. Schön, Phys. Rev. B **54**, 16820 (1996).
17. A.Rosch, J. Paaske, J. Kroha and P. Wölfle, cond-mat/0202404.
18. L.I. Glazman, private communication.
19. A.A. Abrikosov, Physics **2**, 21 (1965).
20. O. Parcollet, and C. Hooley, cond-mat/0202425.
21. H. Schoeller and J. König, Phys. Rev. Lett. **84**, 3686 (2000); see also M. Keil, Ph.D. thesis, U. Göttingen (2002).
22. D.C. Ralph and R.A. Buhrman, Phys. Rev. Lett. **72**, 3401 (1994).

QUANTUM FLUCTUATIONS AND ELECTRONIC TRANSPORT THROUGH STRONGLY INTERACTING QUANTUM DOTS

Proceedings of the ARW NATO Workshop Hvar, Croatia, October 2002

T. A. COSTI
Universität Karlsruhe, Institut für Theorie der Kondensierten Materie 76128 Karlsruhe, Germany

Abstract. We study electronic transport through a strongly interacting quantum dot by using the finite temperature extension of Wilson's numerical renormalization group (NRG) method. This allows the linear conductance to be calculated at all temperatures and in particular at very low temperature where quantum fluctuations and the Kondo effect strongly modify the transport. The quantum dot investigated has one active level for transport and is modeled by an Anderson impurity model attached to left and right electron reservoirs. The predictions for the linear conductance are compared to available experimental data for quantum dots in heterostructures. The spin-resolved conductance is calculated as a function of gate voltage, temperature and magnetic field strength and the spin-filtering properties of quantum dots in a magnetic field are described.

Key words: Quantum Dot, Numerical Renormalization Group, Anderson (Impurity) Model, Kondo Effect, Linear Conductance, Spin-Filtering

1. Introduction

Recent experimental work (1, 2, 3, 4, 5) has demonstrated the importance of correlations in determining the low temperature transport properties of nanoscale size quantum dots. These dots consists of a confined region of electrons, of typical diameter $100nm$, "weakly" coupled to leads via tunnel barriers. A "weak" coupling, Γ, means that the quantized levels of the dot are broadened into resonances, but are not completely washed out. A gate voltage, V_g, controls the position of the quantized levels relative to the chemical potentials of the leads and thereby the total number of electrons on the dot. The charging energy U for adding electrons to the dot from the surrounding electron reservoirs (leads) is typically the largest energy scale, and the dot is strongly correlated provided $U/\Gamma \gg 1$. Typical values for U

A.C. Hewson and V. Zlatić (eds.), Concepts in Electron Correlation, 247–256.

and Γ for the dots in (1, 5) are $U \sim 0.5 - 2.0$ meV and $\Gamma \sim 0.2 - 0.3$ meV, so these dots are strongly correlated.

For temperatures $\Gamma \ll T \ll U$ quantum fluctuations are small, and transport is dominated by charging effects. This regime is well understood (6). The conductance, G, exhibits a series of approximately equidistant peaks as a function of V_g, with spacing U. The peaks correspond to a fractional number of electrons on the dot and alternate "Coulomb blockade" valleys to either an even or an odd number of electrons.

In this paper we shall be interested in the regime $T \lesssim \Gamma \ll U$, where the strong quantum fluctuations can lead to a dramatic modification of the above picture of Coulomb blockade. In particular, for an odd number of electrons on the dot, the quantum dot can have a net spin 1/2, and a Kondo effect can develop. It has been predicted (7, 8), that, at low temperature, this enhances the conductance in the odd electron valleys, turning them instead into plateaus of near perfect transmission.

The outline of the paper is as follows. Sec. 2 describes the model and Sec. 3 the NRG method used to solve it. The results (9, 10, 11) for the conductance are described in Sec. 4-5 and these are used to compare with experiment over a range of gate voltages and temperatures in Sec. 6. Experimental investigation of the regime $T \lesssim \Gamma$ has only recently become possible as a result of better control of the electrode geometry, allowing parameters like Γ to be tuned to values of 1-3K so that $T \lesssim \Gamma$ is accessible (1, 2, 3, 4, 5). Sec. 7 describes the effect of a magnetic field on the transport through a quantum dot. Influencing transport via spin effects is of great current interest (12). The results presented in Sec. 7 should prove useful for interpreting magnetotransport experiments on strongly interacting quantum dots.

2. Model

At sufficiently low temperature, transport through a strongly interacting quantum dot will be mainly determined by the partially occupied level of the dot, denoted ε_d, which lies closest to the chemical potential of the leads. Its occupancy, n_d, can be varied from $n_d = 0$ to $n_d = 2$ by varying the gate voltage V_g, where $-eV_g = \varepsilon_d$. The resulting model of a single correlated level ε_d with Coulomb repulsion U, coupled to left and right free electron reservoirs can be written as

$$\begin{aligned} \mathcal{H} &= \sum_{\sigma} \varepsilon_d d_\sigma^\dagger d_\sigma + U d_\uparrow^\dagger d_\uparrow d_\downarrow^\dagger d_\downarrow + g\mu_B H S_z^d \\ &+ \sum_{k,\sigma,i=L,R} \varepsilon_{k,i} c_{k,i,\sigma}^\dagger c_{k,i,\sigma} + \sum_{k,\sigma,i=L,R} V_i (c_{k,i,\sigma}^\dagger d_\sigma + d_\sigma^\dagger c_{k,i,\sigma}). \end{aligned} \tag{1}$$

We assume energy independent lead couplings $\Gamma_{L,R} = 2\pi\rho_{L,R}(\epsilon_F)V_{L,R}^2$, where $\rho_{L,R}(\epsilon_F)$ is the Fermi level density of states (per spin) of the left/right electron reservoir. The first two terms in $\mathcal{H}$ represent the quantum dot, the third term is a magnetic field coupling only to the dot's spin $S_z^d = \frac{1}{2}(d_\uparrow^\dagger d_\uparrow - d_\downarrow^\dagger d_\downarrow)$ (we set $g = \mu_B = 1$), the fourth term represents the free electron reservoirs and the last term is the coupling between the dot and the reservoirs. This model can be reduced to the standard Anderson impurity model of a single reservoir attached to the dot with strength $\Gamma = \Gamma_L + \Gamma_R$ (7). Note that $\Gamma = 2\Delta$, where Δ is the hybridization strength as usually defined in the Anderson model (13). We use Γ throughout. In experiments (1, 5), Γ is extracted by analyzing the high temperature ($T \gg \Gamma$) behaviour of the conductance peaks (see figure: 2a below).

We assume, from here on, symmetric coupling to the leads, $\Gamma_L = \Gamma_R$. The linear magnetoconductance, $G(T,H) = \sum_\sigma G_\sigma(T,H)$, is written as a sum of spin-resolved magnetoconductances, G_σ, where

$$G_\sigma(T,H) = \frac{e^2}{\hbar}\Gamma \int_{-\infty}^{+\infty} d\omega\, A_\sigma(\omega,T,H)\left(-\frac{\partial f(\omega)}{\partial\omega}\right). \tag{2}$$

$A_\sigma(\omega,T,H)$ is the equilibrium spectral density and is expressed in terms of the local level Green function, $\mathcal{G}_{d,\sigma} = 1/(\omega - \varepsilon_d + i\Gamma/2 - \Sigma_U)$, with Σ_U the correlation part of the self-energy, by

$$A_\sigma(\omega,T,H) = -\frac{1}{\pi}\mathrm{Im}\,\mathcal{G}_{d,\sigma}(\omega + i\epsilon, T, H), \tag{3}$$

3. Method

We calculate $A_\sigma(\omega,T,H)$ by using Wilson's NRG method (14) extended to finite temperature dynamics (9, 15), with recent refinements (16, 17) which improve the high energy features. The NRG procedure for finite temperature dynamics is described in (9, 15). Here, we make a few remarks concerning the above refinements.

The first refinement (16) uses the correlation part of the self-energy, Σ_U, to evaluate the spectral density A in the Anderson impurity model (16). This improves the spectra around the single-particle excitations ε_d and $\varepsilon_d + U$ since the single particle broadening $\Gamma/2$ is put into the Green function, $\mathcal{G}_d$, exactly, thereby making the excitations at ε_d and $\varepsilon_d + U$ slightly sharper than in the earlier procedure (9) which evaluated A directly from its spectral representation. The description of the low energy Kondo resonance is equally accurate within both approaches and close to exact (16).

The second refinement uses the reduced density matrix in the NRG procedure for dynamical quantities (17) in place of the grand canonical

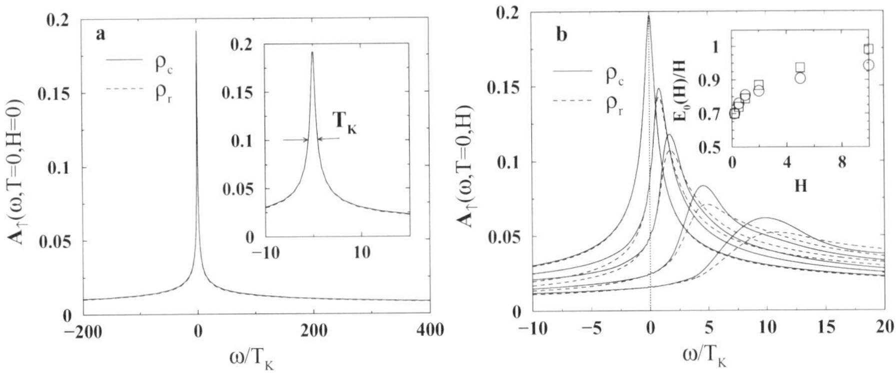

Figure 1. Comparison of methods for $A_\uparrow(\omega, 0, H)$, of the Kondo model, using canonical (ρ_c, solid line) (9) and reduced density matrices (ρ_r, dashed line) (17). $T_K = 3.97\times10^{-5}$ (in units of half bandwidth $D = 1$) is the HWHM of the $T = H = 0$ Kondo resonance : (a) at zero field, with the inset showing the Kondo resonance in more detail, and, (b) in finite field, with the curves from left to right corresponding to $H = 0, 1, 2, 5, 10$ in units of T_K. The inset shows the peak position, $E_0(H)$, as a function of H for the canonical (circles) and reduced density matrix (squares) approaches. Both E_0 and H are measured in units of T_K. The limit (18) $\lim_{H\to0} E_0(H)/H = 2/3$ is recovered in both approaches.

density matrix of the usual procedure (9). This reduces finite-size effects in the spectra. The latter are usually small on all energy scales in the case of zero magnetic field, but can be large in the case of a finite magnetic field, particularly for the high energy parts of the spectra (since these are calculated from the shortest chains and are therefore subject to the largest finite-size effects). Figure: 1a shows that both procedures give equally accurate results for the Kondo model in zero magnetic field on all energy scales. The same holds for the Anderson model in zero magnetic field. At finite magnetic field, the reduced density matrix is required for a correct description of the features at ε_d and ε_d+U in the spin-resolved spectra (17). In contrast, the low energy Kondo resonance in a magnetic field is well described by either procedure for $H \lesssim 10T_K$, as shown in figure: 1b for the Kondo model.

The position, $E_0(H)$, of the spin-resolved Kondo resonance in a weak magnetic field, $H \ll T_K$, is known from an exact Fermi liquid result due to Logan (18) (see also (19) showing the low field quasiparticle resonance). This states that $\lim_{H\to0} E_0(H)/H = 2/3$. This indicates that correlations reduce $E_0(H)/H$ from its expected high field limit of 1 to 2/3. The inset to figure: 1b shows that both procedures recover this result. The use of the reduced density matrix has also proven useful for studying magnetic states in the Hubbard model within the dynamical mean field theory (20).

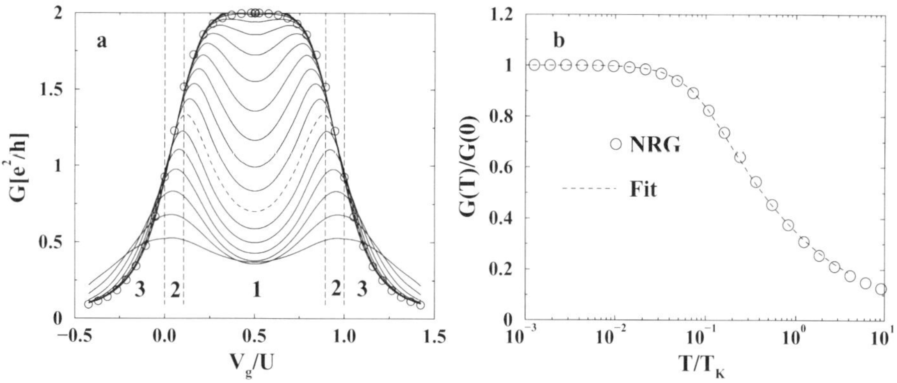

Figure 2. (a) $G(T, H = 0)$ versus V_g in the regime, $T \leq \Gamma$, of strong quantum fluctuations, for $U/\Gamma = 4.712$. The symbols are the $T = 0$ limit, Eq. (4). Temperatures decrease from bottom to top and correspond to $T_N = 0.64\Gamma\Lambda^{-N}$, $N = 0, 1, 2, \ldots, \Lambda = 1.5$. The dashed curve has $T = 1.2T_K$, where $T_K \approx 0.05\Gamma$, is the HWHM of the $T = 0$ Kondo resonance at mid-valley gate voltage. The regions marked $1, 2$ and 3 and separated by vertical dashed lines correspond to the Kondo ($\varepsilon_d \lesssim -0.5\Gamma$, i.e. $n_d \approx 1$), mixed valent ($|\varepsilon_d| \lesssim 0.5\Gamma$, i.e. $n_d \approx 0.5$, or by particle-hole symmetry, $|\varepsilon_d + U| \lesssim 0.5\Gamma$, i.e. $n_d \approx 1.5$), and empty (full) orbital regimes ($\varepsilon_d \gtrsim 0.5\Gamma$, i.e. $n_d \approx 0$, or, by particle-hole symmetry, $\varepsilon_d + U \lesssim -0.5\Gamma$, i.e. $n_d \approx 2$). (b) The universal conductance curve $G(T)/G(0)$ for a quantum dot in the Kondo regime at mid-valley, $V_g/U = 0.5$, (circles) (9, 10). The dashed line is the fit formula $G(T)/G(0) = (T_K'^2/(T^2 + T_K'^2))^s$ with $s = 0.22$ and $T_K' = T_K/\sqrt{2^{1/s} - 1}$ used in (1) to interpolate the NRG results up to $10T_K$ (with T_K as in figure: 2a).

4. Linear conductance for $H = 0$

The gate voltage dependence of the linear conductance of a quantum dot is shown in figure: 2a for a decreasing set of temperatures. There are three distinct physical regimes, corresponding to three ranges of the gate voltage controlling ε_d relative to the Fermi level (see also figure: 3a-c). In the empty (full) orbital regime (region 3) the conductance shows the expected activated behaviour as a function of temperature. In the mixed valent regime (region 2), the behaviour of G versus T is approximately that corresponding to tunneling through a resonant level close to the Fermi level. The most dramatic behaviour is in the Kondo regime (region 1), where one sees an anomalous enhancement of the conductance with decreasing temperature. The conductance continues to increase and eventually reaches the unitarity limit of $2e^2/h$ (see below) at mid-valley ($V_g/U = 0.5$ or $n_d = 1$). This remarkable enhancement of the conductance results from the formation, with decreasing temperature, of the many-body Kondo resonance at the Fermi level of the leads. The three types of behaviour have been observed in experiments on quantum dots carried out at $T \lesssim \Gamma$ (1, 5, 21). The unitarity limit of the conductance, $2e^2/h$, has also recently been observed (5). The $T \to 0$ conductance curve depends only on n_d via the Friedel sum rule (7, 8, 22)

$$G(T = 0, H = 0) = \frac{2e^2}{h} \sin^2(\pi n_d/2), \tag{4}$$

and is shown in figure: 2a (circles), with n_d deduced from the spectral densities as a check on the calculations.

5. Scaling of the linear conductance at $H = 0$

In the Kondo regime, $G(T)/G(0)$, is a universal function of T/T_K starting from low temperatures and extending up to some high temperature which depends on other details such as the precise value of Γ and ε_d. The latter energy scales cut off the universal behaviour of the conductance on the high temperature side. The universal conductance curves for the Kondo and Anderson models have been calculated for the experimentally interesting temperature range $0 \leq T \leq 10T_K$ via the NRG (9, 10, 11). This is shown for the Kondo model in figure: 2b together with an approximate interpolating formula used in (1). The scaling of $G(T)/G(0)$ with T/T_K shown in figure: 2b persists in the Anderson model throughout the Kondo regime (region 1) as long as $T \ll \Gamma$. At higher temperatures and in the other regimes, the conductance curves deviate from this universal shape (see Sec. 6 below). The Kondo scaling of the conductance in figure: 2b agrees well with measurements for both quantum dots in heterostructures (1, 5) and and carbon nanotubes (21).

6. Comparison with experiment

For a comparison of theory with experiment in all regimes, the complete set of conductance curves for the Anderson model is required. These are shown in figure: 3a-c. $G(T)$ in the empty orbital and mixed valent regimes has been calculated in (23) and used to compare with experimental data in (1). Here we make a parameter free comparison to similar data from Reference (5) for all three regimes of interest. The results are shown in figure: 3d. In making the comparisons, we estimated $G(T = 0)$ from $G(T = 30mK)$, close to the lowest effective electron temperature $T = 40$mK (5), and used it to determine n_d from Eq:4 and hence the appropriate ε_d to use in the NRG calculation for $G(T)$ at all T. The calculations also used the values of $\Gamma = 231.4\mu eV$ and $U \approx 0.5$meV from the experiments. It is remarkable that this zero parameter comparison yields the agreement seen. The Kondo scale automatically comes out correctly as seen for the $V_g = -414$mV

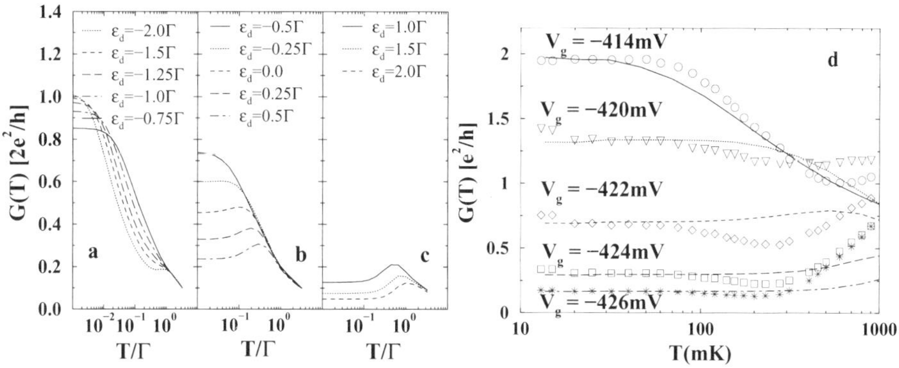

Figure 3. (a-c) $G(T)$ for gate voltages $V_g = -\varepsilon_d/e$ in, (a), the Kondo (K) regime, (b), the mixed valent (MV) regime ($\varepsilon_d = -0.5\Gamma, .., +0.5\Gamma$), and, (c), the empty orbital (EO) regime ($\varepsilon_d = +\Gamma, .., +2.0\Gamma$), and parameters as in figure: 2a. (d) Comparison of theoretical (lines) to experimental (symbols) conductance curves, $G(T)$, for the quantum dot in (5) with $U = 0.5$meV, $\Gamma = 0.231$meV and 13mK $\leq T \leq$ 900mK. The values used for ε_d in the calculations were $\varepsilon_d = -0.85\Gamma$ ($V_g = -414$mV, K), $\varepsilon_d = -0.25\Gamma$ ($V_g = -420$mV, MV), $\varepsilon_d = +0.375\Gamma$ ($V_g = -422$mV, MV), $\varepsilon_d = +1.00\Gamma$ ($V_g = -424$mV, EO), $\varepsilon_d = +1.50\Gamma$ ($V_g = -426$mV, EO). We find a linear dependence $V_g/e = -\alpha\varepsilon_d$ for $V_g \leq -420$mV, which is a consistency check for determining ε_d from $G(T \to 0)$ using Eq. 4.

comparison, and the agreement with the theoretical conductance curve is very good up to 700mK. The conductance curve used here includes the non-universal corrections discussed above and therefore differs slightly from that used in (5) (notably a slope change at 500mK due to corrections from charge fluctuations). The general trends of the experimental data in going to the mixed valent ($V_g = -420$mV, $V_g = -422$mV) and empty orbital cases ($V_g = -424$mV, $V_g = -426$mV) are well reproduced by the calculations. In particular the expected finite T peak in $G(T)$ (23) develops and becomes increasingly more pronounced on entering the empty orbital regime. As described earlier, transport in this regime is likely to involve additional neighbouring levels and a larger conductance at higher temperatures, as observed in the experiments. In this light, the main discrepancy remaining between theory and experiment appears to be the dip in $G(T)$ at 200-300mK in the measurements. No signature of this is present in our model calculations. This could be due to interference effects associated with more than one level. It would therefore be interesting to investigate this possibility further.

7. Effect of a magnetic field

A magnetic field suppresses the Kondo effect on a scale of T_K (13) and consequently it is expected to have a drastic effect on the conductance of

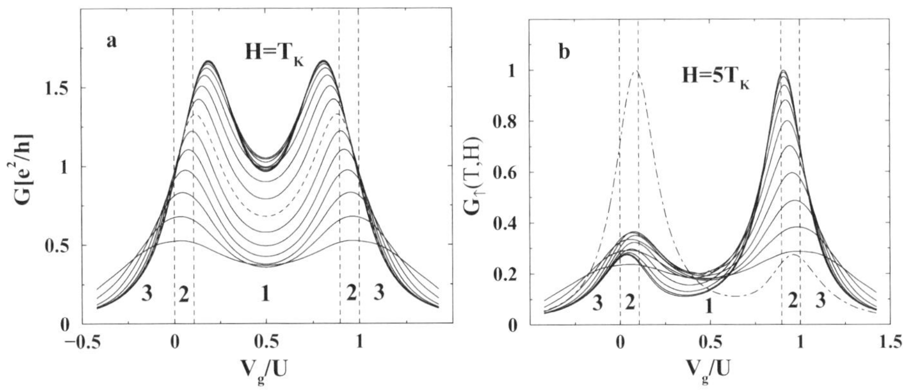

Figure 4. (a) T and V_g dependence of $G(T, H)$ for $H = T_K$ and for temperatures and parameters as in figure:2a. (b) T and V_g dependence of the spin-resolved conductance $G_\uparrow(T, H)$ for $H = 5T_K$ and for temperatures and parameters as in figure:2a. The dot-dashed curve is for the down spin conductance, $G_\downarrow(T, H)$, at the lowest temperature.

a quantum dot (10, 11). Indeed, figure: 4a shows that a field of $H = T_K$ suffices to reduce the $T = 0$ conductance at mid-valley to nearly half its value at $H = 0$ (cf. figure: 2a). This can be understood from the splitting of the Kondo resonance in a magnetic field and its strong reduction at the Fermi level (see figure: 1b and (10)). The suppression of the conductance is large in the whole Kondo regime (region 1) as long as $T \lesssim T_K$. At higher temperatures and in the other regimes, the effect of a field on the total conductance is less pronounced. However, in these regimes too, there can be a large effect in the *spin-resolved* conductances (see below).

The effects of a magnetic field become even more apparent in spin-resolved quantities such as $G_\uparrow$ in figure: 4b (by particle-hole symmetry $G_\downarrow$ is the mirror reflection of $G_\uparrow$ about $V_g/U = 0.5$). As in G, a magnetic field has a large effect on the spin-resolved conductance in the Kondo regime. In addition, there is quite a dramatic spin-filtering effect in the mixed valent regime, with $|G_\uparrow(0, H) - G_\downarrow(0, H)|$ being largest in this regime. In contrast, both up and down spin conductances are approximately equally suppressed in the other regimes. A quantum dot in a field is seen to act as a spin-filter as discussed in (24) for dots very weakly coupled to leads ($G_\sigma \ll e^2/h$).

The field dependence of the conductance of quantum dots defined in carbon nanotubes has been studied (21) and a preliminary comparison between our results and the experimental $G(B, T \lesssim T_K)$ shows good agreement (25).

8. Conclusions

We considered electronic transport through a strongly interacting quantum dot in zero and finite magnetic fields and in the low temperature regime $T \lesssim \Gamma$ where quantum fluctuations strongly modify the transport for an odd number of electrons on the dot.

The NRG conductance curves, $G(T, V_g)$, were used to make a parameter free comparison with experimental results of Reference (5) and we found good agreement. Discrepancies in the empty orbital regime where attributed to the neglect, within our model, of neighbouring levels

The results for the magnetoconductance in the Kondo regime exhibited the strong suppression of the Kondo effect by a magnetic field. A large spin-filtering effect was found in the mixed valent regime.

The field of non-equilibrium transport through quantum dots remains largely unexplored. Perturbative methods valid in the limit of large transport voltage, and magnetic field, $V, H \gg T_K$, are being developed (26), but non-perturbative techniques, such as extensions of NRG, will be required to address strong coupling.

Acknowledgements

I would like to thank the authors of Reference (5) for sending me the experimental data used in figure: 3b and W. van der Wiel for useful discussions. Financial support from the Deutscheforschungsgemeinschaft through SFB 195, and in part by the National Science Foundation under Grant No. PHY99-07949 during the writing of this paper at the KITP of UCSB, is acknowledged.

References

1. D. Goldhaber–Gordon et al., Phys. Rev. Lett. **81**, 5225 (1998).
2. S. M. Cronenwett, T. H. Oosterkamp and L. Kouwenhoven, Science **281**, 540 (1998).
3. J. Schmid, J. Weis, K. Eberl and K. von Klitzing, Physica B **258**, 182 (1998).
4. F. Simmel, R. H. Blick, J. P. Kotthaus, W. Wegscheider and M. Bichler, Phys. Rev. Lett. **83**, 804 (1999).
5. W. G. van der Wiel et al., Science **289**, 2105 (2000).
6. C. W. J. Beenakker and H. van Houten, Solid State Physics, **44**, 1 (1991).
7. L. I. Glazman and M. E. Raikh, Sov. Phys. JETP Lett. **47**, 454 (1988).
8. T. K. Ng and P. A. Lee, Phys. Rev. Lett. **61**, 1768 (1988).
9. T. A. Costi, A. C. Hewson and V. Zlatić, J. Phys. Cond. Matt. **6**, 2519 (1994).
10. T. A. Costi, Phys. Rev. Lett. **85**, 1504 (2000).
11. T. A. Costi, Phys. Rev. B **64**, 241310(R) (2001).
12. Y. Ohno et al., Nature **402**, 790 (1999).

13. A. C. Hewson, *The Kondo Problem to Heavy Fermions* (Cambridge University Press, Cambridge, England 1993)
14. K. G. Wilson, Rev. Mod. Phys.**47**, 773 (1975); H. R. Krishna-murthy, J. W. Wilkins and K. G. Wilson, Phys. Rev. B**21**, 1003 (1980).
15. T. A. Costi, in *Density Matrix Renormalization*, edited by I. Peschel, X. Wang, M. Kaulke and K. Hallberg (Springer, Berlin, Germany 1999).
16. R. Bulla, A. C. Hewson and Th. Pruschke, J. Phys. Cond. Matt. **10**, 8365 (1998).
17. W. Hofstetter, Phys. Rev. Lett. **85**, 1508 (2000).
18. D. E. Logan and N. L. Dickens, J. Phys.: Condens. Matter **13**, 9713 (2001).
19. A. C. Hewson, J. Phys.: Condens. Matter, **13**, 10011 (2001).
20. R. Zitzler, Th. Pruschke and R. Bulla, European Phys. Journal, B**27**, 473 (2002).
21. J. Nygard, D. H. Cobden and P. E. Lindelof, Nature **408**, 342 (2000).
22. D. C. Langreth, Phys. Rev. **150**, 516 (1966).
23. J. König and H. Schoeller, Phys. Rev. Lett. **84**, 3686 (2000).
24. P. Recher, E. V. Sukhorukov and D. Loss, Phys. Rev. Lett. **85**, 1962 (2000).
25. J. Nygard, private communication.
26. A. Rosch, J. Paaske, J. Kroha, P. Wölfle, cond-mat/0202404.

PROPERTIES OF CORRELATED NANOSCOPIC SYSTEMS FROM THE COMBINED EXACT DIAGONALIZATION – AB INITIO METHOD

Proceedings of the ARW NATO Workshop Hvar, Croatia, October 2002

J. SPAŁEK, E.M. GÖRLICH, A. RYCERZ, R. ZAHORBEŃSKI and R. PODSIADŁY
Marian Smoluchowski Institute of Physics, Jagiellonian University, ulica Reymonta 4, 30-059 Kraków, Poland
W. WÓJCIK
Institute of Physics, Tadeusz Kościuszko Technical University, ulica Podchorążych 1, 30-084 Kraków, Poland

Abstract. We review briefly our recent exact results concerning the electronic properties of atomic, molecular, and nanoscopic systems by combining the exact diagonalization and ab initio methods (EDABI). Particular emphasis is put on determining the microscopic parameters. The multiatom system behavior is discussed as a function of interatomic distance. A transformation from a nanometal (for small lattice constant) to an atomic system is observed with increasing distance between the atoms.

Key words: Nanoscopic Systems, Correlated Electrons, Exact Diagonalization, Nanochains, Nanoclusters, Ab Initio Approach

1. Introduction

Theoretical determination of the electronic and atomic properties of nanoscopic systems containing $N \sim 10 - 100$ atoms is very important from the two principal perspectives. First, with the advancement of nanotechnology we begin utilizing the electronics on the atomic scale. Second, it is crucial to determine in a precise manner electronic properties of such small systems to see the evolution from a collection of atoms to a piece of a solid with the delocalized (Bloch-type or Fermi-liquid-like) states. The latter aspect is reviewed briefly here after introducing a novel method of approach (EDABI) developed in our group recently (1, 2, 3), in which we combine the exact

A.C. Hewson and V. Zlatić (eds.), Concepts in Electron Correlation, 257–268.

diagonalization providing us with rigorous treatment of interparticle interaction, with an ab-initio single-particle-wave-function optimization in the system ground state. In this manner, the interaction and the single-particle terms in the Hamiltonian (or Lagrangian) are treated on the same footing. Hence, the method is particularly useful for the strongly correlated systems when the kinetic (band) energy of electrons propagating throughout the system is at best comparable to the interaction among the particles. One can also say that the approach combines the particle and the wave aspects of the electron-state characterization into a single formalism, since the interaction is treated in the occupation number representation, whereas the shape of matter wave for a single particle is found from the renormalized (self-adjusted) wave equation or from a related to it variational procedure. In Sec.2 we summarize briefly our method of approach and its relation to multiconfiguration-interaction (MCI) approach from quantum chemistry. In Secs.3-5 we provide the examples of applications ranging from atomic physics to correlated chain and cluster systems. Sec.6 contains conclusions.

2. The Method

2.1. RENORMALIZED (SELF-ADJUSTED) WAVE EQUATION, SWE

We start with the Hamiltonian in the standard second-quantized form

$$H = \int d^3x \widehat{\Psi}^\dagger(\mathbf{x}) H_1(\mathbf{x}) \widehat{\Psi}(\mathbf{x})$$

$$+\frac{1}{2}\int\int d^3x d^3x' \widehat{\Psi}^\dagger(\mathbf{x})\widehat{\Psi}^\dagger(\mathbf{x}') V(\mathbf{x}-\mathbf{x}')\widehat{\Psi}(\mathbf{x}')\widehat{\Psi}(\mathbf{x}), \tag{1}$$

where $\widehat{\Psi}(\mathbf{x}) = \sum_{i\sigma} w_i(\mathbf{x})\chi_\sigma a_{i\sigma}$ is the field operator, $H_1(\mathbf{x})$ and $V \equiv V(\mathbf{x}-\mathbf{x}')$ are respectively the single-particle and the pair-interaction terms, and $\{w_i(\mathbf{x})\}$ is a complete single-particle basis (arbitrary at this point). A complete solution for N indistinguishable particles would involve determination of $\widehat{\Psi}(\mathbf{x})$ via the Heisenberg equation of motion. Such a solution does not seem to be executable for N-fermion system, so we proceed as follows. We substitute the expression for $\widehat{\Psi}(\mathbf{x})$ into (1) to obtain

$$H = \sum_{ij\sigma} t_{ij} a^\dagger_{i\sigma} a_{j\sigma} + \frac{1}{2}\sum_{ijkl\sigma_1\sigma_2} V_{ijkl} a^\dagger_{i\sigma_1} a^\dagger_{j\sigma_2} a_{l\sigma_2} a_{k\sigma_1}, \tag{2}$$

with the microscopic parameters defined by

$$t_{ij} \equiv < w_i|H_1|w_j >= \int d^3x w_i^*(\mathbf{x}) H_1(\mathbf{x}) w_j(\mathbf{x}), \tag{3}$$

and

$$V_{ijkl} \equiv < w_i w_j | V | w_k w_l > = \int d^3x d^3x' w_i^*(\mathbf{x}) w_j^*(\mathbf{x}') V(\mathbf{x} - \mathbf{x}') w_k(\mathbf{x}) w_l(\mathbf{x}'). \tag{4}$$

In this form, the single and many-particle aspects of the problem are separated in the sense that the calculation of the parameters t_{ij} and V_{ijkl}, both containing the single-particle basis $\{w_i(\mathbf{x})\}$, is separated from the diagonalization procedure of the Hamiltonian in the Fock space (occupation-number representation). This two-step procedure can be seen explicitly when we calculate the system ground state energy

$$E_G \equiv < H > = \sum_{ij\sigma} t_{ij} < a_{i\sigma}^\dagger a_{j\sigma} > + \frac{1}{2} \sum_{ijkl\sigma_1\sigma_2} V_{ijkl} < a_{i\sigma_1}^\dagger a_{j\sigma_2}^\dagger a_{l\sigma_2} a_{k\sigma_1} >, \tag{5}$$

where the averaging $< ... >$ is for the ground $|\Psi_G >$ or excited states; we also take into account all accessible occupancies of the states $\{|i\sigma >\}$, $\{|j\sigma' >\}$, etc. for fixed values of t_{ij} and V_{ijkl} regarded as parameters (if we consider grand canonical ensemble, we diagonalize $H - \mu N$, with $N = \int d^3x \widehat{\Psi}^\dagger(\mathbf{x}) \widehat{\Psi}(\mathbf{x}) = \sum_{ij\sigma} \int d^3x w_i^*(\mathbf{x}) w_j(\mathbf{x}) < a_{i\sigma}^\dagger a_{j\sigma} >$).

To close the solution we treat the expression (5) as a functional $\{w_i(\mathbf{x})\}$. In the most general case of the grand canonical ensemble and with a nonorthogonal basis $\{w_i(\mathbf{x})\}$ this functional can be written as

$$F\{w_i(\mathbf{x})\} = E_G\{w_i(\mathbf{x})\} - \sum_{ij\sigma} (\mu + \lambda_{ij}) \int d^3x w_i^*(\mathbf{x}) w_j(\mathbf{x}) < a_{i\sigma}^\dagger a_{j\sigma} > . \tag{6}$$

In effect, the Lagrange-Euler equation, which plays the role of the stationary *self-adjusted wave equation* (SWE) for $w_i(x)$, takes the form

$$\frac{\delta E_G}{\delta w_i^*(\mathbf{x})} - \nabla \cdot \frac{\delta E_G}{\delta \nabla w_i^*(\mathbf{x})} - \sum_{j \geq i\sigma} (\lambda_{ij} - \mu) w_j(\mathbf{x}) < a_{i\sigma}^\dagger a_{j\sigma} > = 0, \tag{7}$$

where λ_{ij} and μ play the role of Lagrange multipliers (if we use explicitly orthogonal basis and particle-conserving diagonalization procedure in the Fock space, then $\lambda_{ij} = \mu \equiv 0$ and Eq.(7) reduces to the usual Euler equation). Also, when the class of $\{w_i(\mathbf{x})\}$ is selected variationally (as in the following), Eq.(6) is then minimized with respect to trial function parameters (orbital radii a_i in the case when the orthogonal basis $w_i(\mathbf{x})$ is composed of superposed atomic-like wave functions).

2.2. RELATION TO THE MULTICONFIGURATION-INTERACTION (MCI) METHOD

With the help of the field operator $\widehat{\Psi}(\mathbf{x})$ we can define N-particle wave function $\Psi_\alpha(\mathbf{x}_1, .., \mathbf{x}_N)$ for particles distributed among M states as follows

$$\Psi_\alpha(\mathbf{x}_1, .., \mathbf{x}_N) = \frac{1}{\sqrt{N!}} < 0|\widehat{\Psi}(\mathbf{x}_1)...\widehat{\Psi}(\mathbf{x}_N)|\Phi_\alpha >, \tag{8}$$

where $|\Phi_\alpha >$ is the N-particle state in the Fock space

$$|\Phi_\alpha >= \frac{1}{\sqrt{N!}} \sum_{j_1...j_N=1}^{M} C_{j_1...j_N} a^\dagger_{j_1}...a^\dagger_{j_N}|0 >, \tag{9}$$

where $C_{j_1...j_N}$ are the expansion coefficients (note that: $j_1 \equiv j_1\sigma_1$). Within our method we determine $C_{j_1...j_N}$ with the help of Lanczos- or other exact-diagonalization method (4). Substitution (9) to (8) and using the decomposition of $\widehat{\Psi}(\mathbf{x}_i)$ in terms of $\{w_i(\mathbf{x})\}$ we obtain that

$$\Psi_\alpha(\mathbf{x}_1, .., \mathbf{x}_N) = \frac{1}{\sqrt{N!}} \sum_{i_1...i_N=1}^{M} C_{i_1...i_N}\{A, S\}[w_{i_1}(\mathbf{x}_1)...w_{i_N}(\mathbf{x}_N)], \tag{10}$$

where (A, S) are respectively $(-1)^P$, where P is the permutation sign of the indices $(i_1, i_2, ..., i_N)$ for fermions and a symmetrization operation in case of bosons. Therefore if we are able to perform the diagonalization in the Fock space, we have an exact many-particle wave function within the subspace selected by the single-particle basis $\{w_i(\mathbf{x})\}$. In the next Sections we discuss properties of different atomic and nanoscopic systems.

3. Application to atomic and molecular systems

We construct a nanoscopic system out of atoms. Therefore, as a first step we apply our method first to simple atomic and molecular systems.

3.1. LIGHT ATOMS AND IONS

He is the lightest many-electron atom. In the simplest approximation the Hilbert space is defined only by a single orbital $\Phi(\mathbf{r})$. Under these conditions $\widehat{\Psi}(\mathbf{x}) = \Phi(\mathbf{r})\chi_\uparrow a_\uparrow + \Phi(\mathbf{r})\chi_\downarrow a_\downarrow$. The ground state energy is $E_G = \epsilon_a + U$, with

$$\epsilon_a = \int d^3r\Phi^*(\mathbf{r})[-\frac{\hbar^2}{2m}\nabla^2 - \frac{2e^2}{r}]\Phi(\mathbf{r}) \tag{11}$$

TABLE I. Optimized Bohr-orbit radii of 1s, 2s, and 2p orbits (in units of a_0), the overlap S between renormalized 1s and 2s states, and the ground state energy for the lightest atoms and ions.

	a_{1s}	a_{2s}	a_{2p}	S	E_G (Ry)
H	1	2	2	0	-1
H^-	0.9696	1.6485	1.017	-0.1	-1.0487
He	0.4274	0.5731	0.4068	-0.272	-5.79404
He^-	1.831	1.1416	0.4354	-0.781	-5.10058
Li	0.3725	1.066	0.2521	0.15	-14.8334
Be^+	0.2708	0.683	0.1829	0.109	-28.5286

being the atomic energy, and

$$U = \int d^3r d^3r' |\Phi(\mathbf{r})|^2 \frac{e^2}{|\mathbf{r} - \mathbf{r}'|} |\Phi(\mathbf{r}')|^2 \tag{12}$$

being the magnitude of the Coulomb interaction between electrons. Considering E_G as a functional of $\Phi(\mathbf{r})$ we obtain Eq.(7) in the form ($\mu = 0$):

$$-[\frac{\hbar^2}{2m}\nabla^2 + \frac{2e^2}{r}]\Phi(\mathbf{r}) + \Phi(\mathbf{r}) \int d^3r' \frac{e^2}{|\mathbf{r} - \mathbf{r}'|} |\Phi(\mathbf{r}')|^2 = \lambda\Phi(\mathbf{r}). \tag{13}$$

This means $\Phi(\mathbf{r})$ is determined from the Hartree equation. Assuming that $\Phi(\mathbf{r})$ has 1s-like form $\Phi(\mathbf{r}) = (\pi\alpha^3)^{1/2}\exp(-\alpha r)$, and adjusting variationally the size $a = \alpha^{-1}$ of the renormalized 1s orbit we obtain the well known results (5): $E_G = -5.695Ry$ and $a = (16/27)a_0$, where a_0 is the Bohr radius. We can systematically improve on this result by enriching our single-particle basis. Including 1s- and 2s-wave functions, we obtain 6-dimensional Fock space and by repeating both the diagonalization procedure and the variational optimization of E_G with respect to 1s- and 2s-like state radii we obtain an improved value of $E_G = -5.7549Ry$. Finally, taking 1s-, 2s-, and 2p-like states (with the corresponding adjustable radii a_{1s}, a_{2s}, a_{2p}) we obtain the value $E_G = -5.794Ry$, which is close to the accepted "exact" value $-5.8074Ry$. The values of radii are displayed in Table I for the light atoms and ions. The ground-state 2-particle wave function has in the Fock space the form for He atom:

$$|G> \simeq \{0.8a^\dagger_{1s\uparrow}a^\dagger_{1s\downarrow} - 0.4(a^\dagger_{1s\uparrow}a^\dagger_{2s\downarrow} + a^\dagger_{2s\uparrow}a^\dagger_{1s\downarrow}) + 0.135a^\dagger_{2s\uparrow}a^\dagger_{2s\downarrow}$$

TABLE II. Microscopic parameters (in Ry) of the selected atoms and ions all quantities are calculated for the orthogonalized atomic states.

	t	U_1	U_2	U_3	U_p	K_{12}	K_{13}	K_{23}
H^-	0.057	1.333	0.369	0.77	0.728	0.519	0.878	0.457
He	1.186	3.278	1.086	1.924	1.821	1.527	2.192	1.289
He^-	-1.1414	1.232	0.764	1.798	1.701	0.929	1.421	1.041
Li	-0.654	3.267	0.533	3.105	2.938	0.749	3.021	0.743
Be^+	-0.929	4.509	0.869	4.279	4.049	1.191	4.168	1.175

$$-0.035(a^{\dagger}_{p0\uparrow}a^{\dagger}_{p0\downarrow} + a^{\dagger}_{p1\uparrow}a^{\dagger}_{\bar{1}\downarrow} + a^{\dagger}_{p\bar{1}\uparrow}a^{\dagger}_{p1\downarrow})\}|0> . \tag{14}$$

Thus He atom is indeed the smallest atom in the Universe, and we have a substantial (0.32) probability of having the two electrons in the singlet 1s-2s configuration. Nota bene, we can determine the first excited states in this case: the 1s-2s triplet with energy $E_t = -2.3707Ry$, and the next singlet of primary 1s-2s character with $E_{2s} = -1.3737Ry$.

Important are the values of the microscopic parameters, which are displayed for those systems in Table II. They comprise the 1s-2s overlap integral S, the 1s-2s hopping integral t, the magnitudes of 1s (U_1), 2s (U_2), 2p (U_3 for $m = 0$ and U_p for $m = \pm 1$), and 2p (U_p) Coulomb interaction. Additionally , K_{12}, K_{13}, K_{23} represent respectively 1s-2s, 1s-2p, and 2s-2p Coulomb interactions.

3.2. H_2 AND H_2^- SYSTEMS

The next step in building up step by step a model of nanoscopic system is to consider H_2 and H_2^- systems. Here we consider the situation with one orbital per atom, which will be taken as 1s-like. For this case the Fock space for H_2 is spanned on $\binom{4}{2} = 6$ states, whereas that for H_2^- contains $\binom{6}{3} = 4$ basis states. In both cases one can perform the diagonalization in Fock space analytically, whereas the 1s orbit readjustment represents a simple numerical minimization of E_G. In Fig.1 we display the interatomic-distance (R/a_0) dependence of the ground-state energy for those systems. Obviously, within one-orbital-per-atom basis H_2^- state is unstable. However, consideration of those systems provides us also with the values of microscopic parameters such as the hopping integral t, intraatomic (and interatomic) Coulomb interaction $U(K)$, as well as the magnitudes of the correlated hopping parameter V, Heisenberg (J) and the kinetic exchange

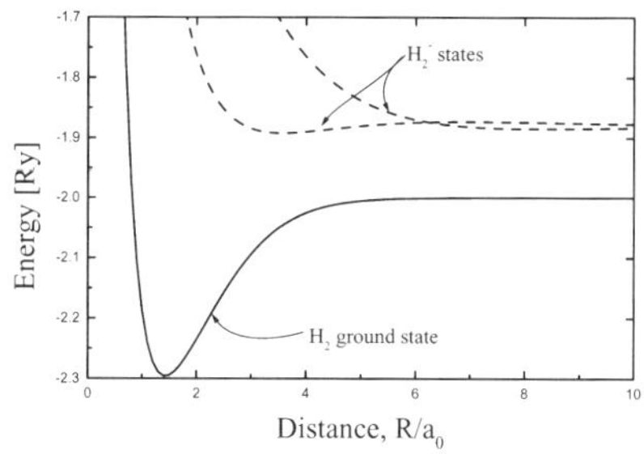

Figure 1. Energies of the ground state for H_2 molecule (solid line), as well as the two lowest laying states for H_2^- ions, all quantities as a function of interatomic distance R/a_0 (a_0 - Bohr radius).

TABLE III. Microscopic parameters (in Ry) of H_2 system vs. R/a_0.

R/a_0	E_G/N	t	U	K	V[mRy]	$J[mRy]$	$\frac{4(t+V)^2}{U-K}$[mRy]
1.	-1.0937	-1.1719	1.8582	1.1334	-13.55	26.254	7755.52
1.5	-1.1472	-0.6784	1.6265	0.9331	-11.687	21.252	2747.41
2.	-1.1177	-0.4274	1.4747	0.7925	-11.577	16.921	1130.19
2.5	-1.0787	-0.2833	1.3769	0.6887	-12.054	13.149	507.209
3.	-1.0469	-0.1932	1.3171	0.6077	-12.594	9.8153	238.939
3.5	-1.0254	-0.1333	1.2835	0.5414	-12.812	6.9224	115.143
4.	-1.0127	-0.0919	1.2663	0.4854	-12.441	4.5736	55.8193
4.5	-1.006	-0.0629	1.2579	0.4377	-11.441	2.8367	26.9722
5.	-1.0028	-0.0426	1.2539	0.3970	-9.9894	1.6652	12.9352
5.5	-1.0012	-0.0286	1.2519	0.3623	-8.3378	0.9334	6.1455
6.	-1.0005	-0.01905	1.251	0.3327	-6.7029	0.5033	2.8902
6.5	-1.00024	-0.0126	1.2505	0.3075	-5.2242	0.2626	1.3452
7.	-1.0001	-0.0083	1.2503	0.2856	-3.9685	0.1333	0.6197

amplitudes which are listed in Table III. Those values will be compared with those for the chains containing up to $N = 14$ atoms to draw some conclusions concerning a similar behavior of both atomic and nanoscopic systems.

4. Nanochains

We now consider a nanoscopic linear chain of up to $N = 14$ atoms, each containing a single valence electron (hydrogenic-like atoms) and including *all* overlap integrals, *all* hopping integrals, as well as *all* interactions (including

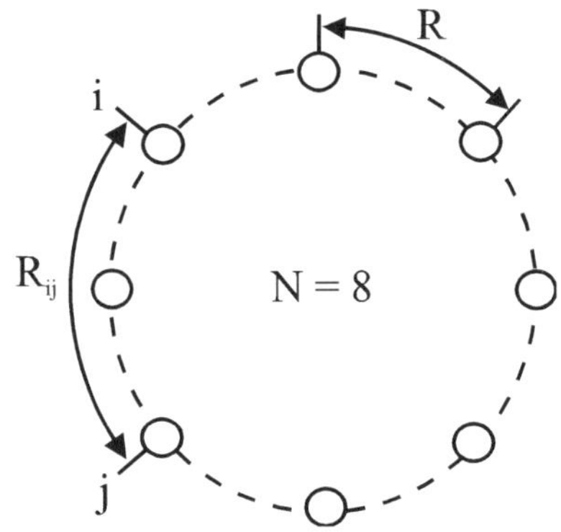

Figure 2. Schematic representation of the finite chain with periodic boundary conditions (or planar cluster) used in the calculations.

TABLE IV. Microscopic parameters for a linear-chain configuration and the Slater orbitals taken.

R/a_0	t_1 [Ry]	t_2 [mRy]	U [Ry]	K_1 [Ry]	K_2 [Ry]	K_3 [Ry]	V_1 [mRy]	V_2 [mRy]
2.0	-0.5851	86.98	2.301	1.077	0.676	0.450	-18.07	33.58
2.5	-0.3302	44.08	1.949	0.843	0.499	0.331	-17.45	19.58
3.0	-0.2000	23.60	1.717	0.692	0.391	0.259	-16.08	11.95
4.0	-0.0825	7.56	1.452	0.508	0.269	0.179	-12.92	4.49
5.0	-0.0366	4.29	1.327	0.403	0.206	0.138	-9.64	1.56

3- and 4-site terms in the Gaussian STO-3G basis). The method and the principal results for the case of Slater orbitals have been published before [2]. Here we present also the first results obtained for the Gaussian (STO-3G) basis, which we will try to apply for realistic systems containing Li and Na atoms in the near future. In Fig.2 we present schematically a ring of N atoms. We analyze its basic properties in a linear-chain configuration with periodic boundary conditions (i.e. overlaps, hoppings, interactions *only* along the ring), whereas in the next Section we discuss a multidimensional ($d = 2, 3$) configurations. By adopting EDABI to this situation we can calculate *all* hopping integrals and therefore, calculate the band energy $\epsilon_{\mathbf{k}}$, as it evolves as a function of interatomic distance R/a_0 (2). For $R/a_0 \gtrsim 4$ the tight-binding approximation works very well. The bandwidth $W = 2z\sum_{j(i)} |t_{ij}|$ falls with growing R systematically reflecting reduced overlap. This illustrates the famous Mott remark (6) that by looking at the single-particle (band) energy one can never observe a transition from atomic to metallic behavior, only a systematic evolution towards the atomic

TABLE V. Microscopic parameters (in Ry) vs. R/a_0 for $N = 4$ planar cluster. The star marks the distance R with the lowest system energy.

R/a_0	t_1	t_2 [mRy]	U	K_1	K_2	V_1 [mRy]	J_1 [mRy]
0,5	-3,8651	1510	2,7792	1,6847	1,2008	-37,30	49,04
1,0	-1,4691	304,4	2,1831	1,2618	0,9237	-12,54	32,54
1,5	-0,7657	87,17	1,8310	1,0020	0,7354	-8,151	24,03
1,8	-0,5563	41,40	1,6877	0,8918	0,6532	-7,884	20,28
1,9*	-0,5043	31,74	1,6482	0,8604	0,6296	-7,960	19,17
2,0	-0,4587	23,91	1,6128	0,8314	0,6077	-8,098	18,12
2,5	-0,2959	1,930	1,4701	0,7101	0,5156	-9,134	13,57
3,0	-0,1991	-5,710	1,3767	0,6187	0,4461	-10,29	9,910
3,5	-0,1368	-7,640	1,3186	0,5467	0,3917	-11,12	6,930
4,0	-0,0945	-7,140	1,2845	0,4879	0,3477	-11,31	4,584

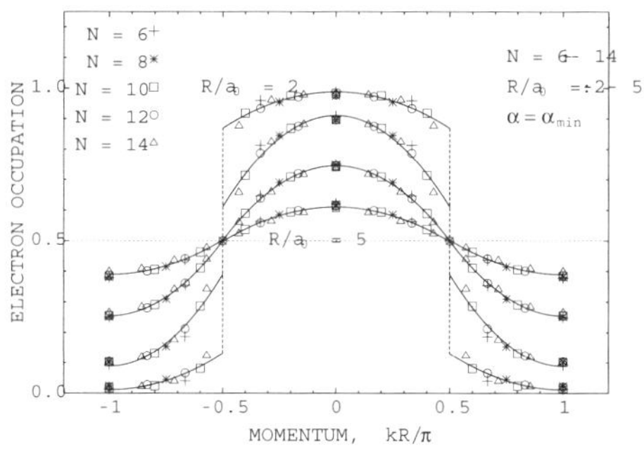

Figure 3. Momentum distribution $n_{k\sigma}$ for electrons in linear chain of $N = 6 \div 12$ atoms; the interatomic distance is specified in units of a_0. The solid line represents the parabolic fit, which is of the same for both $k < k_F$ and $k > k_F$. The discontinuity at $k_F R/\pi = 0.5$ is also marked (STO-3G basis is taken).

states with the increasing R. In Fig.3 we display the statistical distribution function $n(k) =< a^{\dagger}_{\mathbf{k}\sigma} a_{\mathbf{k}\sigma} >$ obtained by using the STO-3G basis to represent the starting atomic wave function, out of which we build the Wannier functions $\{w_i(\mathbf{r})\}_{i=1,\ldots,N}$. These functions are optimized to minimize the ground state energy, as before. We see that for the interatomic distance $R \simeq 3.8 a_0$ the *Fermi* discontinuity Δn_F at the Fermi level vanishes; this is regarded as a quantitative criterion for the electron localization. Further

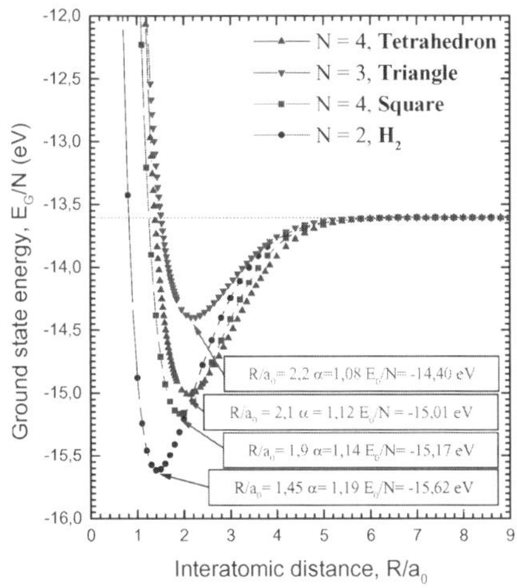

Figure 4. Total ground state energy (per atom, including the ionic contribution) of several cluster systems vs. R/a_0. The optimal values of the parameters for the minimal energy are provided in the insets.

features of this Mott localization crossover are discussed elsewhere (7). To compare the values of the microscopic parameters with those for H_2 and H_2^- ions we have shown in Table IV the interatomic distance dependence of those parameters (for $N = 10$) when using the Slater orbitals also for the linear-chain calculations.The conclusion is that the values of the parameters obtained for the optimized basis are only weakly dependent on the system size.

5. Nanoclusters

We have also applied our method to study various spatial arrangements of clusters containing up to $N \leq 4$ hydrogenic-like atoms and a planar (honeycomb) cluster containing $N = 6$ such atoms. In Fig.4 we display the interatomic distance dependence of the ground-state energy (per atom), together with the values characterizing the minimal point. As before, the size of the atomic-like orbitals ($\alpha^{-1}a_0$) is smaller than that for the hydrogen atom. Hence, the repulsive Coulomb interaction always causes the atomic orbit shrinking. Also, the planar configuration for $N = 4$ is the stable one with respect to tetrahedron configuration. However, it is unstable with respect to the dissociation into two H_2 molecules. Therefore, these systems will be stable only when adsorbed on the surface of other materials.

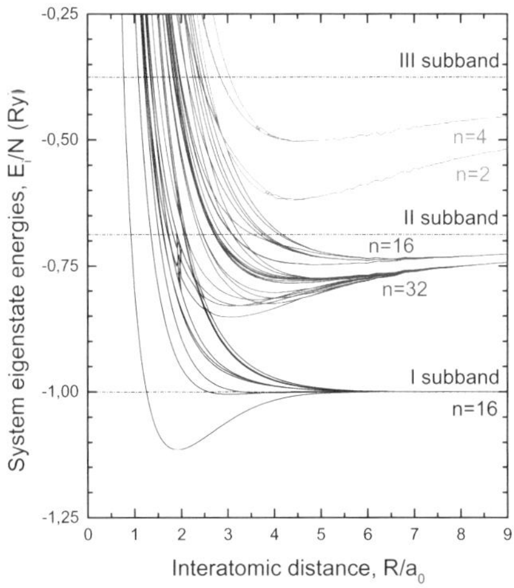

Figure 5. Energies of the excited states for the square configuration of H_4 cluster. The states group into the *Hubbard subbands* with increasing R.

Nevertheless, interesting for us is the nature of the correlated states in given configuration. The microscopic parameters for this case are listed in Table V. They can be compared with those for H_2 systems and the linear chains discussed earlier. A clear consistence of their values is observed. Probably the most interesting result at this stage is the R dependence of the spectrum of excited states displayed in Fig.5. The excitation spectrum for $N = 4$ separates into well defined *Hubbard subbands* located at $U/4$ and $U/2$ in the atomic limit (note that we display the energy per atom). The lowest manifold contains $n = 2^4 = 16$ spin arrangements of roughly singly occupied sites. The second subband reflects $n = 48 = 32 + 16$ states with a single double occupancy, whereas the topmost subband represents $n = 2$ states with two doubly occupied sites across the square diagonal and $n = 4$ states with doubly occupancies located at the nearest neighboring sites. Note a substantional manifold-energy reduction within the topmost subband. The additional splitting within the subband is caused by the difference between n.n and n.n.n. values of intersite Coulomb interaction K_1 and K_2, respectively. The parameters have dependence $K_n \sim R^{-1}$ at large distances and this leads to a slow convergence of the results to the atomic-limit values marked by horizontal lines.

6. Concluding remarks

In this article we have sketched our method of approach (EDABI) to the correlated electronic states that combines the exact diagonalization in the Fock space and a simultaneous *ab initio* readjustment of single-particle wave functions within a selected subspace of their full Hilbert space. The method is particularly useful for small systems, as illustrated on the examples ranging from simple atoms, ions and molecules to nanoscopic chains, rings, and clusters. The method is related to the multiconfiguration-interaction (MCI) method known in quantum chemistry. However, here we put emphasis on questions related to the physics of a systematic evolution from an atomic to a condensed (solid-state) type of state. Additionally, we provide the values of microscopic parameters for the hydrogenic-like systems, as well as set up a self-adjusted wave equation (SWE) for single-particle states in the milieu of correlated particles.

We believe that such an exact analysis within a subspace containing selected orbitals is valuable for a systematic development of reliable methods of approach to nanoscopic systems which represent the systems intermediate between the molecules and the solids.

Acknowledgements

The work was supported by the Committee for Scientific Research (KBN) of Poland, Grant No. 2P03B 050 23. The discussions with Profs. J.E. Hirsch (UCLA) and G. Czycholl (Bremen) at the NATO workshop in Hvar are appreciated.

References

1. J. Spałek, R. Podsiadły, W. Wójcik, and A. Rycerz, Phys. Rev. B **61**, 15676 (2000).
2. A. Rycerz and J. Spałek, Phys. Rev. B**63**, 073101 (2001); *ibid.* B **65**, 035110 (2002).
3. For a brief overview see: J. Spałek et al., Acta Phys. Polonica B **32**, 3189 (2001); *ibid.* B**31**, 2879 (2000).
4. Within the MCI method one determines the $C_{i_1 \dots i_M}$ coefficients variationally, see e.g. R.A. Shavitt, in H. Schaeffer (editor), *Methods of Electronic Structure Theory* (Plenum Press, New York, 1977).
5. See e.g. H.A. Bethe and E.E. Salpeter, *Quantum Mechanics of One- and Two-Electron Atoms* (Academic Press, New York, 1957), pp.154-156.
6. N.F. Mott, *Metal-Insulator Transitions* (Taylor & Francis, London, 1990).
7. J. Spałek and A. Rycerz, Phys. Rev. B **64**, 161105 (2001)(R); cf. also: A. Rycerz, J. Spałek, R. Podsiadły, and W. Wójcik, in *Lectures on the Physics of Highly Correlated Electron Systems VI: Sixth Training Course*, edited by F. Mancini, AIP Conf. Proc. No. 629 (2002).

ON THE MULTICHANNEL-CHANNEL ANDERSON IMPURITY MODEL OF URANIUM COMPOUNDS

Proceedings of the ARW NATO Workshop Hvar, Croatia, October 2002

N. ANDREI and C. J. BOLECH
Center for Materials Theory, Serin Physics Laboratory, Rutgers University, Piscataway, New Jersey 08854-8019

Abstract. In this talk we will present the solution of the two-channel Anderson impurity model, proposed in the context of the heavy fermion compound UBe_{13}, and discuss briefly the more general multi-channel case. We will show results for the thermodynamics in the full range of temperature and fields and make the connection with the current experimental situation.

Key words: Heavy Fermions, Kondo Model, Anderson (Impurity) Model, Multi-Channel Models, Non-Fermi Liquid, Uranium Compounds, UBe_{13}, Bethe Ansatz, Integrability, Integrable Models, Mixed Valence, Crystal Field, Thermodynamic Bethe Ansatz, Impurity Entropy, Impurity Specific Heat

1. Models of Uranium ions

During the last two decades, there has been growing interest in materials whose desciption falls outside the framework of Landau's Fermi Liquid Theory. Examples are the high T_c superconductors, heavy fermions and quasi-1D conductors.

Uranium heavy fermion compounds provide very interesting examples, exhibiting unusual specific heat temperature dependence, $\gamma = C_V/T \sim \ln T$. This behavior is often related to multichannel Kondo physics, occasioned in the case of uranium by the ground state configuration which is degenerate and non-magnetic, so that one is led to consider a quadrupolar Kondo scenario (1, 2).

The starting point is to consider the energies of the uranium ion in its different states of valence. They fall on a parabola depicted in Fig. 1. The ground state corresponds to a U^{4+} ionization state, with a $5f^2$ shell-configuration. Considering the spin-orbit Hund's coupling, the lowest mul-

A.C. Hewson and V. Zlatić (eds.), Concepts in Electron Correlation, 269–276.

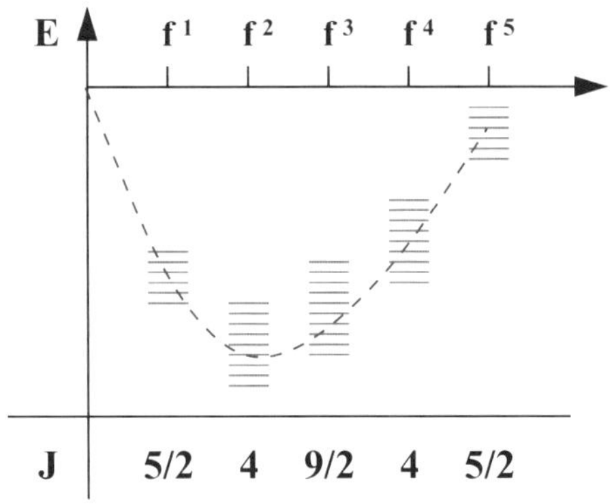

Figure 1. Schematics of the energy and lowest angular momentum multiplet for different states of valence of a uranium ion.

tiplet corresponds to $J = 4$. Taking, further, into account the splitting due to a cubic crystal field (as is the case for the heavy fermion compound UBe_{13}) one finds that the lowest multiplet corresponds to a non-Kramers Γ_3 doublet (see Fig. 2).

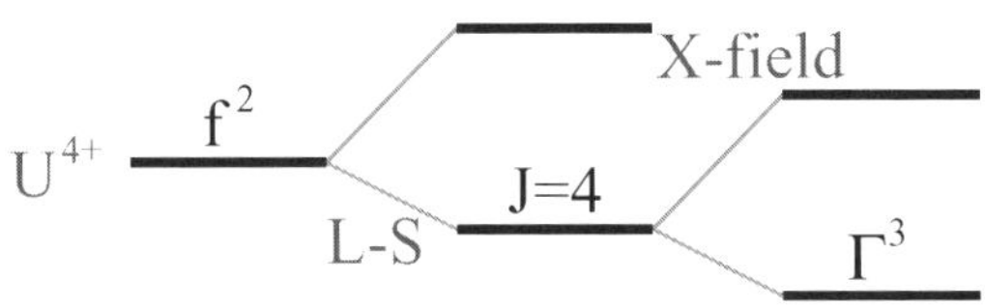

Figure 2. Schematic depiction of the spin-orbit and crystal field splittings for the ground state valence configuration of uranium in a cubic medium.

A similar analysis for the neighboring states of valence indicates that they corresponds to: (**i**) U^{3+}: a $5f^3$ shell-configuration that splits giving a $J = 9/2$ lowest multiplet configuration, that further splits giving a lowest Γ_6 Kramer's doublet and (**ii**) U^{5+}: a $5f^1$ shell-configuration that splits giving a $J = 5/2$ lowest multiplet configuration, that further splits giving a lowest Γ_7 Kramer's doublet. These three valence states are in fact very close in energy and their relative positions are not completely resolved, since the experimental evidence is not conclusive and sometimes contradictory (3, 4). The most accepted scenario is that of a Γ_3 quadrupolar doublet ground state coming from the tetravalent state and next a Γ_6 magnetic doublet coming from the trivalent state of uranium. Neglecting the contributions from the pentavalent state of the uranium as well as those from excited crystalline electric field states we can write down the following Hamiltonian, the *two-channel Anderson impurity model* (see Fig. 3 for a schematic depiction).

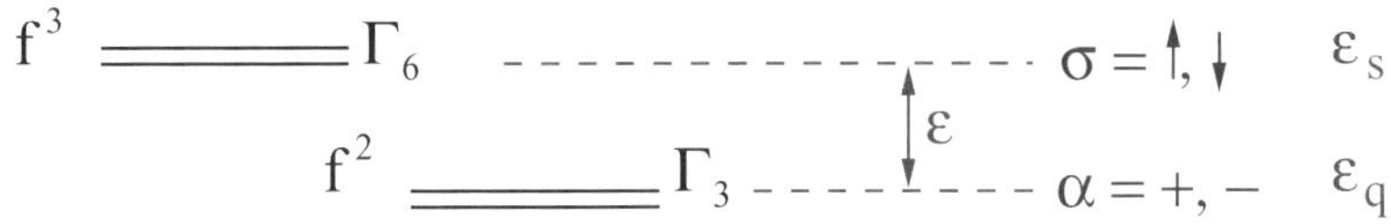

Figure 3. Schematics of the electronic configurations and symmetry multiplets retained for the Hamiltonian of UBe_{13}.

$$
\begin{aligned}
H_{\Gamma_3,\Gamma_6} = H_{\text{host}} &+ \varepsilon_q \sum_{\alpha} b^{\dagger}_{\bar{\alpha}} b_{\bar{\alpha}} + \varepsilon_s \sum_{\sigma} f^{\dagger}_{\sigma} f_{\sigma} \\
&+ V \sum_{k,\alpha,\sigma} \left[f^{\dagger}_{\sigma} b_{\bar{\alpha}} c_{k,\alpha,\sigma} + c^{\dagger}_{k,\alpha,\sigma} b^{\dagger}_{\bar{\alpha}} f_{\sigma} \right] \\
\text{with} \qquad \mathrm{Q} &= \sum_{\alpha} b^{\dagger}_{\bar{\alpha}} b_{\bar{\alpha}} + \sum_{\sigma} f^{\dagger}_{\sigma} f_{\sigma} = 1
\end{aligned}
$$

where the bar on top of the index α indicates that it transforms according to the complex conjugate representation. We used *slave-boson* language and denoted $\varepsilon_q \equiv E_{\Gamma_3}$ and $\varepsilon_s \equiv E_{\Gamma_6}$ (let us also define $\varepsilon \equiv \varepsilon_s - \varepsilon_q$ and $\Delta \equiv V^2/2$). Notice that the model has full $SU(2) \otimes SU(2)$ internal symmetry. In the parameter regimes when $|\varepsilon| \gg \Delta$, we can map it via a Schrieffer-Wolff transformation into a quadrupolar two-channel Kondo model (with a localized quadrupolar moment in the case when $\varepsilon_q \ll \varepsilon_s$), or a magnetic two-channel Kondo model (with a localized magnetic moment in the case when $\varepsilon_q \gg \varepsilon_s$), in both cases with a coupling constant given by $J = V^2/|\varepsilon_s - \varepsilon_q|$ (assuming zero chemical potential).

We will also consider a generalized version of the Hamiltonian, the *multi-channel Anderson impurity model.* In *slave-boson* language the Hamiltonian looks exactly as in the two-channel case, but this time the indices take values according to: (**i**) $\sigma = 1, \ldots, N$ and (**ii**) $\alpha = 1, \ldots, M$. If we follow the standard nomenclature we would call σ the spin index and α the channel index, and the model will be the $SU(N) \otimes SU(M)$ version of the multi-channel Anderson model (5).

As in the two-channel case we can perform Schrieffer-Wolff transformations to derive effective models in any of the two local moment regimes of the problem. When $\varepsilon_q \ll \varepsilon_s$ (assuming always zero electronic chemical potential) we obtain an 'N-channel $SU(M)$ Coqblin-Schrieffer model', whereas when $\varepsilon_s \ll \varepsilon_q$ we obtain an 'M-channel $SU(N)$ Coqblin-Schrieffer model'.

This model has been extensively studied over the past fifteen years by a variety of methods: Large-N methods (6), NCA (4), conserving T-matrix approximation (7), Monte Carlo (4) and Numerical Renormalization Group

(8) among others. All these techniques have difficulties accessing the mixed valence regime.

We showed that this model is integrable and carried out a full analysis by means of the Bethe Ansatz technique. That allowed us to obtain exact results characterizing the ground state of the model as a function of its parameters and very accurate numerical results for the thermodynamics of the model for all temperatures and fields, as well as all valence regimes, including the mixed valence. We stress that is in the mixed valence regime where the Bethe Ansatz solution is most crucial, since all the other techniques have difficulties accessing it.

2. The Thermodynamics of the Model

In this section we present the results of the numerical solution of the Thermodynamic Bethe Ansatz equations for the two-channel case (i.e. $N = M = 2$). This is the first model proposed in the context of UBe_{13}(1). We will study the physics of the model on its own right and also briefly discuss its virtues and inadequacies to describe the physics of the uranium compound that motivated it. The conclusion that we will reach is that one needs to go beyond the simplest scenario given by the two-channel model.

In Fig. 4 we show the behavior of the impurity entropy as a function of temperature for different values of $(\varepsilon - \mu)$, and in Fig. 5 the reader can see the effect of switching on an external field.

At high temperatures the entropy is $S_{\text{imp}} = k_{\text{B}} \ln 4$ in agreement with the size of the impurity Hilbert space. For $|\varepsilon - \mu| \gg \Delta$ the impurity entropy is *quenched* in two stages. The degrees of freedom corresponding to the higher energy doublet are frozen first. The entropy becomes $S_{\text{imp}} = k_{\text{B}} \ln 2$ and the system is in a localized magnetic or quadrupolar moment regime depending on the sign of $(\varepsilon - \mu)$. As the temperature is further decreased, the remaining degrees of freedom undergo frustrated screening leading to entropy $S_{\text{imp}} = k_{\text{B}} \ln \sqrt{2}$. On the other hand, for values of $|\varepsilon - \mu| \ll \Delta$, the quenching process takes place in a single stage. The initial stage of the quenching process happens when the system makes the transition from a high-energy state of valence $n_c = \frac{1}{2}$ (i.e. when both states of valence are equally likely) to the state of valence that will be present at low-energies, $n_c^0(\varepsilon - \mu)$. This is further illustrated in Fig. 6.

As $(\varepsilon - \mu)$ is varied, the behavior interpolates continuously between the magnetic ($n_c^0 = 1$) and the quadrupolar ($n_c^0 = 0$) scenarios. The zero temperature entropy is found to be independent of ε in accordance with our analytic results (9). External fields, either magnetic or quadrupolar, constitute relevant perturbations that drive the system to a Fermi Liquid fixed point with zero entropy.

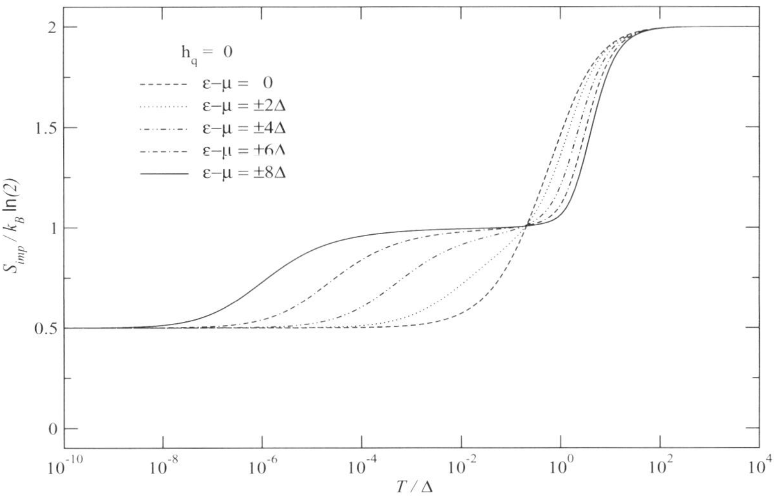

Figure 4. Impurity contribution to the entropy at zero field and as a function of temperature for different values of $\varepsilon - \mu$. As the temperature goes to zero all curves approach the universal value $k_B \ln \sqrt{2}$. Positive and negative values of $\varepsilon - \mu$ fall on top of each other.

In Fig. 7 we show the impurity contribution to the specific heat. The two stage quenching process gives rise to two distinct peaks. The lower temperature peak is the Kondo contribution centered around a temperature $T_l(\varepsilon - \mu)$, whereas the higher temperature peak – often referred to as the Schottky anomaly – is centered around a temperature $T_h(\varepsilon - \mu)$. Approximate expressions for $T_{h,l}$ can be read off from the curves:

$$T_{h,l}(\varepsilon) \approx \frac{\Delta}{a\pi^2} \ln(1 + 2a\, e^{\pm \frac{\pi}{2\Delta}|\varepsilon - \mu|})$$

with $1 < a < 4$. For large $|\varepsilon - \mu|$ the two peaks are clearly separated and the *area*[1] under the Kondo peak is $k_B \ln \sqrt{2}$ while that under the Schottky peak is $k_B \ln 2$.

As mentioned earlier, the model was proposed as a description for the uranium ion physics of UBe_{13}. It is expected to describe the lattice above some coherence temperature. We provide in the inset of Fig. 7 the experimental data for the 5f-derived specific heat of the compound. It is obtained by subtracting from its total specific heat, the specific heat of the isostructural compound $ThBe_{13}$ containing no 5f electrons (10, 11). This way one is subtracting the phonon contribution as well as the electronic

[1] Here by *area* we mean the following integral: $\int C_v(T)\ d(\ln T)$. In other words, the area as plotted, i.e. in a logarithmic scale.

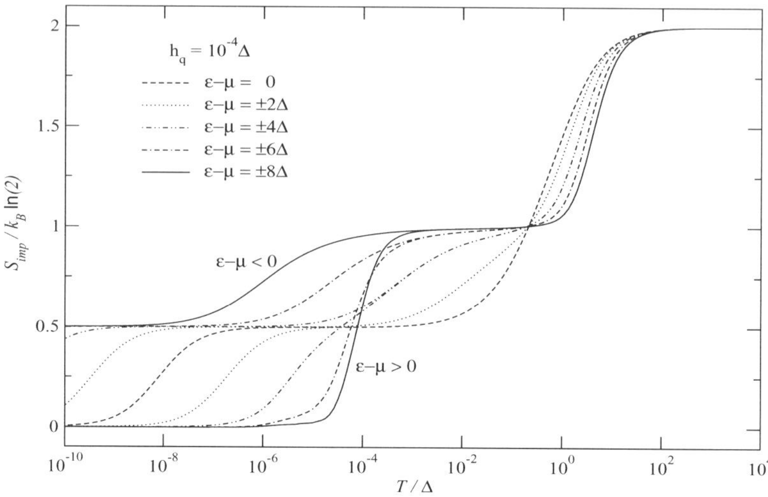

Figure 5. Impurity contribution to the entropy in the presence of an applied quadrupolar field and as a function of temperature for different values of $\varepsilon - \mu$. As the temperature goes to zero all curves approach zero. Positive and negative values of $\varepsilon - \mu$ fall on top of each other only at high temperatures and, asymptotically, for low temperatures.

contribution from electrons in s, p and *d* shells (the procedure is quite involved and we refer the reader to the cited articles for full details). The sharp feature at $\sim 0.8K$ signals the superconducting transition of UBe_{13} and falls outside the range where this compound might be described by considering a single impurity model.

In order to be certain of having eliminated lattice effects it would be better to carry out the measurements on $U_{1-x}Th_xBe_{13}$. For $x > 0.1$ the compound has no longer a superconducting transition and the lattice coherence effects are largely suppressed. Further, there are several experimental indications that support the idea of an impurity model description of the thoriated compound for a wide range of temperatures (12).

Concentrating on the temperature range containing the Kondo and Schottky peaks we conclude that no values of ε and Δ of the 2-channel model yield a good fit. Further, the entropy obtained by integrating the weight under the experimental curve falls between $k_B \ln 4$ and $k_B \ln 6$. This suggests that a full description of the impurity may involve another high energy multiplet (possibly a Γ_4 triplet, cf. with the work of Koga and Cox (8)) close to the Γ_3 to yield the peak for the Schottky anomaly, with the Γ_6 doublet falling outside the range of measurements. The nature of the multiplet could be deduced from further specific heat measurements. For an n-plet degenerate with the Γ_3, one has an $SU(2) \otimes SU(n+2)$ Anderson model and the area under C_V^{imp}/T is then given by $k_B \ln[\frac{n+4}{2} \sec \frac{\pi}{n+4}]$, while

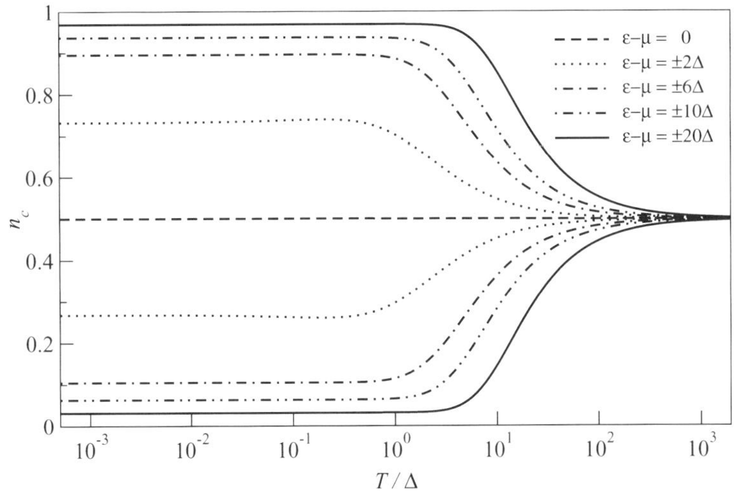

Figure 6. Charge content of the impurity site as a function of temperature for different values of $\varepsilon - \mu$. At high temperatures all curves start from $n_c = 1/2$ and flow, as the temperature is lowered to specific values $n_c^0\,(\varepsilon - \mu)$.

if the n-plet is slightly split off the doublet, the area is given by $k_B \ln[\frac{n+4}{\sqrt{2}}]$ (13). Also the scale of energy involved is then determined by crystal field splitting and corresponds to the scale observed. The split off non-magnetic triplet (i.e. $n = 3$ in the f^2 configuration seems, indeed, to be the most reasonable scenario for UBe_{13}. We are currently working in this direction.

Acknowledgements

We thank the organizers for the invitation to participate in the NATO Advanced Research Workshop on Concepts in Electron Correlation.

References

1. D. L. Cox, Phys. Rev. Lett. **59**, 1240 (1987).
2. A. P. Ramirez *et al.*, Phys. Rev. Lett. **73**, 3018 (1994).
3. F. G. Aliev *et al.*, Europhys. Lett. **32**, 765 (1995).
4. A. Schiller, F. B. Anders and D. L. Cox, Phys. Rev. Lett. **81**, 3235 (1998).
5. D. L. A. Cox and A. Zawadowski, *Exotic Kondo Effects in Metals: Magnetic ions in a crystaline electric field and tunnelling centres* (Taylor & Francis, London UK, 1999).
6. D. L. Cox and A. E. Ruckenstein, Phys. Rev. Lett. **71**, 1613 (1993).
7. J. Kroha and P. Wölfle, Act. Phys. Pol. B **29**, 3781 (1998).
8. M. Koga and D. L. Cox, Phys. Rev. Lett. **82**, 2575 (1999).

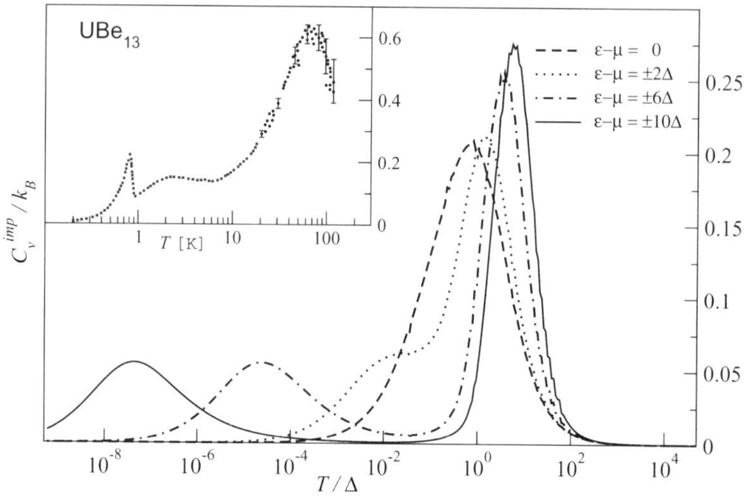

Figure 7. Main plot: impurity contribution to the specific heat as a function of temperature for different values of $\varepsilon - \mu$. As this parameter approaches zero the Kondo contribution (left) and the Schottky anomaly (right) collapse in a single peak. Inset: experimental data for the $5f$-derived specific heat of UBe_{13}.

9. C. J. Bolech and N. Andrei, Phys. Rev. Lett. **88**, 237206 (2002).
10. R. Felten *et al.*, Europhys. Lett. **2**, 323 (1986).
11. R. Felten, G. Weber and H. Rietschel, J. Magn. Magn. Mat. **63 & 64**, 383 (1987).
12. F. G. Aliev, S. Viilar and V. V. Moshchaikov, J. Phys. Cond. Matt. **8**, 9807 (1996).
13. A. Jerez, N. Andrei and G. Zaránd, Phys. Rev. B **58**, 3814 (1998).

ANOMALOUS BEHAVIOR IN RARE-EARTH AND ACTINIDE SYSTEMS.

Proceedings of the ARW NATO Workshop Hvar, Croatia, October 2002

B. COQBLIN
Laboratoire de Physique des Solides, Bat. 510, Universite Paris-Sud, 91405-Orsay, France

Abstract. Many different anomalous behaviours (intermediate valence, heavy fermions, competition between magnetic order and Kondo effect) have been experimentally observed in alloys and compounds containing anomalous rare-earths, Ce, Yb, Eu and also Pr, Sm and Tm, as well as in systems with non-magnetic actinides U, Np and Pu.

This paper reviews theoretical models describing several interesting cases : the transitions accompanied by a valence change such as for example in cerium metal and $Eu(Pd_{1-x}Au_x)_2Si_2$ alloys; the "normal" Kondo effect and the heavy fermion behaviour; the "underscreened" Kondo effect which can be applied to actinide compounds; the occurrence of antiferromagnetic, Kondo, Non Fermi Liquid and spin glass phases in cerium or uranium disordered alloys.

Key words: Kondo Effect, Heavy Fermions, Rare-Earth, Uranium, Kondo Lattice, Underscreened Kondo Effect, Spin Glass

1. Introduction

In the series of rare-earth metals, most of them are magnetic with a magnetic moment corresponding to the $4f^n$ (with integer n values) configuration and a valence of 3 corresponding to three electrons in the conduction band; these metals are called "normal rare-earth metals" (1). There are three exceptions : cerium which has a peculiar phase diagram with two face-centered cubic phases γ and α which have different valences (respectively almost 3 and clearly intermediate between 3 and 4)(2), europium and ytterbium which are divalent at normal pressure. On the other hand, many systems containing rare-earths have been extensively studied and, in addition to cerium, ytterbium and europium, three other rare-earths have been found to have "anomalous" behavior in alloys and compounds, namely Pr, Sm

A.C. Hewson and V. Zlatić (eds.), Concepts in Electron Correlation, 277–286.

and Tm. For example, $PrSn_3$ (3) and TmS (4) present a Kondo effect, while SmS (5) and TmSe (6) present an intermediate valence behavior. Moreover, volume-collapse transitions have been observed at extremely high pressures (7) in praseodymium at 20 GPa (8)and gadolinium at 59 GPa (9); thus, we can imagine that gadolinium changes valence at 600 kbar !

On the other hand, the 5f electrons are less localized at the beginning of the actinide series than the 4f ones in rare-earths and it results that uranium, neptunium and plutonium metals are not magnetic, while a strong magnetism occurs for curium and berkelium. Thus, a strong heavy fermion character or even magnetism can occur in compounds or alloys with uranium or also in a few known cases with neptunium and plutonium.

It is well known that different anomalous behaviors have been observed and theoretically studied in mainly Ce, Yb, Eu, U but also in other anomalous rare-earth systems (10, 11). In the present paper, we will review briefly the intermediate valence or the Kondo behavior associated with the heavy fermion character and we will present some recent features of these two topics, with a special emphasis on the underscreened Kondo effect and the effect of disorder in ternary cerium or uranium compounds.

2. Intermediate Valence transitions

The first extensively studied transition accompanied by a valence change is the γ-α transition of cerium metal (2). The phase diagram of cerium metal has two face-centered cubic phases α and γ and, if one extrapolates the valence from the atomic radius, one obtains a valence very close to 3 for $\gamma - Ce$ and a valence of roughly 3.6 for $\alpha - Ce$. The γ-α transition is first-order at low pressures and temperatures; but the discontinuity in the different physical properties decreases along the transition and the transition becomes a second-order one above a critical point located at roughly 20 kbar and 600 K. At normal pressure and low temperatures, cerium undergoes a transition to the double hexagonal close-packed β phase which orders antiferromagnetically below 13 K and finally cerium becomes superconducting at very low temperatures above 20 kbar (10).

A similar transition has been observed in two systems containing europium : $Eu(Pd_{1-x}Au_x)_2Si_2$ (12) and more recently $Eu(Ge_{1-x}Si_x)_2Ni_2$ (13). For example, in the first case, when gold concentration decreases, there is firstly an antiferromagnetic order and then a first-order transition between an almost divalent Eu to an almost trivalent Eu, which ends up also at a critical point.

We can also cite the case of the compound $YbInCu_4$, which undergoes at normal pressure an isostructural valence transition from a high temperature magnetic phase to a low temperature non magnetic one; ytterbium is trivalent in the magnetic phase, but the valence change to the non magnetic one is very small, of order 0.1 (14). This transition has been studied extensively under pressure (15) or in related materials (14, 16).

These isostructural valence transitions have been extensively studied theoretically. The first theoretical explanation starts from the orbitally degenerate Anderson Hamiltonian and accounts for the magnetic-nonmagnetic transition accompanied by a large valence change and the existence of the critical point (17). Then, the Ramirez-Falicov model starts from an ionic description of the 4f electrons and can also account for the phase diagram of cerium (18). The main point of these two first theoretical models is the explanation of a large valence change. Another theoretical explanation of the γ-α transition of cerium was based on the Kondo effect and assumes a large change of the Kondo temperature when going from the non magnetic phase to the magnetic one (19, 20). More recently, many calculations involving detailed band calculations and based in particular on the dynamical mean-field theory (DMFT) were performed to account for the phase diagram of cerium and to give in particular a better description of the photoemission experiments (21, 22, 23). Several theoretical models, including DMFT calculations within the Falicov-Kimball model (24), have been developed to account for the experimental phase diagrams of Eu or Yb systems such as for example $YbInCu_4$ compound; but in addition, there often exists an antiferromagnetic phase in these phase diagrams and at present there is not a full theoretical model for explaining such a complete phase diagram.

In actinide metals at the beginning of the series, the 5f electrons are less localized than the equivalent 4f electrons in rare-earth metals and the situation is clearly different : there is a strong d-f hybridization and it results that U, Np and Pu metals are not magnetic, while these actinides can be magnetic in alloys or compounds where this hybridization is smaller. Then the 5f electrons become more localized in the middle of the series and a strong magnetism occurs for curium and berkelium. The phase diagrams of the first actinides are complicated and in particular plutonium has five phases at normal pressure (25, 26). The curve of the atomic volumes lies between the corresponding one for 5d transition metals up to Pu and that of lanthanides starting at Am, which is a good indication of the rapid increase of the 5f localization between Pu and Am. On the other hand, there is a huge volume expansion ($26percent$) between the low temperature α phase of plutonium and the high temperature δ phase, which is of the same order as the volume expansion occurring at the γ-α transition of cerium. The volume expansion experiments, as well as the analysis of transport and magnetic properties suggest a strong localization of 5f electrons and a Kondo effect in $\delta - Pu$ (27, 28). On the other hand, plutonium is close to be magnetic at low temperatures and the spin fluctuation model has been used a long time ago in the 70's to account fairly well for the resistivity which is large and passing through a maximum as a function of temperature (29) and for the thermoelectric power (30) of $\alpha - Pu$.

3. The Kondo effect

The long story of the Kondo effect starts 40 years ago with the third-order calculation of J. Kondo (31) who explains theoretically the resistivity minimum observed in dilute alloys such as Cu-based alloys with Fe or Mn impurities and lanthanum-based alloys with Ce impurities. The properties of many cerium, ytterbium or also uranium compounds or alloys are well understood in the theoretical framework of the Kondo effect and we will describe here some features of this difficult problem. Cerium and ytterbium systems are well described by the Kondo effect, when the number of 4f-electrons is close to 1 in the case of cerium or to 13 (or 1 hole) in the case of ytterbium. First of all, the single-impurity Kondo effect has been exactly solved(32): at low temperatures, the system has a "Fermi Liquid" behavior with a T^2 law for the electrical resistivity and very large values of both the electronic constant of the specific heat and the magnetic susceptibility, which were at the origin of the name "heavy fermions" given to these systems (33, 34, 35). At high temperatures with respect to the Kondo temperature, the magnetic contribution to the electrical resistivity passes generally through a maximum corresponding to the overall crystal field splitting and decreases then as Log T (36). The other transport and magnetic properties have been extensively studied in the low temperature and high temperature regimes.

The second extensively studied problem is the Kondo-lattice one. In the case of a lattice, there exists a strong competition between the Kondo effect and magnetic ordering, arising from the RKKY (Ruderman-Kittel-Kasuya-Yosida) interaction between rare-earth atoms at different lattice sites. This situation is well described by the Doniach diagram(37), which gives the variation of the Néel temperature and of the Kondo temperature with increasing antiferromagnetic intrasite exchange interaction $J_K (> 0)$ between localized spins and conduction-electron spins. Usual theories of the one-impurity Kondo effect and of the RKKY interaction yield a Kondo temperature T_{Ko}, that is proportional to $\exp(-1/\rho J_K)$, and an ordering temperature (Néel or, in some cases, Curie), T_{No}, proportional to ρJ_K^2, ρ being the density of states for the conduction band at the Fermi energy. Thus, for small ρJ_K values, T_{No} is larger than T_{Ko} and the system tends to order magnetically, with often a reduction of the magnetic moment due to the Kondo effect; on the contrary, for large ρJ_K, T_{Ko} is larger than T_{No} and the system tends to become non magnetic. The actual ordering temperature T_N, therefore, increases initially with increasing ρJ_K, then passes through a maximum and tends to zero at a critical value ρJ_K^c corresponding to a "quantum critical point" (QCP) in the Doniach diagram. Such a behavior of T_N has been experimentally observed with increasing pressure in many Kondo compounds, such as for example in $CeAl_2$ (Ref. (38)) or in $CeRh_2Si_2$.(39) Thus, we can conclude that the variation of the Néel temperature predicted by the Doniach diagram is well observed experimentally in many cerium compounds. We also know that the Néel temperature starts from zero at a given pressure, and increases rapidly with pressure in $YbCu_2Si_2$ (Ref. (40)) or in related ytterbium compounds, in agreement with the Doniach diagram.

The one-impurity model predicts an exponential increase of the Kondo temperature with ρJ_K. This means that the Kondo temperature should increase with increasing pressure in cerium compounds and with decreasing pressure in ytterbium compounds, which agrees well with many observations. However, deviations seem to occur in some cerium compounds, such as $CeRh_2Si_2$ (Ref. (39)), $CeRu_2Ge_2$(41, 42), or $Ce_2Rh_3Ge_5$ (43), where the actual Kondo temperature observed in a lattice can be significantly different from the one derived for the single-impurity case. Thus, in order to account for such an effect, we have studied in detail the Kondo-lattice model within a mean-field approximation and with both intrasite Kondo exchange and intersite antiferromagnetic exchange, treating successively the half-filled case (corresponding to a number of conduction electrons $n = 1$) (44) and then the general case $n < 1$ (45, 46, 47, 48) which gives a much better description of the metallic cerium systems. We have studied in detail this problem in Ref.(48) and we have shown that the Kondo temperature tends to decrease with an increasing intersite exchange interaction and a decreasing number of conduction electrons corresponding to the so-called "exhaustion" limit (49). This calculation addresses several questions : first, it would be very interesting to understand better the conditions yielding a Kondo temperature for the lattice much different than the single-impurity one and further experiments are certainly needed. Second, the discussion on the Kondo-lattice problem addresses again the difficult issue of the nature of the ground state and screening in the Kondo-lattice problem. We have shown here that, as the number of conduction electrons is reduced, exhaustion may be compensated by formation of intersite singlets of localized spins and exact calculations for small clusters would be interesting (50, 51). The third interesting question concerns the derivation of a correlation temperature below which short-range magnetic correlations appear, in good agreement with neutron scattering experiments in cerium compounds. Finally, it is interesting to notice that taking into account lattice effects seems to be important for describing the properties of cerium or other anomalous rare-earth compounds at low temperatures, as shown, for example, in photoemission experiments (52, 53). Thus, further theoretical and experimental developments are certainly necessary to better understand the Kondo-lattice problem.

Then, up to now, we have described the Kondo compounds within the "Fermi liquid" picture. But, it is now well established that an anomalous effect, called "Non Fermi Liquid" (NFL), occurs very often at or around the QCP in many cerium compounds. This NFL behavior is characterized by laws for the physical properties different from those derived within the classical Fermi Liquid case : for example, the electronic specific heat constant is no more constant with temperature and the electrical resistivity is no more behaving as T^2. Moreover, there are clearly different NFL behaviors according to the different compounds studied. We can distinguish at least three possible origins for these NFL behaviors : the first one is the proximity to the QCP; for example, some cerium compounds are antiferromagnetic, have then an NFL just above the QCP before recovering a Fermi Liquid behavior, when the pressure (for example in Ce_7Ni_3 (54)) or the relative concentration is varied (for example in the well studied case of $Ce(Cu_{1-x}Au_x)_6$ alloys (55)). Then, the disorder can produce an NFL behavior especially in ternary alloys (56) and finally the multichannel Kondo effect can also be at the origin of another NFL behavior. An extensive review discusses the present status of knowledge of NFL, including a compilation of data in over 50 systems containing mainly cerium or uranium citeStewart.

Moreover, the Kondo effect is simply defined in the case of cerium or Ytterbium systems, where there is one 4f electron or hole screened by one conduction electron. But, in the case of uranium systems, the situation is even more complicate, because there are several f-electrons, corresponding to the so-called "multi-channel" Kondo effect. If we call S the f-electron spin and n the number of channels for conduction electrons, the normal Kondo effect corresponds to $n = 2S$, while the "underscreened" Kondo effect corresponds to $n < 2S$ (where there are too many f-electrons with respect to the conduction electrons) and the "overscreened" case to $n > 2S$ (where there are not enough f-electrons). The multichannel model for a single impurity has been completely solved analytically (58). The multichannel is more appropriate for uranium systems where clearly there is a number of 5f electrons larger than 1. In fact, in uranium compounds such as UPt_3 (59), UNi_2Al_3 or UPd_2Al_3 (60), there exist both a large heavy-fermion character and a magnetic ordering and the underscreened model can account for this "coexistence" between the Kondo effect and magnetic order, in contrast to the "competition" observed in cerium systems(61). A remarkable example of underscreened Kondo effect is the case of uranium monochacogenides, as evidenced by Schoenes (62): for example, the compound UTe is ferromagnetic with a large Curie temperature of 102 K and has a Kondo behavior characterized by a LogT decrease of the resistivity at higher temperatures. Thus, there is clearly here coexistence between the Kondo effect and a ferromagnetic ordering, which can be accounted for by the underscreened Kondo effect.

Then, an unusual interplay between Kondo effect and spin-glass behavior has been recently observed in several other disordered cerium alloys such as $CeNi_{1-x}Cu_x$ (63, 64), $CeCoGe_{3-x}Si_x$,(65) $Ce_2Au_{1-x}Co_xSi_3$ (66) or $Ce_2Pd_{1-x}Co_xSi_3$ alloys (67). In particular, a spin glass-Kondo transition occurs in $CeNi_{1-x}Cu_x$ for $x = 0.2$ and for $x > 0.2$ there exists a ferromagnetic phase at low temperatures and a spin glass phase at higher temperatures. In the case of $Ce_2Au_{1-x}Co_xSi_3$ alloys(66), the sequence of SG-AF-Kondo phases is obtained with increasing x at low temperatures. NFL and spin glass phases have been also observed in disordered uranium alloys (68), as for example a sequence of AF-NFL-SG phases with increasing x in $UCu_{5-x}Pd_x$ alloys (69) or the opposite sequence AF-SG-NFL in $U_{1-x}La_xPd_2Al_3$ alloys (70). We have studied theoretically firstly the Spin Glass-Kondo transition (71) by considering a Kondo lattice with the classical intrasite Kondo exchange interaction and a randomly distributed Ising intersite exchange interaction and we have obtained theoretically a SG-Kondo transition as a function of the Kondo exchange parameter J_K, in agreement with the experiment in $CeNi_{1-x}Cu_x$ alloys. Then, we have taken a random Ising intersite exchange interaction with a non zero mean value in order to describe a ferromagnetic phase and we have obtained a Ferromagnetic-Spin Glass-Kondo phase sequence (72), in better agreement with some experimental results. However, we have found a first-order transition without any QCP between the SG and Kondo phases and then we have not obtained any "mixed" SG-Kondo phase. The existence of a QCP in the phase diagram can be solved by considering an additional transverse field which mimics the spin-flipping in the spin glass exchange interaction(73), but the unsolved problem of the absence of a "mixed" SG-Kondo phase is certainly connected to the mean-field approximation used to describe the Kondo intersite exchange interaction.

Finally, a very fascinating subject concerns the occurrence of superconductivity in some cerium or uranium compounds at low temperatures. The first evidence of superconductivity in cerium compounds has been obtained in $CeCu_2Si_2$ at normal pressure below 0.6 K by Steglich et al (74). At present, there are relatively many cerium compounds which become superconducting generally at high pressure close to the QCP(75, 76). There are also several uranium compounds such as UPt_3 (59) or UPd_2Al_3 (60), which become superconducting at low temperatures generally at normal pressure;

these uranium compounds present, therefore, the particularity of having an heavy-fermion character and becoming magnetically ordered and superconducting at low temperatures. We can finally cite the case of the newly studied compound UGe_2 which is ferromagnetic at normal pressure and becomes superconducting at high pressure just before the QCP (76). This competition between superconductivity, the heavy-fermion character and a magnetic order is clearly a new challenge which is not solved theoretically at present.

4. Conclusion

Thus, we have reviewed some features of the anomalous behaviors found in many rare-earth and actinide systems. In the case of intermediate valence systems, it is interesting to note that the isostructural phase transition with a valence change observed in cerium metal has been also found in Eu or Yb alloys or compounds. On the other hand, we have reviewed several features of the Kondo effect which are extensively studied at present, namely the Kondo-lattice problem, the multichannel Kondo effect, the spin glass-Kondo competition and the superconductivity in cerium or uranium compounds.

References

1. B. Coqblin, *The electronic structure of rare-earth metals and alloys : the magnetic heavy rare-earths*, Academic Press, (1977).
2. A. Jayaraman, Phys. Rev., **137**, A179 (1965).
3. P. Lethuillier and P. Haen, Phys. Rev. Lett., **35**, 1391 (1975).
4. F. Lapierre, P. Haen, B. Coqblin, M. Ribault and F. Holtzberg, Proceedings of LT 16, Los Angeles (Aug. 1981).
5. M.B. Maple and D. Wohlleben, Phys. Rev. Lett., **27**, 511 (1971).
6. M. Ribault, J. Flouquet, P. Haen, F. Lapierre, J.M. Mignot and F. Holtzberg, Phys. Rev. Lett., **45**, 1295 (1980).
7. A.K. McMahan, C. Huscroft, R.T. Scalettar and E.L. Pollock, Journal of Computer-Aided Materials Design, **5**, 131 (1998).
8. G.S. Smith and J. Akella, J. Appl. Phys. **53**, 9212 (1982).
9. H. Hua, Y.K. Vohra, J. Akella, S. Weir, R. Ahuja and B. Johansson, Rev. High Press. Sci. Technol., **4**, 233 (1998).
10. B. Coqblin, in *Magnetism of metals and alloys*, ed. by M. Cyrot, (North-Holland, 1982).
11. A. C. Hewson, *The Kondo problem to Heavy Fermions*, (Cambridge University Press, Cambridge, 1992).
12. C.U. Segre, M. Croft, J.A. Hodges, V. Murgai, L.C. Gupta and R.D. Parks, Phys. Rev. Lett., **49**, 1947 (1982).
13. H. Wada, M.F. Hundley, R. Movshovich and J.D. Thompson, Phys. Rev. B, **59**, 1141 (1999).
14. J.L. Sarrao, Physica B, **259-261**, 128 (1999) and references therein.

15. A. Uchida, M. Kosaka, N. Mori, T. Matsumoto, Y. Uwatoko,J.L. Sarrao and J.D. Thompson, Physica B, **312-313**, 339 (2002)
16. M. Ocko and J.L. Sarrao, Physica B, **312-313**, 341 (2002).
17. B. Coqblin and A. Blandin, Adv. in Phys., **17**, 281 (1968).
18. R. Ramirez and L.M. Falicov, Phys. Rev. B, **3**, 2425 (1971).
19. M. Lavagna, C. Lacroix and M. Cyrot, J. Phys.F, **13**, 1007 (1983).
20. J.W. Allen and R.M. Martin, Phys. Rev. Lett., **49**, 1106 (1982).
21. A. Svane, Phys. Rev. B, **53**, 4275 (1996).
22. M.B. Zolfl, I.A. Nekrasov, Th. Pruschke, V.I. Anisimov and J. Keller, Phys. Rev. Lett., **87**, 276403 (2001).
23. K. Held, A.K. McMahan and R.T. Scalettar, Phys. Rev. Lett., **87**, 276404 (2001).
24. V. Zlatic and J.K. Freericks, Acta Phys. Pol. B, **32**, 3253 (2001).
25. D.R. Stephens, J. Phys. Chem. Solids, **24**, 1197 (1963).
26. *Challenges in Plutonium Science*, Los Alamos Science, **26** (2000).
27. S. Meot-Reymond and J.M. Fournier, J. of Alloys and Compounds, **232**, 119 (1996).
28. S.Y. Savrasov, G. Kotliar and E. Abrahams, Nature, **410**, 793 (2001).
29. R. Jullien, M.T. Beal-Monod and B. Coqblin, Phys. Rev. B, **9**, 1441 (1974).
30. J.R. Iglesias-Sicardi, R. Jullien and B. Coqblin, Phys. Rev. B, **17**, 2366 (1978).
31. J. Kondo, Progr. Theoret. Phys. (Kyoto), **32**, 37 (1964).
32. K.G. Wilson, Rev. Mod. Phys., **47**, 773 (1975).
33. P. Nozières, J. Low Temp. Phys., **17**, 31 (1974).
34. K. Andres, J.E. Graebner and H.R. Ott, Phys. Rev. Lett., **35**, 1779 (1975).
35. B. Coqblin, J. Arispe, A.K. Bhattacharjee and S.M.M. Evans, *Selected Topics in Magnetism*, Frontiers in Solid State Sciences, vol. 2, ed. by L.C. Gupta and M.S. Murani, (World Scientific, Singapore 1993), p.75.
36. B. Cornut and B. Coqblin, Phys. Rev. B, **5**, 4541 (1972).
37. S. Doniach, *Proceedings of the Int. Conf. on Valence Instabilities and Related Narrow-band Phenomena*, ed. by R. D. Parks, Plenum Press, p.168 (1976).
38. B. Barbara, H. Bartholin, D. Florence, M.F. Rossignol and E. Walker, Physica B, **86-88**, 177 (1977).
39. T. Graf, J. D. Thompson, M. F. Hundley, R. Movshovich, Z. Fisk, D. Mandrus, R. A. Fischer and N. E. Phillips, Phys. Rev. Lett., **78**, 3769 (1997).
40. K. Alami-Yadri, H. Wilhelm and D. Jaccard, Physica B, **259-261**, 157 (1999).
41. S. Sullow, M.C. Aronson, B.D. Rainford and P. Haen, Phys. Rev. Lett., **82**, 2963 (1999).
42. H. Wilhelm, K. Alami-Yadri, B. Revaz and D. Jaccard, Phys. Rev. B, **59**, 3651 (1999).
43. K. Umeo, T. Takabatake, T. Suzuki, S. Hane, H. Mitamura and T. Goto, Phys. Rev. B, **64**, 144412 (2001).
44. J. R. Iglesias, C. Lacroix and B. Coqblin, Phys. Rev. B, **56**, 11820 (1997).
45. A. R. Ruppenthal, J. R. Iglesias and M. A. Gusmão, Phys. Rev. B, **60**, 7321 (1999).
46. B. Coqblin, M. A. Gusmão, J. R. Iglesias, C. Lacroix, A. Ruppenthal and Acirete S. Da R. Simoes, Physica B, **281-282**, 50 (2000).
47. C. Lacroix, J. R. Iglesias and B. Coqblin, Physica B, **312-313**, 159 (2002).
48. B. Coqblin, C. Lacroix, M. A. Gusmão, J. R. Iglesias, to be published.
49. P. Nozières, Eur. Phys. J. B, **6**, 447 (1998).
50. H. Tsunetsugu, M. Sigrist, and K. Ueda, Rev. Mod. Phys., **69**, 809 (1997).
51. C. Busser, E. V. Anda, J. R. Iglesias, and B. Coqblin, J. Magn. Magn. Mater., **226-230**, 134 (2001).
52. A. N. Tahvildar, M. Jarrell and J. K. Freericks, Phys. Rev. B, **55**, R3332 (1997).

53. A.J. Arko, J.J. Joyce, A.B. Andrews, J.D. Thompson, J.L. Smith, D. Mandrus, M.F. Hundley, A.L. Cornelius, E. Moshopoulou, Z. Fisk, P.C. Canfield and Alois Menovsky, Phys. Rev.B, **56**, R7041 (1997).
54. K. Umeo, T. Takabatake, H. Ohmoto, T. Pietrus, H. von Lohneyssen, K. Koyama, S. Hane and T. Goto, Phys. Rev. B, **58**, 12095 (1998).
55. H. von Lohneyssen, J. Mag. Mag. Mater.,**200**, 532 (1999).
56. E. Miranda, V. Dobrosavljevic and G. Kotliar, Phys. Rev. Lett., **78**, 290 (1997).
57. G.R. Stewart, Rev. Modern Phys., **73**, 797 (2001).
58. P. Schlottman and P. D. Sacramento, Adv. in Phys., **42**, 641 (1993).
59. G.R. Stewart, Z. Fisk, J.O. Willis and J.L. Smith, Phys. Rev. Lett., **52**, 679 (1984).
60. C. Geibel et al., Z. Phys. B, **83**, 305, 1991 and **84**, 1 (1991).
61. Karyn Le Hur and B. Coqblin, Phys. Rev. B, **56**, 668 (1997).
62. J. Schoenes, private communication and J. Schoenes, Journal Less-Common Met., **121**, 87 (1986).
63. J. C. Gómez Sal, J. Garcia Soldevilla, J.A. Blanco, J.I. Espeso, J. Rodriguez Fernandez, F. Luis, F. Bartolomé and J. Bartolomé, Phys. Rev. B, **56**, 11741 (1997).
64. J. Garcia Soldevilla, J.C. Gomez Sal, J.A. Blanco, J.I. Espeso and J. Rodriguez Fernandez, Phys. Rev. B, **61**, 6821 (2000).
65. D. Eom, M. Ishikawa, J. Kitagawa and N. Takeda, J. Phys. Soc. Japan, **67**, 2495 (1998).
66. S. Majumdar, E.V. Sampathkumaran, St. Berger, M. Della Mea, H. Michor, E. Bauer, M. Brando, J. Hemberger and A. Loidl, Solid State Comm., **121**, 665 (2002).
67. E.V. Sampathkumaran, to be published in the proceedings of SCES02 Conference, Acta Physica Polonica.
68. M.B. Maple, M.C. de Andrade, J. Herrmann, Y. Dalichaouch, D.A. Gajewski, C.L. Seaman, R. Chau, R. Movshovich, M.C. Aronson and R. Osborn, J. Low Temp. Phys., **99**, 223 (1995).
69. R. Chau and M.B. Maple, J. Phys. C: Condens. Matter, **8**, 9939 (1996).
70. V.S. Zapf, R.P. Dickey, E.J. Freeman, C. Sirvent and M.B. Maple, Phys. Rev. B, **65**, 024437 (2002).
71. Alba Theumann, B. Coqblin, S.G. Magalhaes and A.A. Schmidt, Phys. Rev. B, **63**, 054409 (2001).
72. S.G. Magalhaes, A.A. Schmidt, Alba Theumann and B. Coqblin, to appear in Eur. Phys. J. B.
73. Alba Theumann and B. Coqblin, to be published.
74. F. Steglich, J. Aarts, C. D. Bredl, W. Lieke, D. Meschede and W. Franz , Phys. Rev. Lett., **43**, 1892 (1979).
75. F. Steglich, J. Mag. Mag. Mater., **226-230**, 1 (2001).
76. S.S. Saxena, P. Agarwal, K. Ahilan, F.M. Grosche, R.K.W. Haselwimmer, M.J. Steiner, E. Pugh, I.R. Walker, S.R. Julian, P. Monthoux, G.G. Lonzarich, A. Huxley, I. Sheikin, D. Braithwaite and J. Flouquet, Nature, **406**, 587 (2000).

DESCRIBING THE VALENCE-CHANGE TRANSITION BY THE DMFT SOLUTION OF THE FALICOV-KIMBALL MODEL

Proceedings of the ARW NATO Workshop Hvar, Croatia, October 2002

V. ZLATIĆ
Institute of Physics, Bijenička c. 46, 10 001 Zagreb, Croatia
J. K. FREERICKS
Department of Physics, Georgetown University, Washington, DC 20057, U.S.A.

Abstract. The anomalous properties of the $YbInCu_4$ family of intermetallic compounds are discussed and the experimental data are compared with the dynamical mean field theory (DMFT) of the Falicov-Kimball model.

Key words: Valence-Change Transition, Falicov-Kimball Model, Dynamical Mean-Field Theory

1. Introduction to Valence-Change Materials

Yb-based intermetallic compounds have an interesting isostructural valence-change transition (1, 2, 3, 4, 5, 6). Here, we briefly describe the most characteristic features of these systems, taking $Yb_{1-x}Y_xInCu_4$ as an example, and show that the Falicov-Kimball model provides a qualitative description of the experimental data.

At zero doping and ambient pressure the properties of $YbInCu_4$ are dominated by a first-order valence-change transition at about 40 K. The valence of the Yb ions changes from Yb^{3+} above T_v to $Yb^{2.85+}$ below T_v (1, 7), where T_v is the transition temperature, and the lattice expands by about 5 %. The crystal structure remains in the C15(b) class and the volume expansion estimated from the atomic radii of the Yb^{3+} and Yb^{2+} ions is compatible with the valence change estimated from L_{III}-edge data (1, 8). Specific heat data shows a first-order transition at T_v with an entropy change $\Delta S \simeq R \ln 8$ corresponding to a complete loss of magnetic degeneracy in the ground state (4). Neutron scattering does not provide any

A.C. Hewson and V. Zlatić (eds.), Concepts in Electron Correlation, 287–296.

evidence for long range order below T_v (9). Y-doping reduces T_v until a critical concentration of 15 % of Y ions is reached, where the high-temperature phase extends down to T=0 K (5, 6, 10). The experimental results for the resistivity, susceptibility and thermopower (6) of $Yb_{1-x}Y_xInCu_4$ are shown in Fig. 1.

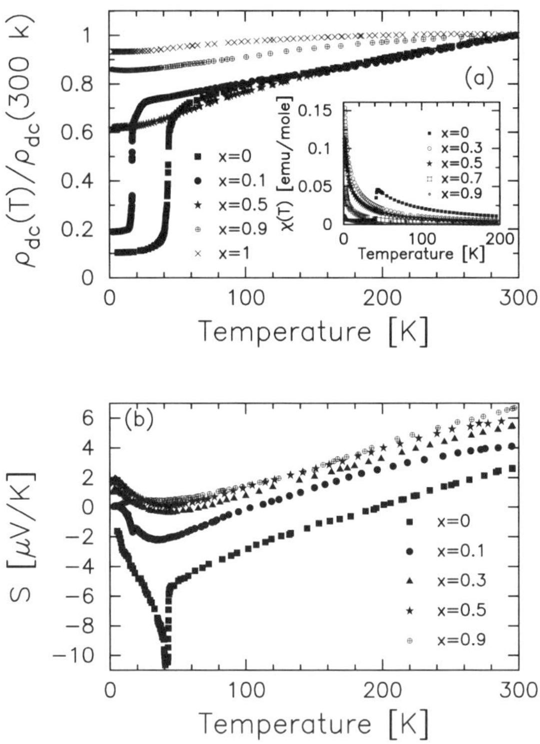

Figure 1. Panel (a) shows the resistivity and the magnetic susceptibility of $Yb_{1-x}Y_xInCu_4$ as function of temperature for various concentrations of Y ions (6). Note, all the "high-temperature" data can be collapsed onto a single universal curve, by normalizing the susceptibility with respect to an effective Yb-concentration (not shown). Panel (b) shows the thermopower of $Yb_{1-x}Y_xInCu_4$ as a function of temperature for various concentrations of Y ions (6).

The low-temperature phase ($T \leq T_v$) shows anomalies typical of an intermetallic compound with a fluctuating valence. The electronic specific heat and the susceptibility are enhanced (4); the ESR data (11) also indicates a large density of states at the Fermi level E_F. The electrical resistance and the Hall constant are small and metallic, and the low-temperature slope of the thermoelectric power is large (6). The optical conductivity is Drude like, with an additional structure in the mid-infrared range which appears quite suddenly at T_v (3), indicating a mixing of the f-states with the conduction band. Neither the susceptibility, nor the resistivity (6), nor the Hall constant (2) show any temperature dependence below T_v, i.e., the system behaves as a fermi liquid with a characteristic energy scale $T_{FL} \gg T_v$. The magnetic moment of the rare earth ions is quenched in the ground state by the f-d hybridization but the onset of the high-entropy phase cannot be explained by the usual Anderson model in which the low- and high-temperature scales are the same and the spin degeneracy is not expected to be recovered below T_{FL}. In these valence-change systems, however, the f-moment is recovered for $T_v \ll T_{FL}$.

The high-T phase that sets in at T_v is also anomalous. In doped systems, the susceptibility data above T_v can be represented by a single universal curve, provided one scales the data by an effective concentration of magnetic f-ions, which is smaller than the nominal concentration of f-ions. The functional form of the magnetic response agrees well with the "single-ion" crystal field (CF) theory for all values of the field. The Yb ions seem to be in the stable 3+ configuration with one f-hole and with the magnetic moment close to the free ion value $g_L\sqrt{J(J+1)}\mu_B = 4.53\mu_B$ ($g_L = 8/7$ is the Landé factor and $J = 7/2$ is the angular momentum of the $4f^{13}$ hole). The dynamical susceptibility obtained from neutron scattering data (12) is typical of isolated local moments, with well resolved CF excitations (13).

The resistivity of $Yb_{1-x}Y_xInCu_4$ alloys exhibits a weak maximum and the thermopower has a minimum above 100 K but neither quantity shows much structure at low temperatures, where the susceptibility drops below the single-ion CF values. The discontinuity of the thermoelectric power at the valence transition is a trivial consequence of the different thermoelectric properties of the two phases: the thermopower of the valence-fluctuating phase has an enhanced slope and grows rapidly up to T_v, where it suddenly drops to values characteristic of the high-temperature phase. The resistivity is not changed much by a magnetic field up to 30 T (14). In typical Kondo systems, on the other hand, one expects a logarithmic behavior on the scale T/T_K and a large negative magnetoresistance. Here, despite the presence of the well defined local moments, there are no Kondo-like anomalies. The Hall constant of the $x = 0$ compound is large and negative in the high-temperature phase, typical of a semi-metal (2); the optical conductivity (3) shows a pronounced maximum of the optical spectral weight at a charge-transfer peak near 1 eV and a strongly suppressed Drude peak.

The hydrostatic pressure and the magnetic field give rise, like the temperature and the doping, to strong and often surprising effects. The critical temperature decreases with pressure (14) but the data cannot be explained with the Kondo volume collapse model (4). We mention also that doping the Yb sites with Lu^{3+} ions (5) reduces T_v despite the fact that Lu has a smaller ionic radius than Yb or Y; doping the In sites by smaller Ag ions enhances T_v in $\mathrm{YbIn}_{1-x}\mathrm{Ag}_x\mathrm{Cu}_4$ (8, 15) without changing appreciably the lattice parameter. Thus, the effects of doping cannot be explained in terms of a chemical pressure. An external magnetic field of a critical strength $H_c(T)$ destabilizes the low-temperature phase and induces a metamagnetic transition which can be seen in the magnetoresistance and the magnetization data (4). The experimental values of $H_c(T=0)$ and $T_v(H=0)$ are of the same order of magnitude.

2. Theoretical description and the DMFT solution

A qualitative description of the properties described above is provided by the Falicov-Kimball (FK) model (16) which takes into account the interaction between a 2–fold degenerate conduction band and a lattice of Yb and Y ions. Each lattice site can be occupied either by a Yb^{2+}, Yb^{3+} or Y^{3+} ion. The Yb^{2+} ion has a full f-shell and is non-magnetic, the Yb^{3+} is magnetic with one f-hole in a J=7/2 spin-orbit state, and the Y^{3+} is non-magnetic, with one additional hole with respect to the Yb^{2+} ion. The number of Y^{3+} ions is fixed in each alloy, while the concentration of Yb^{3+} and Yb^{2+} ions is a thermodynamic variable. The f-holes are localized and the state of a given Yb ion cannot change in time but the relative number of Yb^{2+} and Yb^{3+} changes due to thermodynamic fluctuations. The conduction electrons can hop between nearest-neighbor sites on the D-dimensional lattice, with a hopping matrix $-t_{ij} = -t^*/2\sqrt{D}$; we choose a scaling of the hopping matrix that yields a nontrivial limit in infinite-dimensions (17). We assume that the magnetic f-hole on Yb^{3+} and the spinless hole on Y^{3+} interact with the holes in the conduction band by a Coulomb repulsion U_f and U_Y, respectively. Averaging over all possible random distributions of Yb^{2+} , Yb^{3+} and Y^{3+} ions restores the translational invariance and leads to the Falicov-Kimball model for the lattice of Yb-Y ions,

$$\mathcal{H}_{\mathrm{FK}} = \mathcal{H}_d^0 + \mathcal{H}_f + \mathcal{H}_Y + \mathcal{H}_U, \tag{1}$$

where

$$\mathcal{H}_d^0 = \sum_{ij,\sigma}(-t_{ij} - \mu\delta_{ij})d_{i\sigma}^\dagger d_{j\sigma}, \tag{2}$$

$$\mathcal{H}_f = \sum_{i,\eta}(E_f - \mu)f_{i\eta}^\dagger f_{i\eta}, \tag{3}$$

$$\mathcal{H}_Y = \sum_{i}(E_Y - \mu_Y)c_i^\dagger c_i, \tag{4}$$

and

$$\mathcal{H}_U = U_f \sum_{i,\sigma\eta} d_{i\sigma}^\dagger d_{i\sigma} f_{i\eta}^\dagger f_{i\eta} + U_Y \sum_{i,\sigma} d_{i\sigma}^\dagger d_{i\sigma} c_i^\dagger c_i. \tag{5}$$

Spin-1/2 conduction holes are created or destroyed at site i by $d_{i\sigma}^{\dagger}$ or $d_{i\sigma}$, the 8-fold degenerate localized f-holes are created or destroyed at site i by $f_{i\eta}^{\dagger}$ or $f_{i\eta}$, and the spinless Y-hole is created or destroyed at site i by $c_i^{\dagger}$ or c_i. We use σ and η labels to denote the angular momentum state of the d- and f-holes, respectively. The d-, f- and Y-number operators at each site are $n_d^i = \sum_\sigma n_{d\sigma}^i$, $n_f^i = \sum_\eta n_{f\eta}^i$, and n_Y^i, respectively, and we have the local constraint $n_f^i + n_Y^i \leq 1$. The Y-doping reduces the number of f-holes in the conduction band and provides additional Coulomb scattering for conduction-holes. For a given concentration x of Y ions, the chemical potential μ is employed to conserve the total number of remaining d- and f-particles, $n_d(T) + n_f(T) = n_{tot} - x$. In the presence of a magnetic field the magnetic degeneracy of the f-holes is lifted and the Hamiltonian (1) is supplemented by a Zeeman term. Using the basis that diagonalizes simultaneously the zero-field Hamiltonian and the $J_z^{7/2}$ component of the angular momentum operator, we can write,

$$\mathcal{H}_{\rm Z} = g_d \mu_B H \sum_{i\sigma} \sigma d_{i\sigma}^{\dagger} d_{j\sigma} + g_f \mu_B H \sum_{i\eta} \eta f_{i\eta}^{\dagger} f_{i\eta}, \tag{6}$$

where g_d and g_f are the corresponding Landé g-factors. The numerical calculations are performed for a hyper-cubic lattice with a Gaussian non-interacting density of states $\rho(\epsilon) = \exp[-\epsilon^2/t^{*2}]/(\sqrt{\pi}t^*)$; and t^* is taken as the unit of energy ($t^* = 1$). We consider only the homogeneous phase, where all quantities are translationally invariant.

The DMFT reduces the problem on an infinite-dimensional lattice to the problem of an atomic d-state coupled to an atomic f-state by the same Coulomb interaction as on the lattice (18), and perturbed by an external time-dependent field, $\lambda(\tau, \tau')$. The field is taken in such a way that the local d-electron Green's function of the lattice coincides with the Green's function of the atomic d-state, $G_{loc}^{\sigma}(z) = G_{at}^{\sigma}(z)$. The atomic problem is solved by defining the generating functional (the partition function of the FK atom) in the interaction representation (19),

$$\mathcal{Z}_{at}(\mu, \lambda) = {\rm Tr}_{dfc}\left[T_\tau e^{-\beta \mathcal{H}_{at}} S(\lambda)\right], \tag{7}$$

where the statistical sum runs over all possible quantum states of the system and depends on $\lambda^\sigma(\tau, \tau')$ for $\tau, \tau' \in (0, \beta)$. The Hamiltonian of the FK atom,

$$\begin{aligned}\mathcal{H}_{\rm at} &= -\mu \sum_\sigma d_\sigma^{\dagger} d_\sigma + (E_f - \mu) \sum_\eta f_\eta^{\dagger} f_\eta + (E_Y - \mu_Y) c^{\dagger} c \\ &+ U_f \sum_{\sigma\eta} d_\sigma^{\dagger} d_\sigma f_\eta^{\dagger} f_\eta + U_Y \sum_{\sigma\eta} d_\sigma^{\dagger} d_\sigma c^{\dagger} c,\end{aligned} \tag{8}$$

defines the time evolution of the operators, and the external field defines the time-evolution operator for the state vectors,

$$S(\lambda) = T_\tau e^{-\int_0^\beta d\tau \int_0^\beta d\tau' \sum_\sigma \lambda^\sigma(\tau,\tau') d_\sigma^\dagger(\tau) d_\sigma(\tau')}. \tag{9}$$

In the presence of the magnetic field, we add to (8) a Zeeman term that is obtained from (6) in an obvious way. The Hilbert space can be decomposed into invariant subspaces with respect to n_f and n_c, and the matrix elements in (7) can be calculated within the n_f-invariant subspace by replacing $\sum_\eta f_\eta^\dagger f_\eta$ in $\mathcal{H}_{\rm at}$ by its eigenvalue (0 or 1) and setting $n_c = 0$. Within the n_c-invariant subspace we use $n_f = 0$ and $n_c = 1$. This gives,

$$\mathcal{Z}_{at}(\mu, \lambda) = \mathcal{Z}_0(\mu, \lambda) + \mathcal{Z}_f \mathcal{Z}_0(\mu - U_f, \lambda) + \mathcal{Z}_Y \mathcal{Z}_0(\mu - U_Y, \lambda), \tag{10}$$

where, $\mathcal{Z}_f = \sum_\eta e^{-\beta(E_\eta^{(7/2)} - \mu)}$ and $\mathcal{Z}_Y = e^{-\beta(E_Y - \mu_Y)}$ are the partition functions of the Yb^{3+} and Y^{3+} holes decoupled from the d-states, and $\mathcal{Z}_0(\mu, \lambda)$ is the partition function of the $U_f = U_Y = 0$ atomic d-state coupled to the λ-field only. We have,

$$\mathcal{Z}_0(\mu, \lambda) = \prod_\sigma \mathcal{Z}_0^\sigma(\mu, \lambda), \tag{11}$$

where

$$\mathcal{Z}_0^\sigma(\mu, \lambda) = \mathrm{Tr}_d \left[T_\tau e^{-\beta \mathcal{H}_0^\sigma} S(\mu, \lambda^\sigma) \right]. \tag{12}$$

and

$$\mathcal{H}_0^\sigma = -\sum_\sigma (\mu - g_d \mu_B H) d_\sigma^\dagger d_\sigma. \tag{13}$$

To calculate $\mathcal{Z}_0^\sigma(\mu, \lambda)$ we introduce the d-electron propagator for this simplified atomic problem,

$$G_0^\sigma(\tau, \tau') = -\frac{\delta \ln \mathcal{Z}_0^\sigma}{\delta \lambda^\sigma(\tau', \tau)}, \tag{14}$$

which is determined by the equations of motion (EOM) and the boundary condition $G_0^\sigma(\tau, \tau') = G_0^\sigma(\tau + \beta, \tau')$. The EOM's reduce, in the Matsubara representation, to a set of decoupled linear algebraical equations, such that $[G_{0n}^\sigma]^{-1} = [g_{0n}^\sigma]^{-1} - \lambda_n^\sigma$, where $[g_{0n}^\sigma]^{-1} = i\omega_n + \mu - \sigma \mu_B g_d H$ is the propagator of the trivial $U_f = U_Y = \lambda = 0$ atomic d-state. Using $\mathcal{Z}_0^\sigma(\mu, \lambda) = \det |[G_0^\sigma]^{-1}|$ and $\det |[g_0^\sigma]^{-1}| = 1 + \exp(\mu - \sigma \mu_B g_d H)$, we can write the partition function of the $U_f = U_Y = 0$, $\lambda \neq 0$ atomic model as,

$$\mathcal{Z}_0^\sigma(\mu, \lambda) = (1 + e^{\beta(\mu - \sigma \mu_B g_d H)}) \prod_n (1 - \lambda_n^\sigma g_{0n}^\sigma). \tag{15}$$

The partition functions in the $n_f = 1$ ($n_Y = 0$) and the $n_Y = 1$ ($n_f = 0$) subspaces are obtained from the $n_f = n_Y = 0$ solution by replacing μ in $\mathcal{Z}_0^\sigma(\mu, \lambda)$ by $\mu - U_f$ and $\mu - U_Y$, respectively.

The fully renormalized Green's function of the FK atom is, by definition,

$$G^\sigma(\tau, \tau') = -\frac{1}{\mathcal{Z}_{at}(\mu, \lambda)} \frac{\delta \mathcal{Z}_{at}(\mu, \lambda)}{\delta \lambda^\sigma(\tau', \tau)}, \tag{16}$$

and can be calculated using Eqs.(10) and (14). In the Matsubara representation, where $G_{0n}^\sigma = -\delta \ln \mathcal{Z}_0^\sigma / \delta \lambda_n^\sigma$, we obtain

$$G_{at}^\sigma(i\omega_n) = N_0 G_0^\sigma(i\omega_n) + \frac{N_f}{[G_0^\sigma(i\omega_n)]^{-1} - U_f} + \frac{N_Y}{[G_0^\sigma(i\omega_n)]^{-1} - U_Y}, \tag{17}$$

where $N_0 = \mathcal{Z}_0(\mu, \lambda)/\mathcal{Z}_{at}$, $N_f = \mathcal{Z}_f \mathcal{Z}_0(\mu - U_f, \lambda)/\mathcal{Z}_{at}$, and $N_Y = \mathcal{Z}_Y \mathcal{Z}_0(\mu - U_Y, \lambda)/\mathcal{Z}_{at}$ are the respective average numbers of the Yb^{2+}, Yb^{3+} and Y^{3+} ions in a Y-doped system. The concentration of Y-ions, $N_Y = x$, is kept constant at each temperature, and G_{at}^σ is calculated using $N_0 = 1 - N_f - x$. Since G_0^σ is the solution of the $U_f = U_Y = 0$, $\lambda \neq 0$ problem, the self energy of the full atomic problem is given by the Dyson equation,

$$\Sigma^\sigma = [G_0^\sigma]^{-1} - [G_{at}^\sigma]^{-1}. \tag{18}$$

The DMFT solution for the FK lattice is obtained from the atomic solution for λ-field such that

$$G_{at}^\sigma(i\omega_n) = \int \frac{\rho(\epsilon)}{i\omega_n + \mu + \sigma \mu_B g_d H - \Sigma^\sigma(z) - \epsilon} d\epsilon. \tag{19}$$

The equations (17)–(19), together with the expressions (10) and (15) for the partition function, can be solved by iteration. One starts from some trial self energy and finds G_{at}^σ using Eq.(19). Then, one finds G_0^σ from Eq.(18), finds $\mathcal{Z}_{at}(\mu, \lambda)$ using (10) and (15), calculates G_{at}^σ by functional derivatives, recalculates Σ^σ using (18), and continues until the fixed point is reached. Once the numbers N_0 and N_f are obtained we can iterate (17), (18) and (19) on the real axis and find the retarded quantities. In what follows, we use the DMFT to calculate the thermodynamic and transport properties of the model corrsponding to the $Yb_{1-x}Y_xInCu_4$ alloys.

3. Numerical results

The calculations are performed assuming that doping by Y^{3+} ions removes holes from the conduction band. In an undoped sample the total number of conduction holes and the holes on the Yb^{3+} ions is $n_d + n_f = 1.5$, while for a concentration x of Y ions we assume $n_d + n_f = 1.5 - x$. The parameter space of the model is very large and a quantitative comparison would require a fine-tuning of the parameters. Here, we only show the results for static correlation functions obtained for $U_f = U_Y = 2\,t^*$, and $E_f = -0.6\,t^*$.

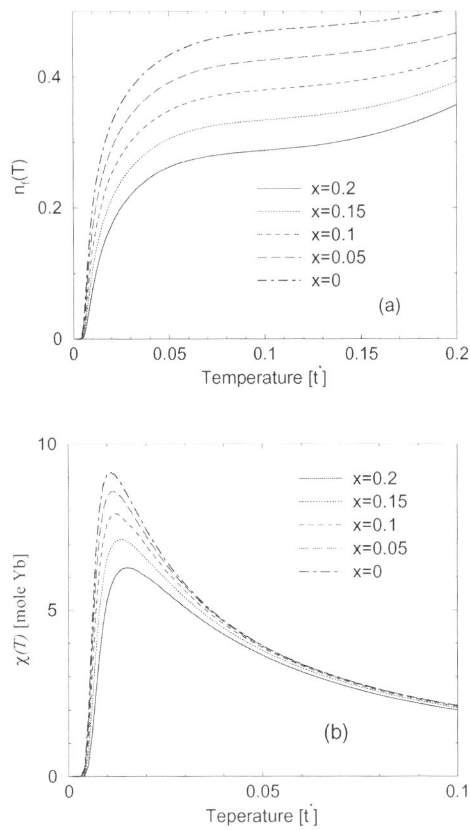

Figure 2. Panel (a) shows localized electron filling for eightfold degenerate doped Falicov-Kimball model with $U_f = U_Y = 2\,t^*$, $E_f = -0.6\,t^*$, and various doping levels x. Panel (b) shows spin susceptibility normalized to the nominal concentration of Yb ions in a $Yb_{1-x}Y_xInCu_4$ alloy.

Panel (a) of Fig.2 shows the effect of temperature and doping on the average concentration of Yb^{3+} ions. Panel (b) shows the f-electron contribution to the spin susceptibility, which vanishes below T_v, and is Curie-like for $T \gg T_v$. Above the transition temperature, defined by the inflection point of the susceptibility curves, the concentration of Yb^{3+} ions becomes significant. The effective Curie constant decreases with doping.

Panel (a) of Fig. 3 shows the result for the DC resistivity, $\rho_{dc}(T)$, normalized by $\rho_0 = 2\hbar d a^{d-2}/e^2\pi$, where a is the lattice spacing and d is the dimensionality of the system ($\rho_0 \approx 2.3 \times 10^{-4}\,\Omega$cm in $d = 3$ with

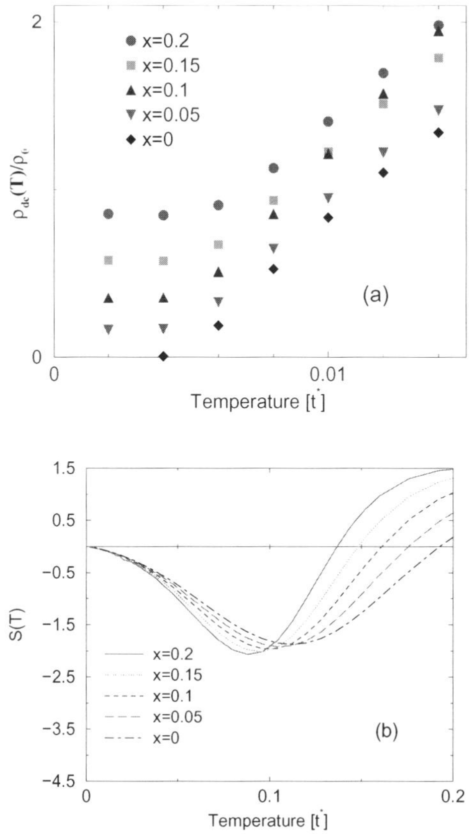

Figure 3. Panel (a) shows low-temeperature DC resistivity of the Falicov-Kimball model for the same parameters as in Fig.(2). Panel (b) shows thermopower on a largere temperature scale.

$a \approx 3 \times 10^{-8}$ cm). In an undoped sample we have $\rho_{dc}(0) = 0$ and the resistivity becomes significant only around T_v, where the concentration of Yb^{3+} ions is large. In doped ($x > 0$) samples, $\rho_{dc}(0) \approx x$, but the relative importance of the doping diminishes for $T \geq T_v$. At high temperatures, the resistivity of all samples has a maximum (not shown here). The thermopower results are shown in panel (b), using a 10 times larger temperature scale than in panel (a). The data reveal the minimum (at about the same temperature as the resisitivty maximum), which is followed by a sign-change. The theoretical results shown in Figs. 2 and 3 have a number of qualitative features that one finds in the experimental data. A detailed comparison will be published elsewhere.

Acknowledgements

We acknowledge useful discussions with C. Geibel and J. Sarrao. We acknowledge the support by the National Science Foundation under grant DMR-0210717. V.Z. acknowledges the support by the Swiss National Science Foundation, grant no. 7KRPJ65554.

References

1. I. Felner and I. Novik, Phys. Rev. B **33**, 617 (1986).
2. E. Figueroa, J. M. Lawrence, J. Sarrao and Z. Fisk, M. F. Hundley and J. D. Thompson, Solid State Commun. **106**, 347 (1998).
3. S. R. Garner, J. N. Hancock, Y.W. Rodriguez, Z. Schlesinger, B. Bucher, Z. Fisk and J. L. Sarrao, Phys. Rev. B **63**, R4778 (2000).
4. J. L. Sarrao, Physica B **259 & 261**, 129 (1999).
5. W. Zhang, N. Sato, K. Yoshimura, A. Mitsuda, T. Goto and K. Kosuge, Phys. Rev. B **66**, 024112-1 (2002).
6. M. Očko and J. L. Sarrao, Physica B **312 & 313**, 341 (2002).
7. C. Dallera, M. Grioni, A. Shukla, G. Vankó, J. L. Sarrao, J. P. Rueff and D. L. Cox, Phys. Rev. Lett. **88**, 196403 (2002).
8. A. L. Cornelius , J. M. Lawrence, J. Sarrao, Z. Fisk, M. F. Hundley, G. H. Kwei, J. D. Thompson, C. H. Booth and F. Bridges, Phys. Rev. B **56**, 7993 (1997).
9. J. M. Lawrence, S. M. Shapiro, J. L. Sarrao and Z. Fisk, Phys. Rev. **55**, 14467 (1997)
10. A. Mitsuda, T. Goto, K. Yoshimura, W. Zhang, N. Sato, K. Kosige and H. Wada, Phys. Rev. Lett. **88**, 137204 (2002).
11. C. Rettori, S. B. Oseroff, D. Rao, P. G. Pagliuso, G. E. Barberis, J. Sarrao, Z. Fisk and M. F. Hundley, Phys. Rev. B **55**, 1016 (1997).
12. E. A. Goremychkin and R. Osborn, Phys. Rev. B **47**, 14280 (1993).
13. A. Murani, D. Richard and R. Bewley, Physica B **312 & 313**, 346 (2002).
14. C. D. Immer, J. Sarrao, Z. Fisk, A. Lacerda, C. Mielke and J. D. Thompson, Phys. Rev. B **56**, 71 (1997).
15. J. M. Lawrence, R. Osborn, J. L. Sarrao and Z. Fisk, Phys. Rev. **59**, 1134 (1999).
16. L. M. Falicov and J. C. Kimball, Phys. Rev. Lett. **22**, 997 (1969).
17. W. Metzner and D. Vollhardt, Phys. Rev. Lett. **62**, 324 (1989).
18. U. Brandt and C. Mielsch, Z. Phys. B **75**, 365 (1989).
19. L.P. Kadanoff and G. Baym, *Quantum Statistical Physics* (W. A. Benjamin, Menlo Park, CA, 1962).
20. J. Freericks and V. Zlatić, Phys. Rev. B **58**, 322 (1998); V. Zlatić and J. Freericks, in *Proc. NATO ARW Conference, Bled 2000*, edited by P. Prelovsek, S. Sarkar, J. Bonca (North-Holland, Amsterdam, 2001).

NEUTRON SPECTROSCOPY OF VALENCE FLUCTUATION COMPOUNDS OF CERIUM AND YTTERBIUM.

Proceedings of the ARW NATO Workshop Hvar, Croatia, October 2002

A. P. MURANI
Institut Laue Langevin, 6 Rue Jules Horowitz, 38042 Grenoble.

Abstract. At low temperatures, the magnetic response of Ce and Yb-based valence fluctuation systems consists of a broad inelastic lorentzian distribution centered on a characteristic energy T_K which varies from ~ 10 meV e.g. $YbAgCu_4$ (1), $CeIn_{1.5}Sn_{1.5}$ (2) to ~ 35 - 55 meV, e.g. $CeSn_3$ (3), $YbAl_3$ (4), $YbInAu_2$ (5), $CePd_3$ (6, 7), going up to ~ 120 - 140 meV, e.g. $CeRh_2$ (8), $CeNi_2$ (9), and ~ 170 - 180 meV as in α-Ce (10) and $YbAl_2$ (11). In strongly hybridized Ce systems such as $CeFe_2$ (12), $CeRu_2$ (13) very broad magnetic responses with characteristic energies of ~ 500 meV have been observed. In all these systems the variation of the spectral intensity with Q follows, or is consistent with, the single ion form factors of Ce^{3+} or Yb^{3+}.

Measurements of excitations to the upper spin-orbit state in Ce show a dramatic shift at the $\gamma \rightarrow \alpha$ transition from ~ 260 meV in the γ phase to around ~ 500 meV in the α phase (10, 14). A similar enhancement of the spin-orbit excitation energy by about ~ 50 ± 30 meV is observed in the compound $YbInCu_4$ as it undergoes a similar transition from a low-T_K (~ 2 meV) heavy fermion state to one with a significantly higher T_K (~ 32 meV) (15). In $CePd_3$ with a ground state T_K of ~ 55 meV, the spin-orbit excitation is observed at ~ 370 meV representing an enhancement of ~ 100 meV (16). Also, a progressive evolution of the spin-orbit excitation energy from ~ 275 meV to ~ 340 meV is observed in $CeIn_{3-x}Sn_x$ as x is increased form 0 to 3 (2). Remarkably, in all these systems the enhancement of the spin-orbit energy relative to the free-ion value is about twice the characteristic energy T_K of the ground state.

Key words: Neutron Spectroscopy, Paramagnetic Scattering, Valence Fluctuation Systems, Heavy Fermions, Cerium Systems, Ytterbium Systems, Kondo Effect

1. Introduction

Metals, alloys and intermetallic compounds containing the end members of the 4f electron series of the periodic table, viz. Ce and Yb, very often show interesting physical properties that deviate significantly from those

A.C. Hewson and V. Zlatić (eds.), Concepts in Electron Correlation, 297–305.

expected for a localized 4f ion having a Russel-Saunders ground state and interacting lightly with the surroundings (crystalline field and inter-ion couplings). The latter is the most common feature amongst the rare earth series, except for a number of deviant Ce and Yb systems as well as a few systems containing Sm, Eu and Tm.

Two principal characteristic properties of the 'anomalous' Ce and Yb based systems are the enhanced magnitude of the linear coefficient of the specific heat, and the constant low temperature (T $\rightarrow$ 0 K) susceptibility $\chi(0)$. Intermetallic compounds in which the Sommerfeld constant $\gamma \gtrsim 100$ mJ mole^{-1}K^{-2}) are commonly classified as heavy fermion systems, while those with $\gamma \lesssim 50$ mJ mole^{-1}K^{-2} are called valence fluctuation systems. The low temperature susceptibility $\chi(0)$ and the linear specific heat coefficient γ are related via a common characteristic energy parameter T_K as expressed by the relations $\chi(0) = \mu^2\langle n_f\rangle/3T_K$ and $\gamma = \pi^2 k^2\langle n_f\rangle/3T_K$, where $\langle n_f\rangle$ (≤ 1) is the occupancy of the 4f state. For heavy fermion systems the occupancy factor $\langle n_f\rangle$ is close to unity while in valence fluctuation systems $\langle n_f\rangle$ can deviate from unity by a significant amount. Some of the early theories have shown that the Wilson ratio $R = \pi^2 k^2\chi(0)/\mu^2\gamma$ is, ideally, close to unity for heavy fermion/valence fluctuation systems (17).

There is no fundamental difference between heavy fermions and valence fluctuation systems except in the degree of hybridization, and the relative strength of inter-site correlations. In a heavy fermion system hybridization is relatively weak and spin-spin correlations are usually quite strong, while hybridisation is much stronger in valence fluctuation systems and inter-site correlations almost negligible. If a heavy fermion system does not order magnetically or become superconducting at low temperatures, its bulk magnetic susceptibility tends to a constant value $\chi(0)$ as T $\rightarrow$ 0 K. In general, hybridization acts to counter the tendency for a system to order magnetically. Increasing hybridization thus favors paramagnetic states with increasing T_K and progressively decreasing susceptibility $\chi(0)$ and specific heat coefficient γ. In the following we review and discuss results of neutron scattering measurements on these relatively strongly hybridized (valence fluctuation) systems in which the inter-site magnetic correlations are relatively weak or negligible compared with T_K.

2. Formalism

In the following we show that if a **characteristic** temperature T_K is defined via the low temperature susceptibility relation $\chi(0) = \mu^2/3T_K$ then T_K can be identified with the energy ω_0 representing the centroid

of the broad lorentzian spectral distribution observed in neutron scattering measurements on valence fluctuation systems.
Following Marshall and Lowde (18) we express $\chi''(\omega)$ in terms of a lorentzian spectral function as,

$$\chi''(\omega) = \frac{A\omega}{\pi}[\frac{\Gamma}{\Gamma^2 + (\omega - \omega_0)^2}] \tag{1}$$

where A $= \pi\chi(0) = \pi^2\mu^2/6\Gamma$, and Γ is assumed Q-independent.
Using the Kramers-Kronig relation

$$\chi(0) = \frac{1}{\pi}\int_{-\infty}^{\infty} \frac{\chi''(\omega)}{\omega}\mathrm{d}\omega = \frac{2}{\pi}\int_{0}^{\infty} \frac{\chi''(\omega)}{\omega}\mathrm{d}\omega \tag{2}$$

and substituting equation 1, we obtain

$$\chi(0) = \frac{\mu^2}{3}\int_{0}^{\infty} \frac{1}{\Gamma^2 + (\omega - \omega_0)^2}\mathrm{d}\omega = \frac{\mu^2}{3\Gamma}[\frac{\pi}{2} - \tan^{-1}(\frac{\omega_0}{\Gamma})] \tag{3}$$

Thus, for $\Gamma/\omega_0 \leq 1$, (generally, Γ/ω_0 lies within the range 1/2 to 2/3) we obtain,

$$\chi(0) \simeq \frac{\mu^2}{3\omega_0} \tag{4}$$

hence, $T_K \simeq \omega_0$. However, for a quasi-elastic spectral distribution we obtain

$$\chi(0) = \frac{\pi}{2}\frac{\mu^2}{3\Gamma} \tag{5}$$

hence, $T_K = \frac{2}{\pi}\Gamma$. Note that the relation (5) is implicit via the constant A in the definition of $\chi''(\omega)$, in relation (1). Furthermore it can be easily shown that relation (1) satisfies the moment sum rule

$$\int_{-\infty}^{\infty} S(\omega)\mathrm{d}\omega = \frac{2/\pi}{g^2\mu_B^2}\int_{-\infty}^{\infty} \frac{1}{[1 - exp(-\frac{\omega}{kT})]}\chi''(\omega)\mathrm{d}\omega = \frac{2}{3}\mu_0^2 \tag{6}$$

where $\mu_0 = \frac{\mu}{g\mu_B}$, and g = 2. Unlike the susceptibility integral, the integral of S(ω), of course, diverges logarithmically. Hence a cut-off needs to be applied around $\omega_c \sim 10 - 12\omega_0$, or $\sim 10\Gamma$ for a quasi-elastic distribution.

3. Experimental Results

In heavy fermion systems owing to the relatively weak hybridization one can still observe crystal field excitations which are, nevertheless, energy broadened. As the hybridization increases the crystal field states broaden further and eventually merge into a single broad distribution of lorentzian

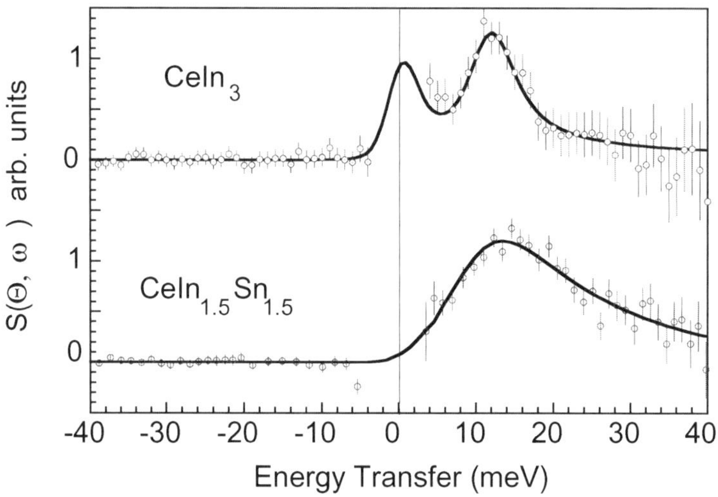

Figure 1. Paramagnetic spectral response from $CeIn_3$ at 20 K and $CeIn_{1.5}Sn_{1.5}$ at 5 K. The continuous curves represent fits to the data using two lorentzian components for $CeIn_3$ corresponding to the Γ_7 ground state and Γ_8 excited state. For $CeIn_{1.5}Sn_{1.5}$ the curve represents a broad lorentzian distribution centered on $\sim$ 10 meV

shape. This phenomenon is well illustrated by the system $CeIn_{3-x}Sn_x$ which evolves continuously from the heavy fermion state in $CeIn_3$ to a valence fluctuation system in $CeSn_3$. The data in fig.1 show that the centroid of the broad distribution observed in $CeIn_{1.5}Sn_{1.5}$ (2) has roughly the same energy as the overall crystal field splitting in the pure $CeIn_3$ sample. The broad spectrum observed for $CeIn_{1.5}Sn_{1.5}$ could also be interpreted in terms of broadened crystal field states as proposed by Zwicknagel, Zevin and Fulde (19). For example, the observed broad spectral distribution with a characteristic energy of $\sim$ 9 meV in $YbAgCu_4$ has been re-interpreted assuming the crystal field scheme observed in iso-structural $YbAuCu_4$ and a Kondo temperature T_o of $\sim$ 6 meV (23). However, if one accepts to define a **characteristic** temperature via the relation $\chi(0) = \mu^2/3T_K$ then, as shown above, the centroid of the broad lorentzian spectral distribution represents T_K directly. The question of whether the broad lorentzian distribution can be decomposed into broadened crystal field states is then less relevant.

A classic example of a valence fluctuation system is the compound $CePd_3$ which shows the characteristic lorentzian inelastic spectral distribution, figure 2 (16). The measurements performed with high energy neutrons (E_i = 900 meV) on the HET spectrometer at ISIS permits a substantial energy range to be covered with, however, a moderately poor energy resolution. The non-magnetic scattering has been subtracted out with reference to the data taken on the iso-structural compound YPd_3. The continuous curve

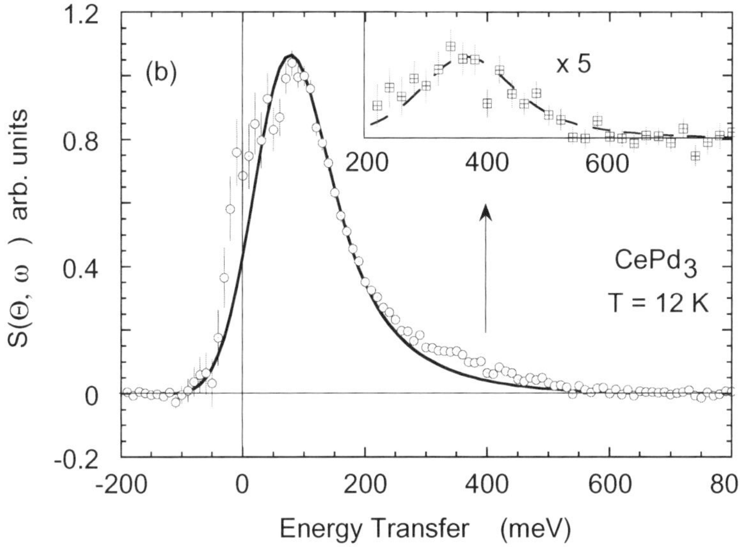

Figure 2. Paramagnetic spectral response from $CePd_3$ at T = 12 K measured with high energy neutrons. The continuous curve represents a fit to the data using the Kuramoto-Müller Hartmann (KMH) spectral form (20) with a characteristic temperature of $\sim$ 55 meV. The inset shows the additional scattering relative to the fitted curve, on an expanded vertical scale and re-binned on the energy scale. It represents the spin-orbit excitation at $\sim$ 370 meV.

shows a fit to the KMH function which is similar to the lorentzian fit over a very wide energy range. In the region above $\sim$ 200 meV one observes some additional scattering, lying above the fitted curve, which is shown in the inset on an expanded vertical scale. It corresponds to the spin-orbit excitation which is centered on $\sim$ 370 meV, representing an enhancement over the free ion value of $\sim$ 100 meV.

The $\gamma \rightarrow \alpha$ transition in Ce which occurs on cooling below $\sim$ 110 K (21) provides one of the best illustrations of the changes in the magnetic spectral response induced via a change in the hybridization with temperature. A similar transition occurs in the system $YbInCu_4$ below 42 K (22). In their high temperature phases both can be described as heavy fermion systems showing a central quasi-elastic distribution together with broad but well defined crystal field excitations. Both systems transform abruptly below their respective transition temperatures and show valence fluctuation character, in particular broad, lorentzian spectral distributions centered on $\omega_0 \sim 32$ meV in $YbInCu_4$ (15) and $\sim$ 170 meV in α-Ce (10).

Following Johansson's interpretation of the $\gamma \rightarrow \alpha$ transition in Ce in terms of the transformation of the localized 4f state into an itinerant 4f

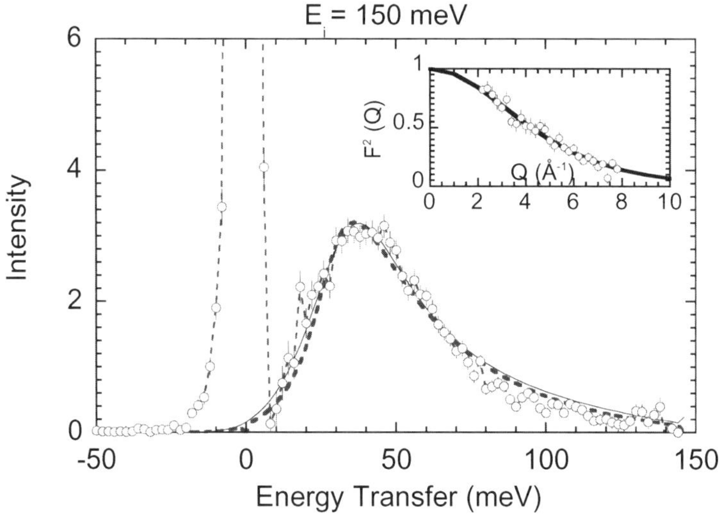

Figure 3. Paramagnetic spectral response from $YbInCu_4$ at 35 K. The continuous and the dashed curves, which are practically indistinguishable, represent fits to the data above 10 meV to a lorentzian distribution centered on $\sim$ 32 meV and the KMH spectral function. The inset shows the variation of the intensity around the peak as a function of Q. The thick solid line represents a fit to the Yb^{3+} form factor.

band (24), as well as the observations of 'coherence' effects in a number of physical properties including heavy band masses on the Fermi surface of $CeSn_3$ (25) and many similar systems, a number of band theoretical calculations (LDA+U) have been performed assuming, at the outset, 4f band states. One such calculation (26) for $CeFe_2$, shows clearly that while both the spin and orbital angular momenta are quenched substantially due to band formation, the reduction is much greater *in proportion* for the *orbital moment* with dramatic consequences for the magnetic form factor. While the observations in U-based systems as for example UFe_2 (27) appear consistent with theory (28), the data for α-Ce, $YbInCu_4$ shown in figure 3, and many others indicate, on the contrary, a remarkably single ion-like form factor dependence. In the case of $CeFe_2$ inelastic scattering measurements (12) are not of adequate statistical quality to permit an explicit verification of the form factor dependence, but the data can be fitted over a wide angular range Θ to the single ion spectral forms (lorentzian and KMH distributions) assuming the Ce^{3+} form factor variation of intensity with ω, hence Q.

Finally, in figure 4 we present data showing the spin-orbit excitations in γ and α Ce measured at very large Q-vectors, using the Be-filter technique on the IN1 spectrometer at ILL (14). They substantiate our earlier observations of the same at lower Q's using the HET time-of-flight spectrometer

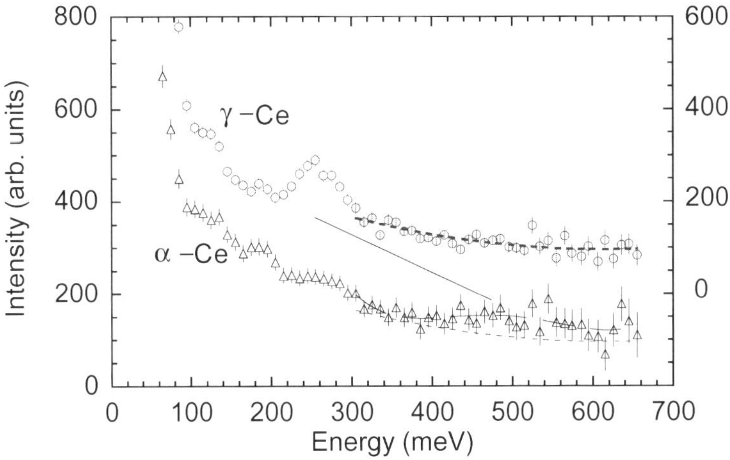

Figure 4. Paramagnetic scattering from γ-Ce showing the spin-orbit excitation at $\sim$ 260 meV which transforms to a broad excitation centered on $\sim$ 500 meV in the α-phase.

at ISIS (10). The well defined excitation seen around 260 meV in the γ phase is found to have broadened and shifted to $\sim$ 500 meV in the α phase, as found earlier. However, the most remarkable aspect of these later measurements is that, with the Be-filter technique, at energy transfers of $\sim$ 500 meV the momentum transfers Q of $\sim$ 15 A^{-1} are attained. Hence, the presence of measurable intensity in the spin-orbit excitations at such large Q-vectors further testifies to the single-ion character of the 4f state in the α phase. If, the enhancement of the excitation energy by about $\sim$ 250 meV can be interpreted as representing the characteristic energy of the upper $^2F_{7/2}$ spin-orbit state, then $T_K{}^{7/2}$ is roughly $\sim$ 1.5 times the characteristic energy $T_K{}^{5/2}$ ($\sim$ 170 meV) of the $^2F_{5/2}$ ground state. As mentioned earlier the data in figure 2 for $CePd_3$ indicate $T_K{}^{7/2} \simeq 2 \times T_K{}^{5/2}$. A very similar ratio for the excited to ground state characteristic energies is observed for the series of compounds $CeIn_{3-x}Sn_x$ over the full range $0 \leq x \leq 3$.

4. Conclusions

We have shown that a lorentzian spectral form centered on an energy ω_0 gives an adequate description of the observed spectra in valence fluctuation systems where ω_0 can be identified with the **characteristic** energy or temperature T_K defined via the relation, $\chi(0) = \mu^2/3T_K$. The question whether the characteristic temperature represents the Kondo temperature or corresponds, in reality, to some spin fluctuation or moment fluctuation temperature needs still to be addressed. However, what is abundantly clear from the experimental observations is that the magnetic response shows the localised, single ion, character via the Q dependence of the spectral intensity as well as its single-ion spectral form.

Observations of the energy-enhanced spin-orbit excitations in both Ce and Yb-based valence fluctuation systems are rather intriguing. A simple interpretation would suggest that the characteristic energy of the upper spin-orbit state is about $\sim$ 1.5 to 2 times that of the ground state. In view of the large range in the T_K's of the ground states from $\sim$ 10 to $\sim$ 170 meV (for systems on which these observations are made) such a small and relatively constant ratio would appear surprising on the basis of the Anderson model which indicates an exponential variation of T_K's with the various physical parameters including the degeneracy as well as the energy of a state relative to the Fermi level (29). Perhaps other hybridization processes such as with the ligand electrons may also be operative in giving rise to the observed phenomena (30). Clearly much further progress is still to be achieved to reach a coherent understanding of these exciting systems.

References

1. A. Severing, A.P. Murani, J.D. Thompson, Z. Fisk, C.-K, Loong, Phys. Rev. B **41**, 1739 (1990).
2. A.P. Murani, A.D.Taylor, R.Osborn and Z.A.Bowden, Phys. Rev. B**48**, 10606 (1993).
3. A.P. Murani, J. Phys. C **33** 6359, (1983); Phys. Rev. B **28**, 2308 (1983).
4. A.P. Murani, Phys. Rev. B **50** 9882, (1994).
5. A.P. Murani and J. Pierre, Physica B **206-207**, 329 (1995).
6. R.M. Galera, A.P. Murani, J. Pierre and K.R.A. Ziebeck, J. Magn. Magn. Mater. **63-64**, 594 (1987).
7. A.P. Murani, A. Severing and W.G. Marshall, Phys. Rev. B **53**, 2641 (1996).
8. A.P. Murani and R.S. Eccleston, Physica B **230-232**, 126 (1997).
9. A.P. Murani and R.S. Eccleston, Phys. Rev. B **53**, 48 (1996).
10. A.P. Murani, Z.A. Bowden, A.D. Taylor, R. Osborn and W.G. Marshall, Phys. Rev. B **48**, 13981 (1993).
11. A.P. Murani, Z.A.Bowden, A.D. Taylor and R. Osborn, Phil. Mag. B**65**, 3092 (1992).
12. A.P. Murani, B. Ouladdif and R.S. Eccleston, Physica B **259-261**, 1167 (1999).
13. A.P. Murani and R.S. Eccleston, Physica B **241-243**, 850 (1998).
14. A.P. Murani, J. Reske, A.S. Ivanov and P. Palleau, Phys. Rev. B **65**, 094416 (2002).
15. A.P. Murani D. Richard and R. Bewley, Physica B **312-313**, 346 (2002).
16. A.P. Murani, R.Raphel, Z.A.Bowden and R.S.Eccleston, Phys.Rev. B**53**, 8188 (1996).
17. H. Lustfeld and A. Bringer, Solid State Comm. **28**, 119 (1978).
18. W. Marshall and R.D. Lowde, Rep. Prog. Phys. **31**, 705 (1968).
19. G. Zwicknagl, V. Zevin and P.Fulde, Physica B **163**, 577 (1990).
20. Y. Kuramoto and E. Müller-Hartmann, J. Magn. Magn. Mater. **52**, 122 (1985).
21. A.W. Lawson and T.-Y. Tang, Phys. Rev. **76**, 301 (1949).
22. I. Felner et al. Phys. Rev. B **35**, 6956 (1987).
23. G. Polatzek and P. Bonville, Z. Phys. B **88**, 189 (1992).

24. B. Johansson, Phil. Mag. **30**, 469 (1974).
25. W.R. Johanson, G.W. Crabtree, A.S. Edelstein and O.D. McMasters, Phys. Rev. Lett. **46**, 504 (1981).
26. O. Eriksson, L. Nordstrom, M.S.S. Brooks and B. Johansson, Phys. Rev. Lett. **60** 2523 (1988).
27. M. Wulff, G.H. Lander, B. Lebech and A. Delapalme, Phys. Rev. B**39**, 4719 (1989).
28. M.S.S. Brooks, O. Eriksson, B. Johansson, J.J.M. Franse and P.H. Frings, J. Phys. F **18**, L33 (1988).
29. N.E. Bickers, D.L. Cox and J. W. Wilkins, Phys. Rev. B **36**, 2036 (1987).
30. Q.G. Sheng and B.R. Cooper, Phil. Mag. Lett. **72**, 123 (1995).

GENERALIZATIONS OF DMFT, CPA AND NCA

Proceedings of the ARW NATO Workshop Hvar, Croatia, October 2002

H. KEITER, K. BAUMGARTNER and DIRK OTTO
Institute of Physics, Dortmund University, D-44221 Dortmund

Abstract. Local and non-local generalizations of dynamical mean-field theory (DMFT) are presented. They come from a self-avoiding walk in a hypercubic lattice, in which the principle of exclusion and inclusion is used. The coherent potential approximation (CPA) can be treated in a similar way. CPA and DMFT can be mapped onto each other. The numerical corrections to the DMFT are relatively weak, with the non-crossing approximation (NCA) used for the single impurity part. NCA-generalizations similar to those of the Wölfle group are set up with direct self-consistent perturbation theory.

Key words: Dynamical Mean-Field Theory, Coherent Potential Approximation, Non Crossing Approximation, Self-Avoiding Walk, Generating Functional, Scattering Matrices, Anderson Lattice Model, Anderson (Impurity) Model

1. Self-avoiding loop and DMFT generalizations

The competition between itinerant and local moment behavior of electrons determined 40 years of research on strongly correlated electron systems, starting from a magnetic impurity which could form a local moment (1) and interact via hybridization with a conduction band. The strength of the latter interaction (hybridization V or a corresponding hopping term) competes with the strong on-site part of the screened Coulomb interaction U. Usually U is the largest energy in the system, so perturbation expansions in the hybridization seemed to be an adequate tool for the impurity case (2) and for the lattice (3). For reviews see (4, 5, 6, 7). In general, the following picture emerges: Physical processes contributing to the (grand canonical) partition function start from a lattice site "0", perform all "on-site" interactions there, then migrate to a (nearest neighbor) site x_1 with on-site interactions there, etc. and finally return to the initial lattice site. In such a process a certain site may be visited several times (3). One of the

A.C. Hewson and V. Zlatić (eds.), Concepts in Electron Correlation, 307–315.

Hamiltonians which give rise to this behavior is the Anderson lattice model

$$\begin{aligned} H_{PAM} &= \sum_{xy,m} t_{xy} c^\dagger_{x,m} c_{y,m} + \sum_{x,m} \epsilon_f f^\dagger_{x,m} f_{x,m} + V \sum_{x,m} (c_{x,m} f^\dagger_{x,m} + h.c.) + H_U \\ H_U &= \sum_{x,m<m'} U f^\dagger_{x,m} f_{x,m} f^\dagger_{x,m'} f_{x,m'} \end{aligned} \tag{1}$$

N-fold degeneracy of the f- and band-electrons, dispersionless hybridization, discrete sums on the (hypercubic) lattice sites x in d dimensions are assumed. U will be infinite in the following, and $t_{xy} = t/\sqrt{d}$ is the scaling suggested in (8) for nearest neighbor hopping. With $\epsilon_k = \sum_x t_{0x} \exp ikx = \sqrt{\frac{2}{d}} t \sum_{j=1}^{d} \cos k_j$ one obtains for the unperturbed propagators

$$g_{xy} = \langle \frac{e^{ik(x-y)}}{z - \epsilon_k} \rangle; \quad g_{00} = \frac{1}{i} \int_0^\infty ds\, e^{isz} (J_0(\sqrt{\frac{2}{d}} t))^d = \frac{1}{z}(1 - \sqrt{2d} t g_{10}(z)) \tag{2}$$

Here, the average denotes the sum on k, normalized by the number of lattice sites. J_0 is the standard Bessel function, the lattice constant is 1. The unperturbed propagators are needed to set up the self-avoiding walk through the lattice:

$$G_{0y}(z) = g_{0y}(z) + \sum_{n=1}^{\infty} \sum_{x_1,\ldots,x_n}{}' g_{0x_1}(z) T(z) g_{x_1 x_2}(z) \cdots T(z) g_{x_n y}(z) \tag{3}$$

The prime on the sum indicates, that all sites have to be pairwise different from each other and (up to y) from the initial site. The scattering matrix $T(z) = V^2 G_f(z)$ is related to the local f-Green's function to be calculated from a single impurity theory. Setting $y = 0$ one obtains a self-avoiding loop (SAL). Corrections to the SAL especially by multiple visits at a site vanish for large spatial dimensions (9). At finite dimensions the SAL is known from the XNCA (10). In (11) some of its mathematical properties were studied. Here we use a self-avoiding *walk* to derive DMFT and generalizations of it. Defining a self-energy via

$$\breve{G}_{0y}(z) = \frac{\delta_{0y}}{T(z)} + \sum_x g_{0x}(z) T(z) \breve{G}_{xy}(z) + \sum_x \Sigma_{0x}(z) \breve{G}_{xy}(z) \tag{4}$$

In order to find the first contribution to Σ_{0x}, start from site 0 and go in a first step to x. But then in going to site y one must exclude the return to site 0. This correlation in the walk generates the self-energy. For, if one includes the site 0 into the walk from x to y, one needs a subtraction of the term which contains site 0. In this term we have a walk from x to 0 to y. Correlations on this walk arise, because in the first part the same site may

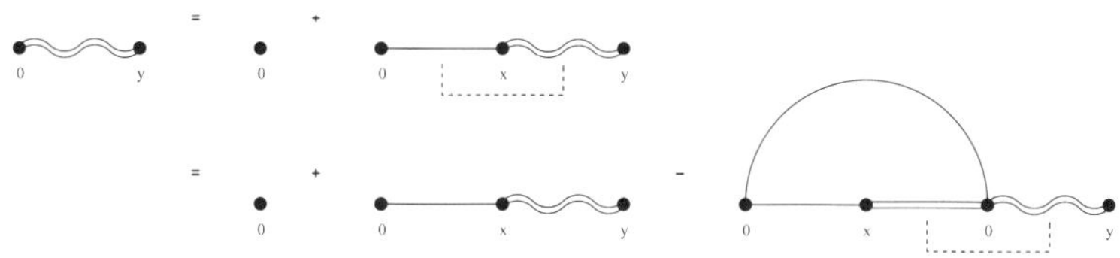

Figure 1. Principle of exclusion and inclusion for $\breve{G}_{0y}(z)$. If the correlation line is left off, one obtains $\Sigma_{xy}^{(1)}$ and a subtraction with 4 correlation lines. If these are left off, the first non-local contribution to $\Sigma_{xy}^{(2)}$ is obtained and another subtraction etc..

be visited as in the second. It is easier to see this from the graphics in fig. 1. Anyway, the first contribution to the self-energy $\Sigma_{xy}^{(1)}(z) = -\delta_{xy} T(z) G_{00}(z)$, if it is inserted into the eq. 4 which is multiplied from the right by g gives the self-consistency equation of DMFT

$$G_{00}(z) = \langle \frac{g_k(z)}{1 - T(z)(g_k(z) - G_{00}(z))} \rangle \tag{5}$$

with $g_k(z) = (z - \epsilon_k)^{-1}$. The second approximation to the self-energy contains all second nearest neighbor contributions and has a diagonal as well as a non-diagonal term

$$\Sigma_{xy}^{(2)} = \frac{T^3(z) \bar{G}_{xy}^2(z) \bar{G}_{yx}(z)}{1 - T^2(z) \bar{G}_{xy} \bar{G}_{yx}(z)} - \delta_{xy} \sum_w \frac{(T(z) \bar{G}_{0w}(z) T(z) \bar{G}_{w0})^2}{1 - T^2(z) \bar{G}_{0w}(z) \bar{G}_{w0}(z)} \tag{6}$$

Here $\bar{G}_{xy}(z) = G_{xy}(z)$ for $x \neq y$ and 0 otherwise. The corrections to the DMFT are displayed in fig. 2.

2. Equivalence of DMFT and CPA

From the matrix equations of standard many-body theory, involving the full single particle Green's function G, the unperturbed one g, the scattering matrix $\tilde{T}$ and the self-energy Σ

$$G = g + g\tilde{T}g = g + g\Sigma G \tag{7}$$

one obtains

$$\tilde{T} = (1 - \Sigma g)^{-1} \Sigma; \quad \Sigma = \tilde{T}(1 + g\tilde{T})^{-1} \tag{8}$$

We next cast the DMFT equation (5) into a similar form by multiplying eq. (5) with $1 + T(z)G_{00}(z)$. This yields

$$G_{00}(z)(1 + T(z)G_{00}(z)) = \langle \frac{g(z)}{1 - \frac{g(z)T(z)}{1+T(z)G_{00}(z)}} \rangle = g_{00}(z - \Sigma(z)) \equiv \tilde{G}(z) \tag{9}$$

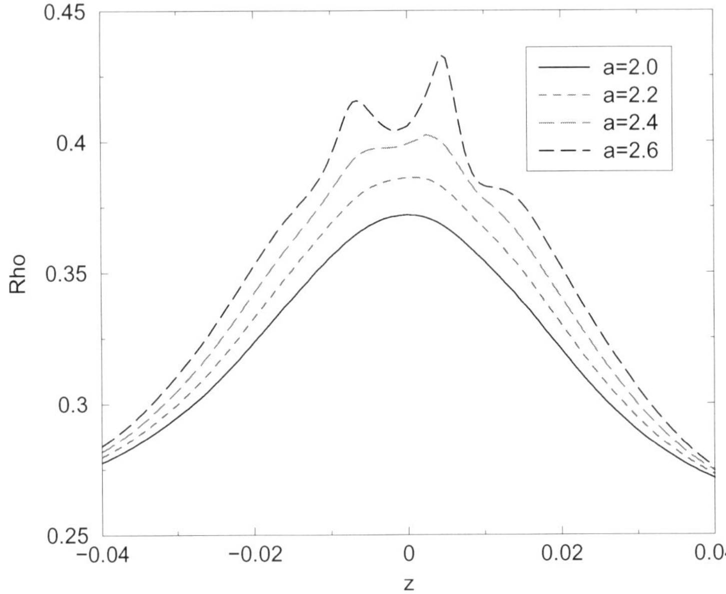

Figure 2. Conduction band density of states with eq. (6). $T(z)$ is approximated by two Lorentzians $T(z) = |V|^2(\frac{a_f b_f}{z-\epsilon_f+ib_f} + \frac{ab}{z+ib})$. Parameters are $|V|^2 = 0.4$, $\epsilon_f = -2$, $a_f = b = 0.2$, $b_f = 1$.

where $\Sigma(z) = T(z)(1 + T(z)G_{00}(z))^{-1}$ was introduced. This allows us to write the DMFT eq. (5) in the equivalent form

$$\tilde{G}(z) = G_{00}(z) + G_{00}(z)T(z)G_{00}(z) = G_{00}(z) + G_{00}(z)\Sigma(z)\tilde{G}(z) \qquad (10)$$

In contrast to the general matrix equations (7) of many-body theory the eq. (5) or the equivalent form eq. (10) are *scalar* equations containing only the diagonal elements of all quantities.

The equation for the *single site* CPA

$$\Sigma(z) = \overline{\frac{V}{1 - \overline{\tilde{G}(z)}^{\mathrm{conf}}(V - \Sigma(z))}}^{\mathrm{conf}} \qquad (11)$$

contains two configuration averages over the random potentials V and over $\tilde{G}(z)$. Then $\overline{\tilde{G}}^{\mathrm{conf}}$ enters into all the equations above, replacing $\tilde{G}$. Within the DMFT the scattering matrix $T(z)$ is calculated via the effective propagator $G_{00}(z)$, which in the CPA describes the influence of other scattering centers. Thus

$$T(z) = \overline{\frac{V}{1 - G_{00}(z)V}}^{\mathrm{conf}} \qquad (12)$$

Given the local quantity $T(z)$ one obtains the self-energy $\Sigma(z)$ via

$$\Sigma(z) = \overline{\frac{T(z)}{1+T(z)G_{00}(z)}}^{\text{conf}} = \frac{V}{1-V(G_{00}(z)+G_{00}(z)T(z)G_{00}(z))+T(z)G_{00}(z)} \tag{13}$$

Using eq. (10) with $\tilde{G}(z) \to \overline{\tilde{G}(z)}^{\text{conf}}$ we obtain the CPA equation (11). In view of this, we may interprete the CPA as a special case of the DMFT for disordered systems. This completes the equivalence of both approaches. We note in passing, that also the non-local scattering matrix $\overline{T}^{\text{conf}}$ of the CPA can be obtained in a closed form from eq. (10)

$$\overline{T}^{\text{conf}}(z) = \frac{T(z)}{1-T(z)(g-G_{00}(z))} \tag{14}$$

3. Improvements of the NCA

For the numerical results obtained in section 1 we used either model scattering matrices $T(z)$ (see fig. 2) or the NCA. The latter, however, is restricted to temperatures above the single impurity Kondo-temperature. Then the corrections to the DMFT results by $\Sigma^{(2)}_{xy}(z)$ from eq. (6) are only a few percent. This is why one has to use another single impurity input. Besides numerical renormalization group methods or density matrix renormalization group, NCA-extensions have been set up. The extension used by the Wölfle group, the CTMA, can be found in (13), with earlier references therein. Their derivation via auxiliary bosons can be avoided, if one uses direct perturbation methods reviewed in (4, 5, 6, 7). In the generating functionals used before in the direct scheme one can also vary with respect to the conduction electrons (i.e. the lines crossing the circles in fig. 3). This can be easily be done, using the formalism in the last appendix of (4). The first diagram for the generating functional in fig. 3 then reads

$$2\beta\Phi^{NCA} = \int \frac{dz}{2\pi i} e^{-\beta z} \sum_{\sigma, i\omega_n} \tilde{F}_\sigma(i\omega_n) P_0(z) P_\sigma(z+i\omega_n) \tag{15}$$

β is the inverse temperature, the contour integral surrounds all singularities of the integrand in a counter-clockwise direction, $\omega_n = (2n+1)\pi/\beta$, $P_0(z) = (z-\Sigma_0(z))^{-1}$, $P_\sigma(z) = (z-\epsilon_f-\Sigma_\sigma(z))^{-1}$ are the auxiliary propagators for the empty f-level (dotted lines in the diagrams) and the occupied one (full lines on the circumference in the diagrams of fig. 3; both are related to their self-energies, of which the lowest orders are depicted in figs. 4 and 5. There is energy- and spin-conservation at every vertex. Varying $2\beta\Phi$ with respect to the bare band-electron Green's function $\tilde{F}(i\omega_n)$ (this amounts to cutting lines crossing the circles in the diagrams for the generating functional) a formal self-energy $\Sigma_{c,\sigma}(i\omega_n)$ is obtained, which is related to the full f-Green-function $G_{f_\sigma f_\sigma^\dagger}(i\omega_n)$ as follows:

$$\mathcal{Z}_f G_{f_\sigma f_\sigma^\dagger}(i\omega_n) = \Sigma_{c,\sigma}(i\omega_n) \tag{16}$$

Here $\mathcal{Z}_f$ is the full partition function divided by the band part. In

$$\tilde{F}_\sigma(i\omega_n) = V^2 \int \frac{A_\sigma(\omega)}{i\omega_n - \omega} d\omega \tag{17}$$

$A_\sigma(\omega)$ denotes the density of states, which is needed in the diagrammatic contributions for $\Sigma_0(z)$, $\Sigma_\sigma(z)$ and $\Sigma_{c,\sigma}(z)$. The rules are similar to the ones in (4, 5, 7) with the following changes: The complex energy z enters into the lowest vertex in the diagrams of figs. 4 and 5. A curved line entering a vertex carries an energy ω and a spin σ and stands for $A_\sigma(\omega)f(\omega)$, where $f(\omega)$ is Fermi's function. A curved line leaving a vertex stands for $A_\sigma(\omega)(1-f(\omega))$. There is a factor $(-1)^{cr}$, where cr is the number of crossings of curved lines ($=3$ for the first diagram in fig. 4). The f-level (straight lines) on the left of the diagrams in figs. 4 and 5 is alternatively occupied (lines with an arrow) or empty (dotted lines). It carries the auxiliary propagators. The essentials can be seen from the analytic expression for the first diagram in fig. 4.

$$\begin{aligned}\Sigma_0^{(a)}(z) &= \sum_\sigma V^4 \int d\omega_1 d\omega_2 d\omega_3 f(\omega_1) f(\omega_2)(1-f(\omega_3)) P_\sigma(z+\omega_1) \times \\ &\quad \times P_0(z+\omega_1-\omega_3) T_1(z-\omega_3,\omega_1,\omega_2) P_0(z+\omega_2-\omega_3) P_\sigma(z+\omega_2)\end{aligned} \tag{18}$$

The one-dimensional integral equation for T_1 is depicted in fig. 6 and reads

$$\begin{aligned}-T_1(z,\omega_1,\omega_2) &= V^2 P_\sigma(z+\omega_1+\omega_3) \\ &+ V^2 \int d\omega_3 A_\sigma(\omega_3) f(\omega_3) T_1(z,\omega_1,\omega_3) P_0(z+\omega_3) P_\sigma(z+\omega_2+\omega_3)\end{aligned} \tag{19}$$

The scattering matrix T_1 can also be used for the self energy $\Sigma_\sigma^{(a)}$, as shown in fig. 6 and for the first part of the conduction band self energy.

$$\begin{aligned}\Sigma_{c,\sigma}^{(a)}(i\omega_n) &= -V^2 \int \frac{dz}{2\pi i} e^{-\beta z} \int d\omega_2 d\omega_3 A_\sigma(\omega_2) A_\sigma(\omega_3) f(\omega_2)(1-f(\omega_3)) P_0(z) \\ &P_\sigma(z+i\omega_n) P_0(z+i\omega_n-\omega_3) T_1(z+\omega_3, i\omega_n, \omega_2) P_0(z+\omega_2-\omega_3) P_\sigma(z+\omega_2)\end{aligned} \tag{20}$$

The case with two spin sums is more complicated. In the 3rd diagram of fig. 4 the curved line entering the lowest vertex carries the energy $i\omega_n$ and the dashed dotted box is replaced by $(\sum_{\sigma'} T_{2,\sigma'}(z+i\omega_n+\omega_4,\omega_1,\omega_2))+$

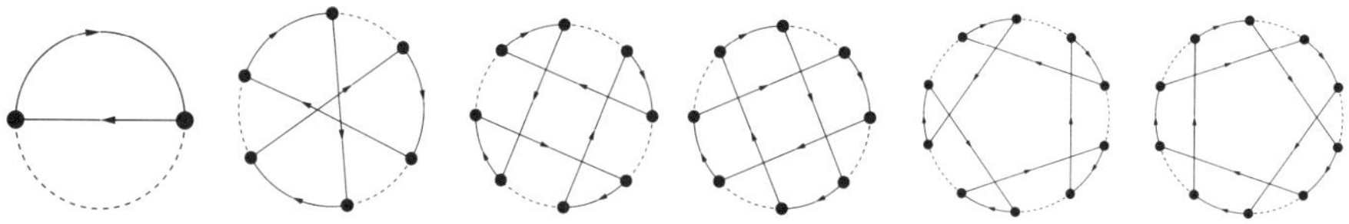

Figure 3. Generating functional for improvements of the NCA. The diagrams have the inverse number of vertices as symmetry factor. All lines are dressed (skeleton diagrams) and the propagators of the corresponding lines are given in the main text.

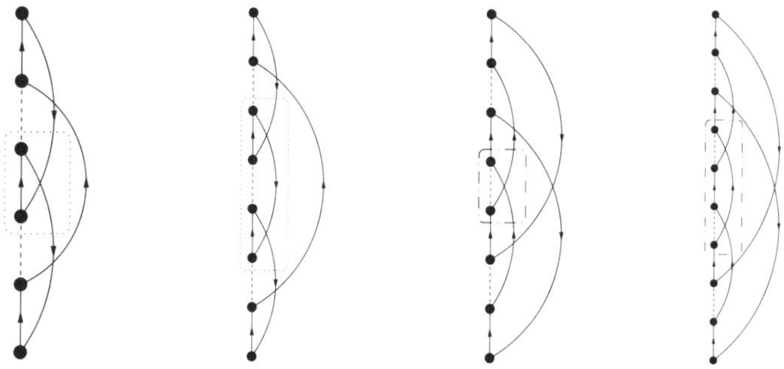

Figure 4. Self-energy contributions to Σ_0, obtained by cutting a dotted line in the generating functional. Note that there is an additional sum on the spins in the 3rd diagram.

$T_3(z+i\omega_n+\omega_4,\omega_1,\omega_2)$. Multiplying by the corresponding spectral functions $A_\sigma(\omega)$, by Fermi's functions and by the auxiliary propagators, times an external $P_0(z)$, and integrating all read off from the 3rd diagram of fig. 4 we obtain the contribution $\Sigma^{(b)}_{c,\sigma}(i\omega_n)$. The final equation then is

$$\Sigma_{c,\sigma}(i\omega_n) = \Sigma^{NCA}_{c,\sigma}(i\omega_n) + \Sigma^{(a)}_{c,\sigma}(i\omega_n) + \Sigma^{(b)}_{c,\sigma}(i\omega_n) \tag{21}$$

together with the contribution from the 1st diagram in fig. 3

$$\Sigma^{NCA}_{c,\sigma} = \int \frac{dz}{2\pi i} e^{-\beta z} P_0(z) P_\sigma(z+i\omega_n) \tag{22}$$

The contributions for the self-energies of the auxiliary propagators follow from the 1st diagram in fig. 3 and from the 3rd diagram of fig. 4 with the dashed dotted box replaced in the same way as for $\Sigma^{(b)}_{c,\sigma}(i\omega_n)$, which then gives $\Sigma^{(b)}_0(z)$. The self-energy $\Sigma^{(b)}_\sigma(z)$ follows with the same substitution from the third diagram in fig. 5. Finally the equations for T_2 and T_3 follow from fig. 6 and will not be written here because of lack of space.

References

1. P.W. Anderson, Phys. Rev. **124**, 41 (1961).

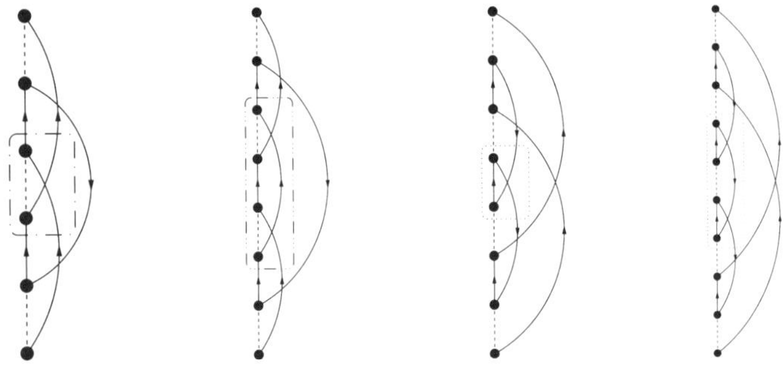

Figure 5. Self-energy contributions to Σ_f, obtained by cutting an arrowed line on the circumference in the generating functional. Note that there is an additional sum on spins in the 2nd diagram.

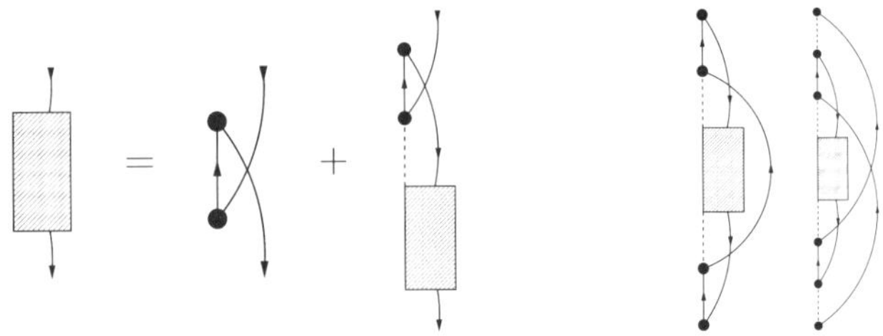

Figure 6. Scattering matrix T_1 and self-energies $\Sigma_0^{(a)}$ and $\Sigma_\sigma^{(a)}$.

2. H. Keiter, J.C. Kimball, Phys. Rev. Lett. **25**, 672 (1970).
3. N. Grewe, H. Keiter, Phys. Rev. B **24**, 4421 (1981).
4. H. Keiter, G. Morandi, Phys. Rep. **109**, 227 (1984).
5. N.E. Bickers, Rev. Mod. Phys. **59**, 845 (1987).
6. A. Georges, G. Kotliar, W. Krauth, M.J. Rozenberg, Rev. Mod. Phys. **68**, 13 (1996).
7. A.C. Hewson, *The Kondo Problem to Heavy Fermions*, Cambridge University Press (1993).
8. W. Metzner, D. Vollhardt, Phys. Rev. Lett. **62**, 324 (1989).

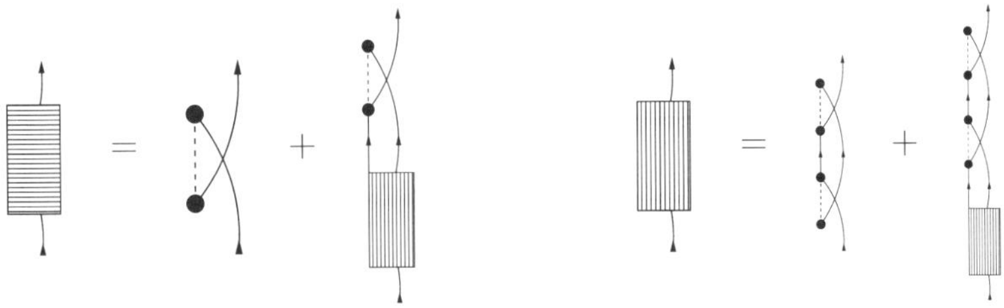

Figure 7. Scattering matrices T_2 and T_3. The complication is related to the additional spin summation.

9. G. Hülsenbeck, Q. Qin, Solid State Comm. **90**, 195 (1994).
10. C. Kim, Y. Kuramoto, T. Kasuya, J. Phys. Soc. Japan **59**, 2414 (1990).
11. H. Keiter, T. Leuders, Europhys. Lett. **49**, 801 (2000).
12. N. Madras, G. Slade, *The self-avoiding walk*, Birkhäuser Verlag, Basel (1993).
13. S. Kirchner, J. Kroha, JLTP **126**, 1233 (2002).

ITINERANT FERROMAGNETISM FOR MIXED VALENCE SYSTEMS

Proceedings of the ARW NATO Workshop Hvar, Croatia, October 2002

J. BONČA and S. EL SHAWISH
J. Stefan Institute, 1000 Ljubljana, Slovenia

C.D. BATISTA
Center for Nonlinear Studies and Theoretical Division, Los Alamos National Laboratory, Los Alamos, NM 87545

J.E. GUBERNATIS
Theoretical Division, Los Alamos National Laboratory, Los Alamos, NM 87545

Abstract. We investigate the Periodic Anderson model in the regime of itinerant ferromagnetism. We compare Quantum Monte Carlo (QMC) results with results obtained using the Gutzwiller approximation (GA). As expected, the energy of the paramagnetic state is overestimated by the GA in comparison with QMC results; however, the partially saturated ferromagnetic (FM) state energies obtained by both methods are surprisingly similar. While the GA gives the ferromagnetic instability too easily, its results for the FM ground state, once its existence is confirmed by more dependable methods, can be used for further examination. This thesis is confirmed by a direct comparison of the mean values of the occupation numbers $n_{\mathbf{k}}$ obtained using both methods.

Key words: Gutzwiller Approximation, Ferromagnetism, Periodic Anderson Model, Quantum Monte Carlo, Mixed Valence Systems

Introduction

Recently, a novel mechanism for itinerant ferromagnetism has been suggested Refs.(1, 2). The simplest model, containing this mechanism, is a simple two band model consisting of a narrow correlated band hybridized with a dispersive, uncorrelated one. The hybridization between bands and the particular band structure play a crucial role in this mechanism by generating a multi-shell structure for the correlated orbitals. This structure, when

A.C. Hewson and V. Zlatić (eds.), Concepts in Electron Correlation, 317–326.

combined with the local Coulomb repulsion, favors a partially saturated ferromagnetic state (FM) which forms as a consequence of a generalized Hund's rule. The most important physical mechanism that leads to FM is associated with the fact that spin-polarized electrons are more successful in avoiding the Coulomb interaction due to Pauli principle than unpolarized ones.

In the past, many Gutzwiller approximation approaches were developed to investigate the ground state properties of the periodic Anderson model (PAM), Refs.(3, 4, 5, 6). Most of these approaches were focused on the Kondo limit of the PAM while the itinerant ferromagnetic state has been found in the mixed valence regime, Refs.(1, 2). In this Paper we focus on a direct comparison between QMC results and GA calculations in the mixed valence regime. We derive GA equations for the spin-polarized PAM by closely following the work by Fazekas and Brandow Ref.(5). We focus on the direct comparison of energies as functions of magnetization using both above mentioned methods. For a closer inspection of the nature of the FM state in comparison to the energetically higher paramagnetic (PM) state we compute average particle densities.

Model

The PAM was originally introduced to explain the properties of the the rare-earth and actinide metallic compounds including the so called heavy fermion compounds. The basic ingredients of this model are a narrow and correlated f band hybridized with a dispersive and uncorrelated d band. The Hamiltonian associated with this model is:

$$H = - \; t \sum_{\langle \mathbf{r},\mathbf{r}' \rangle,\sigma} (d^\dagger_{\mathbf{r}\sigma} d_{\mathbf{r}'\sigma} + d^\dagger_{\mathbf{r}'\sigma} d_{\mathbf{r}\sigma}) + \epsilon_f \sum_{\mathbf{r},\sigma} n^f_{\mathbf{r}\sigma}$$
$$+ \; V \sum_{\mathbf{r},\sigma} (f^\dagger_{\mathbf{r}\sigma} d_{\mathbf{r}\sigma} + d^\dagger_{\mathbf{r}\sigma} f_{\mathbf{r}\sigma}) + \frac{U}{2} \sum_{\mathbf{r},\sigma} n^f_{\mathbf{r}\sigma} n^f_{\mathbf{r}\bar{\sigma}} \,, \qquad (1)$$

where $f^\dagger_{\mathbf{r}\sigma}$ and $d^\dagger_{\mathbf{r}\sigma}$ create an electron with spin σ in f and d orbitals at lattice site $\mathbf{r}$ and $n^f_{\mathbf{r}\sigma} = f^\dagger_{\mathbf{r}\sigma} f_{\mathbf{r}\sigma}$. The hopping t is limited to the nearest-neighbor sites. We will work in units where $t = 1$.

Gutzwiller method

We extend the two-band version of the Gutzwiller $U \to \infty$ method proposed by Fazekas and Brandow Ref.(5) to the spin-polarized ground state. After replacing the determinant expressions by their average values we make no further approximations in optimizing the ansatz with respect to a large number of independent variational parameters. Laborious and exacting derivations not presented here can be found in Refs.(3, 5).

We construct a magnetic trial state by allowing both d- and f-band polarization, as well as spin-dependence of the mixing amplitudes. The average electron density per site n is assumed to be sufficiently small, i.e. $n < 1$, so that only the lower hybridized band needs to be used for either spin, Ref.(5). Our ansatz for a magnetic ground state is thus obtained by projecting out the f^2 component from an optimally chosen free-electron hybridized ground state,

$$|\psi\rangle = \prod_i (1 - n^f_{i\uparrow} n^f_{i\downarrow}) \prod_\sigma \prod_{\mathbf{k}\in \mathrm{FS}_\sigma} (1 + a_\sigma(\mathbf{k}) f^\dagger_{\mathbf{k}\sigma} d_{\mathbf{k}\sigma}) \, |\mathrm{FS}_\sigma\rangle \, . \tag{2}$$

In the spin-up Fermi sea $|\mathrm{FS}_\uparrow\rangle$, the d-band is filled up to the Fermi level $\epsilon_{F\uparrow}$, and similarly for down-spins, leaving all f-states empty. The number of independent variational parameters $a_\sigma(\mathbf{k})$ equals the number of $\mathbf{k}$-states within the Fermi volume.

As in the single-band Gutzwiller method we seek the term dominating the normalization factor $\langle\psi|\psi\rangle$ to set as a replacement for the sum over all possible configurations. After the average treatment of the determinants is carried out, Ref.(5), the combinatorics brings us to the self-consistency condition

$$n_{f\sigma} = \frac{1}{L} \sum_{\mathbf{k}\in \mathrm{FS}_\sigma} \frac{a^2_\sigma(\mathbf{k})}{q_\sigma^{-1} + a^2_\sigma(\mathbf{k})}, \tag{3}$$

where

$$q_\sigma = \frac{1 - n_f}{1 - n_{f\sigma}}, \tag{4}$$

with $n_f = n_{f\sigma} + n_{f\bar{\sigma}}$. With the help of Eq.3 we obtain

$$\langle H\rangle = \sum_{\mathbf{k},\sigma} \varepsilon_{\mathbf{k}} + \frac{(\varepsilon_f - \varepsilon_{\mathbf{k}}) a^2_\sigma(\mathbf{k}) - 2V a_\sigma(\mathbf{k})}{q_\sigma^{-1} + a^2_\sigma(\mathbf{k})}, \tag{5}$$

where $\varepsilon_{\mathbf{k}} = -2t \sum_i \cos(k_{x_i})$ is the $d-$band dispersion relation. Here, as in all subsequent summations over $\mathbf{k}$, the sums are understood to be taken such that $\mathbf{k} \in \mathrm{FS}_\sigma$.

The minimization of the expression Eq.5 for $\langle H\rangle$ is complicated by the fact that the self-consistency equation Eq.3 for $n_{f\sigma}$ connects all variational parameters $a_\sigma(\mathbf{k})$. Taking the partial derivative of Eq.3 with respect to $a_\sigma(\mathbf{k})$ we obtain

$$\frac{\partial n_{f\sigma}}{\partial a_\sigma(\mathbf{k})} = -A_\sigma (1 - n_f)^2 \frac{\partial q_\sigma^{-1}}{\partial a_\sigma(\mathbf{k})} + C_\sigma, \tag{6}$$

$$\frac{\partial n_{f\bar{\sigma}}}{\partial a_\sigma(\mathbf{k})} = -A_{\bar{\sigma}} (1 - n_f)^2 \frac{\partial q_{\bar{\sigma}}^{-1}}{\partial a_\sigma(\mathbf{k})}, \tag{7}$$

where

$$A_\sigma = \frac{1}{L(1-n_f)^2}\sum_{\mathbf{p}}\frac{a_\sigma^2(\mathbf{p})}{(q_\sigma^{-1}+a_\sigma^2(\mathbf{p}))^2},$$
$$B_\sigma = \frac{1}{L(1-n_f)^2}\sum_{\mathbf{p}}\frac{(\varepsilon_f-\varepsilon_{\mathbf{p}})a_\sigma^2(\mathbf{p})-2Va_\sigma(\mathbf{p})}{(q_\sigma^{-1}+a_\sigma^2(\mathbf{p}))^2},$$
$$C_\sigma = \frac{2a_\sigma(\mathbf{k})}{Lq_\sigma(q_\sigma^{-1}+a_\sigma^2(\mathbf{k}))^2}. \tag{8}$$

Similarly, by recalling Eq.4, we arrive at the second set of equations

$$\frac{\partial q_\sigma^{-1}}{\partial a_\sigma(\mathbf{k})} = \frac{\partial n_{f\sigma}}{\partial a_\sigma(\mathbf{k})}\frac{n_{f\bar\sigma}}{(1-n_f)^2}+\frac{\partial n_{f\bar\sigma}}{\partial a_\sigma(\mathbf{k})}\frac{1-n_{f\sigma}}{(1-n_f)^2}, \tag{9}$$
$$\frac{\partial q_{\bar\sigma}^{-1}}{\partial a_\sigma(\mathbf{k})} = \frac{\partial n_{f\bar\sigma}}{\partial a_\sigma(\mathbf{k})}\frac{n_{f\sigma}}{(1-n_f)^2}+\frac{\partial n_{f\sigma}}{\partial a_\sigma(\mathbf{k})}\frac{1-n_{f\bar\sigma}}{(1-n_f)^2}, \tag{10}$$

from which a straightforward expression for all derivatives follows

$$\frac{\partial n_{f\sigma}}{\partial a_\sigma(\mathbf{k})} = C_\sigma\frac{1+A_{\bar\sigma}n_{f\sigma}}{1+A_{\bar\sigma}n_{f\sigma}+A_\sigma n_{f\bar\sigma}+A_\sigma A_{\bar\sigma}(1-n_f)}, \tag{11}$$
$$\frac{\partial n_{f\bar\sigma}}{\partial a_\sigma(\mathbf{k})} = C_\sigma\frac{A_{\bar\sigma}n_{f\bar\sigma}-A_{\bar\sigma}}{1+A_{\bar\sigma}n_{f\sigma}+A_\sigma n_{f\bar\sigma}+A_\sigma A_{\bar\sigma}(1-n_f)}. \tag{12}$$

We next minimize the energy, Eq.5

$$\frac{\partial\langle H\rangle}{\partial a_\sigma(\mathbf{k})} = \frac{2}{q_\sigma(q_\sigma^{-1}+a_\sigma^2(\mathbf{k}))^2}\left[a_\sigma(\mathbf{k})(\tilde\varepsilon_{f\sigma}-\varepsilon_{\mathbf{k}})-V\left(1-q_\sigma a_\sigma^2(\mathbf{k})\right)\right], \tag{13}$$

where the renormalized f-level is given by

$$\tilde\varepsilon_{f\sigma} = \varepsilon_f - \frac{B_{\bar\sigma}+B_\sigma n_{f\bar\sigma}-B_{\bar\sigma}n_{f\bar\sigma}-A_{\bar\sigma}B_\sigma(1-n_f)}{1+A_\sigma n_{f\bar\sigma}+A_{\bar\sigma}n_{f\sigma}-A_\sigma A_{\bar\sigma}(1-n_f)}. \tag{14}$$

Using $\frac{\partial\langle H\rangle}{\partial a_\sigma(\mathbf{k})}=0$ we get from Eq.13 the corresponding variational parameters

$$a_\sigma(\mathbf{k}) = \frac{2V}{\tilde\varepsilon_{f\sigma}-\varepsilon_{\mathbf{k}}+\sqrt{(\varepsilon_{\mathbf{k}}-\tilde\varepsilon_{f\sigma})^2+4V^2q_\sigma}}. \tag{15}$$

Comparing Eq.15 to the $U=0$ solution, Ref.(5), the effect of switching on $U\to\infty$ is: a) to replace ε_f by the renormalized f-level $\tilde\varepsilon_{f\sigma}$ and b) to replace the bare hybridization V by the renormalized $V\sqrt{q_\sigma}$ in the denominator of $a_\sigma(\mathbf{k})$, but not in the numerator. As a simple test to verify the relation Eq.14 we compute the non-magnetic solution, i.e. $\sigma=\bar\sigma$, which leads to the correct result, Ref.(5), $\tilde{\varepsilon}_{f\sigma} = \tilde{\varepsilon}_{f\bar{\sigma}} = \varepsilon_f - B_\sigma/(1+A_\sigma)$. The Eq.5 is just a formal relation, since the determination of $\tilde{\varepsilon}_{f\sigma}$ and $n_{f\sigma}$ still needs to be done. For a general set of model parameters, this can be achieved numerically using iterative methods.

Quasiparticle density
Contrary to the Hubbard model, Refs.(7, 8), the number of the strongly correlated f-electrons in the two-band model is not conserved. In order to control this number in lower and upper band, we calculate the quasiparticle density as a function of $\mathbf{k}$. Following the steps in Ref.(5) we find the d-electron distribution

$$\langle n^d_{\mathbf{k}\sigma}\rangle = \langle d^\dagger_{\mathbf{k}\sigma} d_{\mathbf{k}\sigma}\rangle = 1 - \frac{a^2_\sigma(\mathbf{k})}{q_\sigma^{-1} + a^2_\sigma(\mathbf{k})}, \tag{16}$$

for $\mathbf{k}$'s contained in the FS_σ. The f-band occupation number does not obey a simple relationship like Eq.16. Instead, just as in the Gutzwiller solution of the Hubbard model Refs. (7, 8), correlation causes the f-state occupation to spread over the entire Brillouin zone, rather than being confined to the $\mathbf{k}$-values contained in the FS_σ,

$$\langle n^f_{\mathbf{k}\sigma}\rangle = \langle f^\dagger_{\mathbf{k}\sigma} f_{\mathbf{k}\sigma}\rangle = \begin{cases} (1-q_\sigma)n_{f\sigma} + \dfrac{q_\sigma a^2_\sigma(\mathbf{k})}{q_\sigma^{-1} + a^2_\sigma(\mathbf{k})} & ; \mathbf{k} \in \mathrm{FS}_\sigma \\ (1-q_\sigma)n_{f\sigma} & ; \mathbf{k} \notin \mathrm{FS}_\sigma \end{cases}. \tag{17}$$

The above relation suggests that the f-electron distribution experiences a jump at the Fermi wave vector. Analogous procedure, Ref.(5), brings us to the mixing terms,

$$\langle n^{fd}_{\mathbf{k}\sigma}\rangle = \langle f^\dagger_{\mathbf{k}\sigma} d_{\mathbf{k}\sigma}\rangle = \langle n^{df}_{\mathbf{k}\sigma}\rangle = \langle d^\dagger_{\mathbf{k}\sigma} f_{\mathbf{k}\sigma}\rangle = \frac{a_\sigma(\mathbf{k})}{q_\sigma^{-1} + a^2_\sigma(\mathbf{k})}, \tag{18}$$

which holds only for $\mathbf{k}$'s contained in FS_σ. Instead of presenting the actual dispersion of f- and d-electrons in the $\mathbf{k}$-space, we decide to investigate the quasiparticle density of hybridized bands defined with the following operators

$$\begin{aligned} \alpha^\dagger_{\mathbf{k}\sigma} &= u(\mathbf{k}) f^\dagger_{\mathbf{k}\sigma} + v(\mathbf{k}) d^\dagger_{\mathbf{k}\sigma}, \\ \beta^\dagger_{\mathbf{k}\sigma} &= -v(\mathbf{k}) f^\dagger_{\mathbf{k}\sigma} + u(\mathbf{k}) d^\dagger_{\mathbf{k}\sigma}, \end{aligned} \tag{19}$$

with

$$\begin{aligned} u(\mathbf{k}) &= \frac{E_\mathbf{k} - \varepsilon_f}{\sqrt{(E_\mathbf{k} - \varepsilon_f)^2 + V^2}}, \\ v(\mathbf{k}) &= \frac{-V}{\sqrt{(E_\mathbf{k} - \varepsilon_f)^2 + V^2}}, \end{aligned} \tag{20}$$

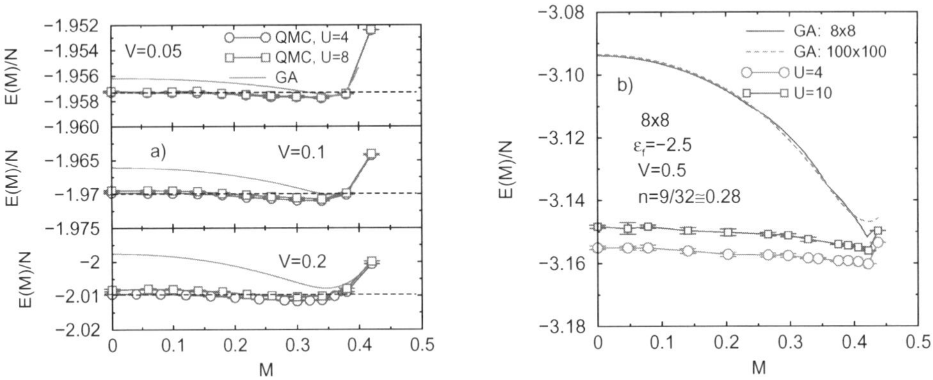

Figure 1. a) E/N for 1D PAM vs. M. Other parameters of the system are $N = 50$, $n = 0.3$, $\varepsilon_f = -1.5$. Results are presented for various $V = 0.05, 0.1, 0.2$ and $U = 4, 8$. Gutzwiller results are presented with full lines. b) E/N for 2D PAM vs. M. Other parameters are presented in the inset. QMC results were calculated on a 8×8 square lattice, Gutzwiller results were calculated on a 8×8 lattice (full line) and 100×100 (dashed lines).

and $E_{\mathbf{k}} = \frac{1}{2}\left(\varepsilon_{\mathbf{k}} + \varepsilon_f + \sqrt{(\varepsilon_{\mathbf{k}} - \varepsilon_f)^2 + 4V^2}\right)$. We finally arrive at desired relations for the upper (α) and lower (β) band, respectively,

$$\begin{aligned} \langle n^{\alpha}_{\mathbf{k}\sigma}\rangle &= u^2(\mathbf{k})\langle n^{f}_{\mathbf{k}\sigma}\rangle + v^2(\mathbf{k})\langle n^{d}_{\mathbf{k}\sigma}\rangle + u(\mathbf{k})v(\mathbf{k})\langle n^{fd}_{\mathbf{k}\sigma} + n^{df}_{\mathbf{k}\sigma}\rangle, \\ \langle n^{\beta}_{\mathbf{k}\sigma}\rangle &= v^2(\mathbf{k})\langle n^{f}_{\mathbf{k}\sigma}\rangle + u^2(\mathbf{k})\langle n^{d}_{\mathbf{k}\sigma}\rangle - u(\mathbf{k})v(\mathbf{k})\langle n^{fd}_{\mathbf{k}\sigma} + n^{df}_{\mathbf{k}\sigma}\rangle. \end{aligned} \tag{21}$$

Results

Our numerical method, the constrained-path Monte Carlo (CPMC) method, is extensively described and benchmarked elsewhere, Refs.(9, 10). For details about the implementation of the method to PAM we refer the reader to our recent publication Ref.(2). Most of our QMC calculations were performed on $N = 50$ and 70 lattice sites for 1D cases and on 2D square lattices with $N = N_x \times N_x$ lattice sites with $N_x = 8$ and 12. Calculations were done at various dopings $n = N_e/N$ where N_e is the number of electrons $N_e = N_\uparrow + N_\downarrow$ and various magnetizations $M = (N_\uparrow - N_\downarrow)/2N$. In Fig.1.a. we show E/N for 1D PAM as a function of the magnetization M for various values of hybridization V and two coupling strengths U. In all cases $E(M)/N$ reaches minimum at finite magnetization $0.3 \leq M \leq 0.35$. Comparing QMC results with GA, we observe that the difference between energies obtained using the two methods systematically decreases with increasing M. This is simple a consequence of the fact that with increasing the magnetization the system becomes less correlated which renders the GA results more accurate. It is less straightforward to understand why lowering of V also leads to a similar effect. The main effect of lowering V is diminishing the crossover region in $\mathbf{k}-$space where the $d-$ and $f-$ bands hybridize. We would also like to stress that even though GA overestimates the paramagnetic energy, it predicts reasonably well the value of magnetization of the ground state. This is particularly evident at small $V = 0.05$. Comparing systems with substantially different $U = 4, 8$ we notice that at small $V = 0.05$ we obtain almost identical results (the two plots for $U = 4$ and 8 nearly overlap) while at larger values of $V = 0.1$, and 0.2 differences are visible nevertheless they remain small. This confirms our hypothesis from Refs.(1, 2) that the relevant energy scale leading to ferromagnetism is the hybridization gap Δ. When $U \gg \Delta$ the system essentially reaches $U \to \infty$ limit.

Results from the 2D system show similar trends as the 1D results, see Fig.1.b. The main difference is that in the contrast to the 1D case, the system size used to obtain QMC results in 2D is not large enough to reach the thermodynamic limit. While finite-size scaling has been performed in our previous work Refs.(1, 2), we only show QMC results for 8×8 system and GA results for 8×8 and 100×100 systems. The most prominent finite-size effect in the GA (comparing 8×8 and 100×100 systems) is a slight increase of the ground state energy around $M \sim 0.42$. As in the 1D case, the GA overestimates the paramagnetic energy in comparison to QMC calculation. We should nevertheless stress that even in the paramagnetic case the relative difference between GA and QMC energies is no more than 2%.

Next we compare the mean value of the occupation numbers $\langle n_{k\sigma}^{(\alpha,\beta)} \rangle$ for the 1D PAM as obtained using GA, Eqs.21 and QMC results for the paramagnetic (Fig.2.a.c.) and ferromagnetic solutions (Fig.2.b.d.) as obtained using $V = 0.05$ and $V = 0.2$. GA calculation captures surprisingly well many details of $\langle n_{k\sigma}^{(\alpha,\beta)} \rangle$ in both the PM and FM cases. Starting with the lower band $\langle n_{k(u,d)}^{\alpha} \rangle$ for both PM solutions, presented in Fig.2.a.c., we first see a downturn around $k \sim 0.2\pi$, which is more pronounced in the $V = 0.05$ case, followed by the plateau-like $k-$dependence. Even though the position of the downturn can be easily predicted by the position of the crossover from the $d-$like to the $f-$like nature of the hybridized α band, the actual hight of the plateau for $k > 0.2\pi$ is set entirely by correlation effect. Even more surprising is the finding that the GA nearly perfectly predicts the size of the jump at the Fermi wave vector $k = k_F = 0.6\pi$ and the magnitude of $\langle n_{k(u,d)}^{\alpha} \rangle$ for $k > k_F$.

Good agreement persists in the FM regime, Fig.2.b.d., where upper band majority $\langle n_{ku}^{\alpha} \rangle$ and minority $\langle n_{kd}^{\alpha} \rangle$ spin occupancies are equally well

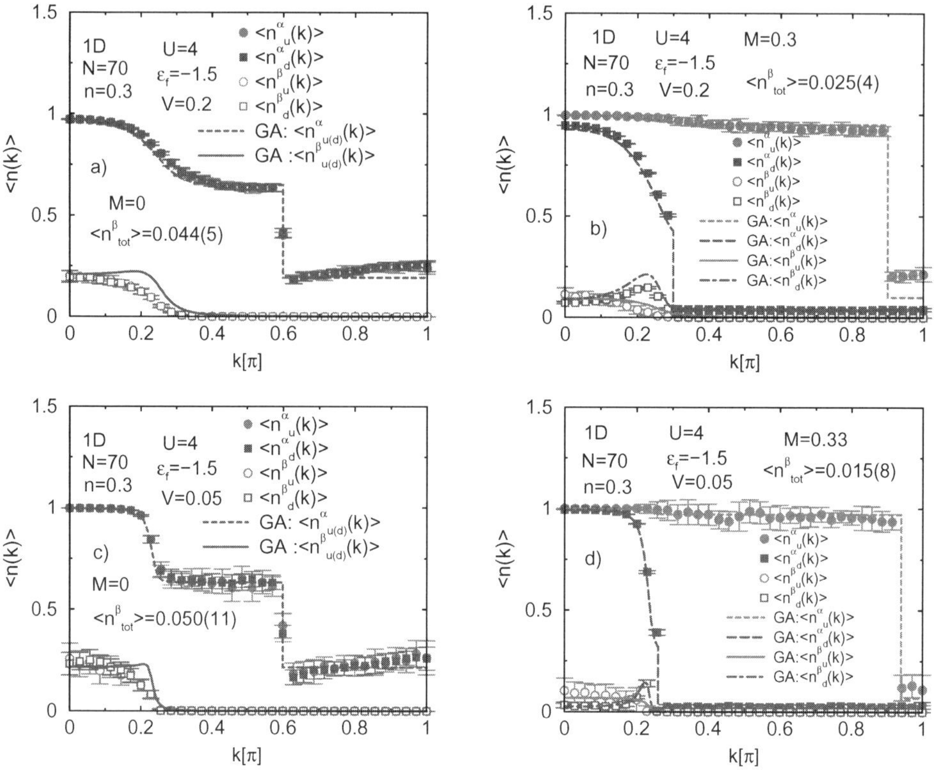

Figure 2. Mean value of the occupation numbers of the non-interacting band states of the PAM in the 1D lattice for the paramagnetic state $M = 0$ a) and c) and for the ground state with partially saturated ferromagnetic solution b) and d). Legends: $\left\langle n_{ku}^{\alpha}\right\rangle$ represents lower-band occupation number of up-spins, $\left\langle n_{kd}^{\beta}\right\rangle$ represents upper-band occupation number of down-spins, etc.

reproduced using both methods. Furthermore, we notice that the minority-spin upper band occupancy $\langle n_{kd}^{\beta}\rangle$ reaches a peak with increasing k that is located around $k \sim 0.2\pi$. On the other hand, the majority $\langle n_{ku}^{\beta}\rangle$ occupancy monotonously decreases with increasing k. It seems as if the spin-up and spin-down occupancies are avoiding each other in the k-space. This effect is captured by the QMC as well as with the GA calculations. Finally, we would like to emphasize yet another interesting result. Comparing the total occupancy of the upper $\beta-$band for the PM and FM solution

$$\langle n_{tot}^{\beta}\rangle = \sum_{k,\sigma}\langle n_{k\sigma}^{\beta}\rangle, \tag{22}$$

we find, that $\langle n_{tot}^{\beta}\rangle$ for the PM solution is always larger than $\langle n_{tot}^{\beta}\rangle$ for the FM solution (actual numbers are given in the inserts of Figs.2). This agrees

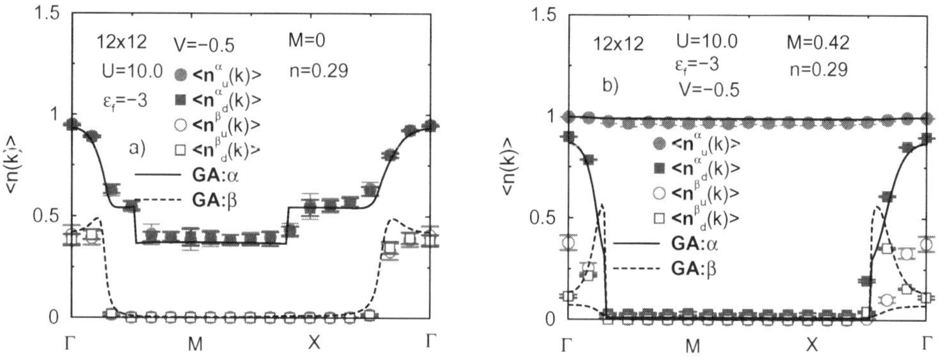

Figure 3. Mean value of the occupation numbers of the non-interacting band states of the PAM in the 2D 12×12 lattice for the a) paramagnetic state $M = 0$ and b) for the ground state with partially saturated ferromagnetic solution with $M = 0.42$.

with our general picture of itinerant ferromagnetism Refs.(1, 2) where a FM ground state develops as a consequence of a generalized Hund's rule where electrons in the partially saturated FM state feel weaker Coulomb interaction in comparison to electrons in the PM state. Since the upper band occupancy develops as a consequence of electron interactions, it is then reasonable to expect the inequality $\langle n^{\beta}_{tot}(M = 0)\rangle > \langle n^{\beta}_{tot}(M > 0)\rangle$.

In Fig.3 we present the occupation numbers for the 2D system. While agreement between QMC and GA results is close in the PM case, it is rather surprising that in the FM case agreement is not as close as in the 1D calculation. The unusual up-turn of the minority upper-band occupancy seems to be well captured by both methods. However the majority upper-band occupation numbers seem to show a large discrepancy. In part we attribute the disagreement to finite-size effects that are much more pronounced in 2D due to a smaller linear dimension of the lattice and in part to different parameters of the model.

Conclusions

We have presented a quantitative comparison of the QMC and GA results in the mixed valence regime of PAM. Following work by Fazekas and Brandow Ref.(5), we have extended their two-band version of the Gutzwiller method to the spin-polarized state. Apart from the so-called Gutzwiller approximation we have made no further simplifications to the model. We have solved Gutzwiller equations numerically using iterative procedures. As naively expected, GA overestimated the energy of the PM solution; however captured reasonably well the energy of the partially saturated FM state. For this reason, GA was not suitable to determine the phase diagram; nevertheless it can be used to investigate physical properties of the partially saturated FM ground state as long as its existence has been established by more reliable methods. Contrary to naive expectations, we have shown, that despite overestimated PM energy, GA predicted well the unsaturated magnetization of the FM ground state and furthermore, it gave the quasiparticle occupation numbers with surprising accuracy in the FM as well as in the PM case.

Acknowledgements

J. B. acknowledges the support of Slovene Ministry of Education Science and Sports, FERLIN, and the hospitality and financial support of the Los Alamos National Laboratory where part of this work has been performed.

References

1. C. D. Batista, J. Bonča and J. E. Gubernatis, Phys. Rev. Lett. **88**, 187203 (2002).
2. C. D. Batista, J. Bonča and J. E. Gubernatis, cond-mat/0208604.
3. T. M. Rice, K. Ueda, Phys. Rev. Lett., **55**, No. 9, 995 (1985); Phys. Rev. B **34**, 6420 (1986).
4. C.M. Varma, W. Weber, and L.J. Randall, Phys. Rev. B **33**, 1015 (1986).
5. P. Fazekas, B. H. Brandow, Physica Scripta, **36**, 809 (1987).
6. P. Fazekas and E. Müller-Hartmann, Z. Phys. B - Condensed Matter **85**, 285 (1991).
7. M. C. Gutzwiller, Phys. Rev., **137**, No. 6A, A1726 (1965).
8. D. Vollhardt, Rev. Mod. Phys., **56**, No. 1, 99 (1984).
9. S. Zhang, J. Carlson and J. E. Gubernatis, Phys. Rev. Lett., **74**, 3652 (1995).
10. S. Zhang, J. Carlson and J. E. Gubernatis, Phys. Rev. B **55**, 7464 (1997); J. Carlson, J. E. Gubernatis, G. Ortiz, and Shiwei Zhang, Phys. Rev. B **59**, 12788 (1999).

TRANSPORT PROPERTIES OF HEAVY FERMION SYSTEMS

Proceedings of the ARW NATO Workshop Hvar, Croatia, October 2002

C. GRENZEBACH and G. CZYCHOLL
Institut für Theoretische Physik, Universität Bremen, D-28334 Bremen, Germany

Abstract. Heavy fermion systems (HFS) are described by the periodic Anderson model (PAM), which is studied within the dynamical mean-field theory (DMFT), mapping the PAM on an effective single-impurity Anderson model (SIAM) to be determined selfconsistently, which becomes exact in the limit of high spatial dimensions, $d \to \infty$. We use the modified perturbation theory (MPT) as approximation for the effective SIAM. The MPT is exact up to second order in the Coulomb correlation U and simultaneously reproduces the exact results for the atomic limit and the lowest moments. Within this approximation we have calculated the temperature dependence of the resistivity, the thermopower and the frequency and temperature dependence of the dynamical conductivity. For all these quantities the typical HFS-behavior is qualitatively well reproduced within our treatment.

Key words: Heavy Fermions, Resistivity, Thermo(electric) Power, Temperature Dependence of Transport Quantities, Periodic Anderson Model, Dynamical Mean-Field Theory, Modified Perturbation Theory, Dynamical Conductivity

1. Introduction

Besides the specific heat (γ-coefficient) and the magnetic susceptibility the unusual transport properties are most characteristic for heavy fermion systems (HFS). Pure metallic HFS, for example $CePd_3$ (1), $CeCu_6$ (2), have a small zero temperature resistivity $R(T = 0)$, a sharp increase of $R(T)$ for low temperature T, a maximum at some characteristic temperature T^*, and a $R(T)$ decreasing with increasing T, i.e. a negative temperature coefficient (NTC), for higher temperatures. There are, however, also some HFS (for instance $UPt_3, YbCuAl$) for which no maximum but only a flattening and a plateau behavior of $R(T)$ is observed for higher T.

Another interesting transport quantity is the thermoelectric power (TEP). In HFS the TEP $S(T)$ may have a sign change at low T (6, 7, 8, 9, 10) and it can assume giant values; much of the recent interest in the heavy fermion

A.C. Hewson and V. Zlatić (eds.), Concepts in Electron Correlation, 327–334.

thermo-electricity is due to the belief that some HFS with a thermopower larger than 150 μV/K might be useful for application.

In the frequency dependent (dynamical) conductivity $\sigma(\omega, T)$ one usually observes a narrow Drude peak at low T and an additional peak at higher frequency. With increasing T the Drude peak quickly disappears and the finite frequency peak becomes less pronounced and there is a crossover to a behavior with one small, but broad peak around zero frequency (11).

Finally, we mention the Kondo insulators like SmB_6 and $Ce_3Bi_4Pt_3$ which show clear evidence of a gap formation at low T resulting in an activation behavior of $R(T)$ for very low T and an (exponentially) vanishing conductivity $\sigma(\omega, T)$ for very low T and low ω, whereas for higher T the behavior is similar as in the metallic HFS. The thermopower can be very large also for the Kondo insulators.

In this contribution we present an oversimplified model and approximation within which all mentioned experimentally observed transport properties being typical for HFS can be reproduced and explained.

2. Model

It is well accepted that the basic electronic properties of HFS are described by the periodic Anderson model (PAM), which contains a conduction band, strongly correlated f-levels localized at the sites of a lattice simulating the rare-earth or actinide ions, and a hybridization between f-states and conduction electron states. For real rare earth or actinide atoms or ions the f-shells are highly degenerate, namely 14-fold degenerate in the bare (hydrogen like) atom, which is split into a 6- and an 8-fold degenerate level due to spin-orbit coupling, and this degeneracy is further reduced in a crystal due to crystalline electric fields so that the lowest f-level may be only two-fold degenerate. But the crystal-field (CF) split level is thermally accessible so that taking into account only a two-fold degenerate f-level is probably not sufficient to account for the full temperature dependence of the relevant physical quantities.

Nevertheless for the actual calculations presented in this paper the standard PAM with only a two-fold degeneracy of the f-levels has been used, i.e. (with $n^f_{\mathbf{R}\sigma} := f^\dagger_{\mathbf{R}\sigma} f_{\mathbf{R}\sigma}$)

$$H = \sum_{\mathbf{k}\sigma} \epsilon_{\mathbf{k}} c^\dagger_{\mathbf{k}\sigma} c_{\mathbf{k}\sigma} + \sum_{\mathbf{R}\sigma} \left[E_f n^f_{\mathbf{R}\sigma} + \frac{U}{2} n^f_{\mathbf{R}\sigma} n^f_{\mathbf{R}-\sigma} + V(c^\dagger_{\mathbf{R}\sigma} f_{\mathbf{R}\sigma} + c.c.) \right] \quad (1)$$

where $\epsilon_{\mathbf{k}} = 2t \sum_{l=1}^{d} \cos(k_l a)$ denotes the band electron dispersion (t nearest neighbor hopping, a lattice constant), for which a tight-binding model on a simple-cubic lattice in dimension d is assumed. To calculate transport quantities one needs also the current operator, and for the current operator being consistent with the PAM the component in x-direction is given by

$$j_x = \sum_{\mathbf{k}\sigma} \frac{\partial \epsilon_{\mathbf{k}}}{\partial k_x} c^\dagger_{\mathbf{k}\sigma} c_{\mathbf{k}\sigma} \Big[+ \underbrace{\frac{\partial V_k}{\partial k_x} (f^\dagger_{\mathbf{k}\sigma} c_{\mathbf{k}\sigma} + c.c.)}_{= 0, \text{ if V k-independent}} \Big] \tag{2}$$

So for k-independent hybridization V one has in site representation

$$j_x = ita \sum_{\mathbf{R}\Delta_x\sigma} (c^\dagger_{\mathbf{R}+\Delta_x\sigma} c_{\mathbf{R}\sigma} - c^\dagger_{\mathbf{R}\sigma} c_{\mathbf{R}+\Delta_x\sigma}) \tag{3}$$

(i imaginary unit). Being interested in transport quantities two-particle Green functions for the PAM have to be calculated. To calculate the conductivity we start from the Kubo formula for the (diagonal element of the) frequency (ω-) dependent conductivity (tensor), which (according to standard linear response theory (12, 13, 14)) is given by the current-current response function:

$$\sigma(\omega) = -\frac{1}{\omega + i0} \mathrm{Im}\chi_{j_x,j_x}(\omega + i0) \tag{4}$$

$$\chi_{j_x,j_x}(z) = \ll j_x; j_x \gg_z = -i \int_0^\infty dt' e^{izt'} < [j_x(t'), j_x(0)]_- > \tag{5}$$

The static conductivity can be obtained as the zero frequency ($\omega \to 0$) limit of the dynamical conductivity and then the TEP follows by means of the standard Mott relation (15). So one must determine two-particle Green functions of the kind $\ll c^\dagger_{\mathbf{k}\sigma} c_{\mathbf{k}\sigma}; c^\dagger_{\mathbf{k}'\sigma'} c_{\mathbf{k}'\sigma'} \gg_z$. Using standard relations of many-body theory (12) the two-particle Green functions can be obtained from (products of) one-particle Green functions and additional vertex corrections. The one-particle Green functions are defined by

$$G^{ab}_{\mathbf{k}\sigma}(z) = \ll a_{\mathbf{k}\sigma}; b^\dagger_{\mathbf{k}\sigma} \gg_z = -i \int_0^\infty dt' e^{izt'} < [a_{\mathbf{k}\sigma}(t'), b^\dagger_{\mathbf{k}\sigma}(0)]_+ > \tag{6}$$

(with $a, b \in \{c, f\}$). The f-electron selfenergy is defined by:

$$\begin{pmatrix} \ll c_{\mathbf{k}\sigma}; c^\dagger_{\mathbf{k}\sigma} \gg_z & \ll c_{\mathbf{k}\sigma}; f^\dagger_{\mathbf{k}\sigma} \gg_z \\ \ll f_{\mathbf{k}\sigma}; c^\dagger_{\mathbf{k}\sigma} \gg_z & \ll f_{\mathbf{k}\sigma}; f^\dagger_{\mathbf{k}\sigma} \gg_z \end{pmatrix} = \begin{pmatrix} z - \epsilon_{\mathbf{k}} & -V \\ -V & z - E_f - \Sigma_{\mathbf{k}f}(z) \end{pmatrix}^{-1} \tag{7}$$

3. Approximations

We study the PAM within the dynamical mean-field theory (DMFT) (19), which becomes correct in the limit of infinite dimension $d \to \infty$, $t \to 0$

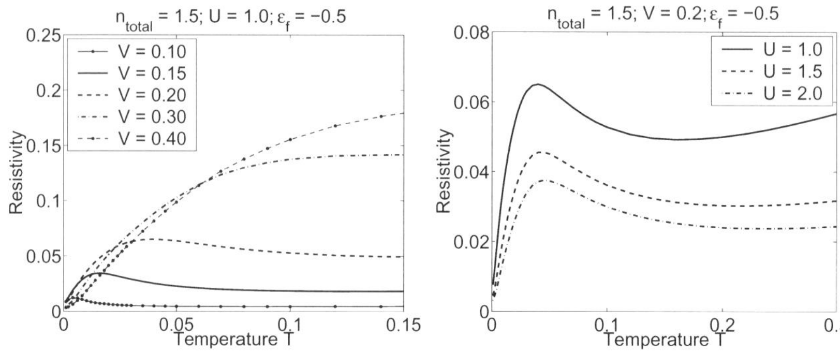

Figure 1. T-dependence of resistivity for $E_f = -0.5, n_{tot} = 1.5$ and for $U = 1$ and different hybridizations V (left panel) and fixed hybridization $V = 0.2$ and different U (right panel)

keeping dt^2 = constant (16). Within DMFT certain simplifications become exact, which extremely simplify practical calculations. It has been shown (17) that these simplifications are already a reasonable additional approximation for realistic dimension $d = 3$, i.e. 3-dimensional systems behave already similar as systems in the mathematical limit $d \to \infty$. The most important simplification is that the selfenergy becomes site-diagonal (k-independent), i.e. the so called "local approximation" becomes exact within DMFT and is a reasonable approximation for $d = 3$. Consistent with the site diagonality of the self energy is the vanishing of vertex corrections for the current-current response functions. Then the frequency dependent conductivity is simply given by (18)

$$\sigma(\omega, T) = \frac{e^2 a^{2-d}}{2\pi\hbar} t^2 \int dE \, \frac{f(E) - f(E+\omega)}{\omega} \, L(E, E+\omega) \tag{8}$$

where $f(E)$ is the Fermi function and the function $L(E_1, E_2)$ is defined by

$$L(E_1, E_2) = \frac{2}{N} \sum_{\mathbf{k}\sigma} \Big(\mathrm{Im} G^c_{\mathbf{k}\sigma}(E_1 + i0) \mathrm{Im} G^c_{\mathbf{k}\sigma}(E_2 + i0) \Big) \tag{9}$$

where

$$G^c_{\mathbf{k}\sigma}(z) = \langle\langle c_{\mathbf{k}\sigma}; c^\dagger_{\mathbf{k}\sigma} \rangle\rangle_z = \frac{1}{z - \frac{V^2}{z - E_f - \Sigma_f(z)} - \varepsilon_k} \tag{10}$$

is the band electron Green function of the PAM and $\Sigma_f(z)$ the (k-independent) f-electron selfenergy. The static conductivity (and the resistivity $R(T)$ as its inverse) follow from the limit $\omega \to 0$, and the TEP is also determined by the function $L(E_1, E_2)$ according to (15)

$$S = \frac{\int dE \left(-\frac{df}{dE} \right) (E - \mu) L(E, E)}{eT \int dE \left(-\frac{df}{dE} \right) L(E, E)} \tag{11}$$

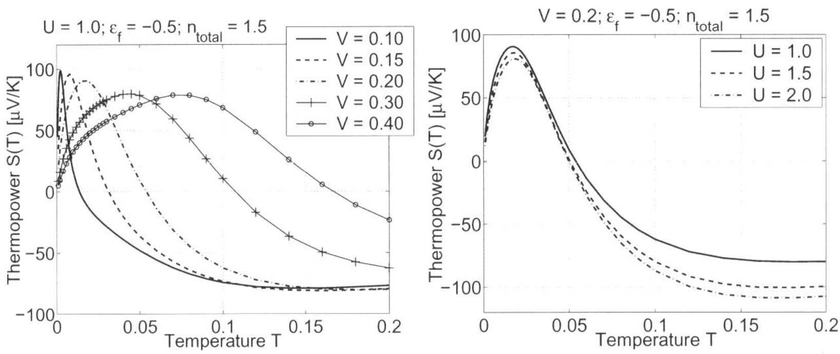

Figure 2. T-dependence of thermopower, otherwise as in Fig.1

Therefore, in the limit $d \to \infty$ the transport quantities of interest can be obtained from the one-particle Green functions or the (k-independent) selfenergy alone.

But a full exact solution of the PAM is not possible even in infinite dimensions and one needs additional approximations. Within DMFT a self-consistent mapping on an effective single-impurity Anderson model (SIAM) is possible. (19) The DMFT selfconsistency relation reads

$$\begin{aligned} G_R^{fSIAM}(z) &= \frac{1}{z - E_f - \Sigma_f(z) - \Delta(z)} = \\ G_R^{fPAM}(z) &= \frac{1}{N} \sum_{\mathbf{k}} \frac{1}{z - E_f - \Sigma_f(z) - \frac{V^2}{z - \epsilon_{\mathbf{k}}}} \end{aligned} \tag{12}$$

Here the "bath" Green function $\Delta(z)$ (i.e. the effective SIAM conduction band Green function) has to be determined selfconsistently. But the selfenergy for the effective SIAM is not known rigorously and an additional suitable approximation for the SIAM has to be used. We have applied the "modified perturbation theory" (MPT), which starts from the following ansatz (20, 21, 22)

$$\Sigma_f(z) = U n_{f-\sigma} + \frac{\alpha \Sigma_f^{SOC}(z)}{1 - \beta \Sigma_f^{SOC}(z)} \tag{13}$$

where $\Sigma_f^{SOC}(z)$ is the second order contribution in U to the SIAM selfenergy relative to the Hartree-Fock solution and the parameters α, β can be determined by the condition that the atomic limit (of vanishing V) and an additional criterion (Fermi liquid sum rule, reproduction of the first four moments) are fulfilled. The MPT has the advantage that it is exact up to order U^2 and the atomic limit is fulfilled simultaneously.

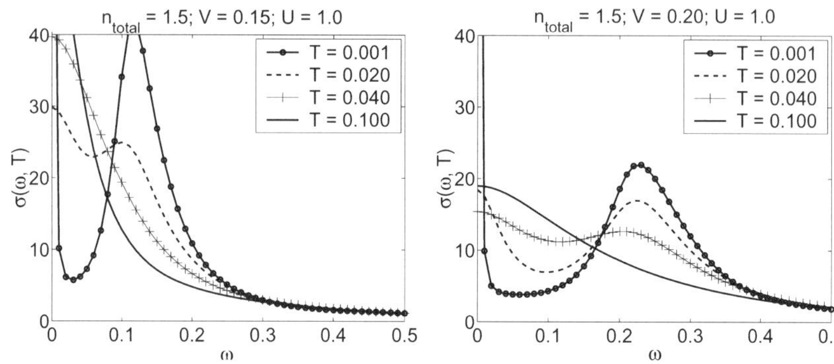

Figure 3. Frequency dependence of dynamical conductivity for different T, $E_f = -0.5, n_{tot} = 1.5, U = 1$ and hybridization $V = 0.15$ (left panel) and $V = 0.2$ (right panel)

4. Results

For practical calculations we choose a semielliptical model density of states for the unperturbed conduction band, i.e. we use the assumption

$$\frac{1}{N}\sum_{\mathbf{k}}\frac{1}{z-\epsilon_{\mathbf{k}}} = 8z\cdot\left(1-\sqrt{1-(4z)^{-2}}\right), \tag{14}$$

which becomes exact for a Bethe lattice in the limit of infinite coordination number. Thereby we have chosen the unperturbed conduction band width as our energy (and temperature) unit. The chemical potential μ has to be determined selfconsistently for a given total number n_{tot} of electrons per site. Choosing $n_{tot} = 1.5$ μ does not fall into a possibly existing hybridization gap, i.e. the model describes a metallic situation. Typical results are shown in Figs. 1-3. The resistivity $R(T)$ is small for $T \to 0$, has a rapid increase with increasing T, a maximum at T^*, and an NTC for higher $T > T^*$. With increasing hybridization V T^* also increases and for larger V one can get a behavior with no maximum but a plateau behavior. Changing U for fixed V only the absolute value of $R(T)$ varies whereas T^* remains fixed. In the thermopower $S(T)$ we observe the large absolute value (of the magnitude $100\mu V/K$), a low T extremum and a sign change at a temperature $\sim T^*$. In the frequency dependent (optical) conductivity $\sigma(\omega, T)$ (Fig.3) we observe a Drude peak and an additional (mid infrared) peak at finite frequency for low T. With increasing T we obtain a crossover to a behavior with only one rather broad peak around zero frequency. The position of the mid infrared peak obviously increases with increasing hybridization V.

Choosing $n_{tot} = 2, E_f = -0.5, U = 1$ we have the "symmetric PAM" and the chemical potential μ falls into a hybridization gap at zero T; this

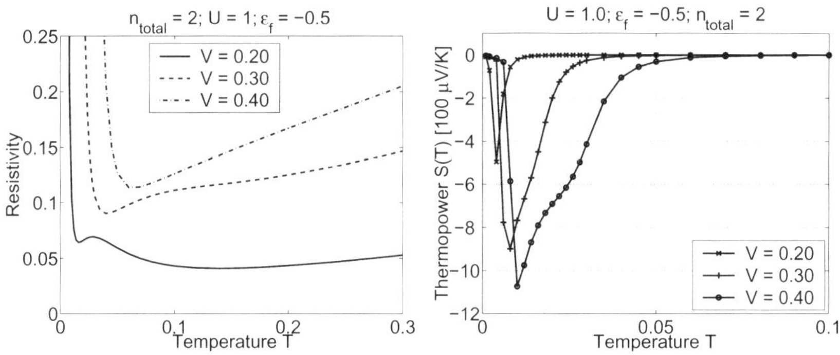

Figure 4. T-dependence of $R(T)$ (left panel) and $S(T)$ (right panel) for $n_{tot} = 2$, $E_f = -0.5, U = 1, V = 0.2$

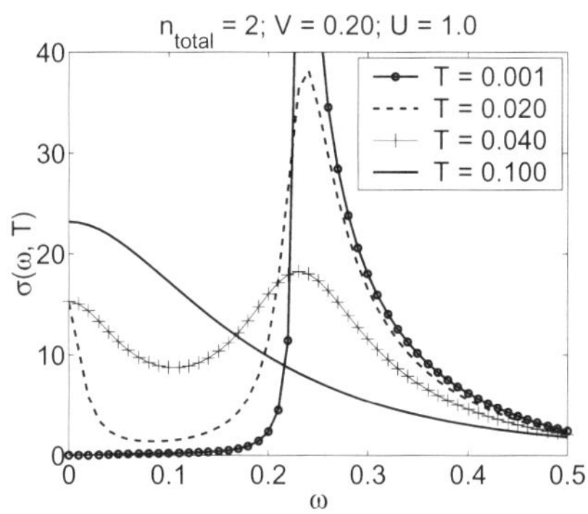

Figure 5. ω- and T-dependence of $\sigma(\omega, T)$ for $n_{tot} = 2$, $E_f = -0.5, U = 1$ and $V = 0.2$

situation may, therefore, model a Kondo insulator. Results for $R(T), S(T)$ and $\sigma(\omega, T)$ are shown in Figs. 4, 5. The hybridization gap (of width Δ) manifests itself in an activation behavior, i.e. an exponential increase of $R(T)$ for very low T and in a vanishing of $\sigma(\omega, T = 0)$ for $\omega < \Delta$. With increasing T the hybridization gap disappears and we get a crossover to a behavior similar as in metallic HFS for sufficiently high T. The thermopower is also extremely large, even one magnitude larger than in the metallic HFS.

5. Conclusion

The most characteristic features observed in the T-dependence of the resistivity and the thermopower and in the T- and frequency-(ω-)dependence of the dynamical (optical) conductivity of HFS can qualitatively be reproduced within simple approximations of the PAM, in particular within the DMFT/MPT. This investigation can be extended into several directions. The effects of impurities and of alloying (and of "chemical pressure") on $R(T), S(T)$ and $\sigma(T, \omega)$ can be modelled using additionally the CPA for disordered systems, similarly as it has been done for $R(T)$ in (23) using the SOPT. The treatment can also be applied to a more realistic version of the PAM with a more realistic (higher) orbital degeneracy of the f-levels and more realistic assumptions for the hybridization (dispersion); then also crystal field effects can be studied and included. Furthermore, using a two-channel Anderson model it should be possible to study also non-Fermi liquid systems within this approach.

References

1. P. Scoboria, J. E. Crow, T. Mihalisin, J. Appl. Phys. **50**, 1895 (1979).
2. Y. Onuki, T. Komatsubara, J. Magn.& Magn. Matter **63** & **64**, 281 (1987).
3. F.J.Blatt et al., *Thermoelectric Power of Metals*, Plenum New York (1962); R.D. Barnard, *Thermoelectricity in Metals and Alloys*, Taylor & Francis, London (1962).
4. F. Steglich, Festkörperprobleme XVII, 319 (1977).
5. D. Jaccard et al., J. Magn. Magn. Mater.**47 & 48**, 23 (1985).
6. E. Bauer, Adv. Phys. **40**, 417 (1991).
7. P. Link, D. Jaccard and P. Lejay, Physica B **225**, 207 (1996).
8. M. Očko, B. Bushinger, C. Geibel and F. Steglich, Physica B **259-261**, 87 (1999).
9. D. Huo, K. Mori, T. Kuwai, S. Fukuda, Y. Isikawa, J. Sakurai, Physica B**281 & 282**, 101 (2000).
10. J. Sakurai, D. Huo, D. Kato, T. Kuwai, Y. Isikawa, K. Mori, Physica B**281 & 282**, 98 (2000).
11. B.C.Webb, A.J.Sievers, T.Mihalisin, Phys. Rev. Letters **57**, 1951 (1986).
12. G. Mahan, *Many-Particle Physics*, Ch.3.8 and 7, Plenum New York (1990).
13. O. Madelung, *Introduction to Solid-State Theory*, Springer Heidelberg (1978).
14. G. Czycholl, *Theoretische Festkörperphysik*, Vieweg Wiesbaden (Springer Heidelberg) (2000).
15. G. Mahan in: Solid State Physics, Vol. **51**, p.82, H.Ehrenreich, F.Spaegen (eds.), Academic Press (1997).
16. W. Metzner and D. Vollhardt, Phys. Rev. Letters **62**, 324 (1989).
17. H.Schweitzer, G.Czycholl, Solid State Commun. **74**, 735 (1990); Z. Physik B **83**, 93 (1991).
18. H. Schweitzer and G. Czycholl, Phys. Rev. Letters **67**, 3724 (1991).
19. A. Georges, G. Kotliar, W. Krauth, M.J. Rozenberg, Rev. Mod. Phys. **68**, 13 (1996).
20. A. Martin-Rodero, E.Louis, F. Flores, C.Tejedor, Phys. Rev. B **33**, 1814 (1986).
21. H. Kajueter, G. Kotliar, Phys. Rev. Letters **77**, 131 (1996).
22. D. Meyer, W. Nolting, Phys. Rev. B **62**, 5657 (2000).
23. S. Wermbter, K. Sabel, G.Czycholl, Phys. Rev. **B 53**, 2528 (1996).

TRANSPORT PROPERTIES OF CORRELATED ELECTRONS IN HIGH DIMENSIONS

Proceedings of the ARW NATO Workshop Hvar, Croatia, October 2002

N. BLÜMER and P. G. J. VAN DONGEN
Institut für Physik, Johannes Gutenberg-Universität,
55099 Mainz, Germany

Abstract. We develop a new general algorithm for finding a regular tight-binding lattice Hamiltonian in infinite dimensions for an arbitrary given shape of the density of states (DOS). The availability of such an algorithm is essential for the investigation of broken-symmetry phases of interacting electron systems and for the computation of transport properties within the dynamical mean-field theory (DMFT). The algorithm enables us to calculate the optical conductivity fully consistently on a regular lattice, e.g., for the semi-elliptical (Bethe) DOS. We discuss the relevant f-sum rule and present numerical results obtained using quantum Monte Carlo techniques.

Key words: High Dimensions, Infinite Dimensions, Correlated Electrons, Optical Conductivity, Quantum Monte Carlo, Transport, Density of States, General Dispersion Method, F-Sum Rule

1. Introduction

Microscopic studies of strongly correlated electron systems require methods which take the Coulomb repulsion between electrons explicitly into account. Many features of such systems can be modeled using the single-band Hubbard model (1)

$$\hat{H} = \hat{H}_0 + \hat{H}_{\text{int}} = \sum_{ij,\sigma} t_{ij}\, \hat{c}^{\dagger}_{\boldsymbol{R}_i\sigma} \hat{c}_{\boldsymbol{R}_j\sigma} + U \sum_i \hat{n}_{\boldsymbol{R}_i\uparrow} \hat{n}_{\boldsymbol{R}_i\downarrow}\,, \tag{1}$$

where the operators $\hat{c}^{\dagger}_{\boldsymbol{R}_i\sigma}$ and $\hat{c}_{\boldsymbol{R}_i\sigma}$ create and destroy electrons of spin σ on site $\boldsymbol{R}_i$, respectively; $\hat{n}_{\boldsymbol{R}_i\sigma}$ measures the corresponding occupancy. A general nonperturbative treatment of this model is only possible in the limit of infinite dimensionality where the dynamical mean-field theory becomes exact: due to a local self-energy the model reduces for $d \to \infty$ to a

A.C. Hewson and V. Zlatić (eds.), Concepts in Electron Correlation, 335–343.

single impurity Anderson model plus a self-consistency equation. For homogeneous phases, local properties then only depend on the lattice via the noninteracting DOS $\rho(\epsilon) := \frac{1}{N}\sum_{\boldsymbol{k}}\delta(\epsilon-\epsilon_{\boldsymbol{k}})$. In the simplest case of uniform nearest-neighbor hopping on a hypercubic (hc) lattice, the dispersion reads $\epsilon_{\boldsymbol{k}}^{\rm hc} = -2t\sum_{\alpha=1}^{d}\cos(k_\alpha)$ which for the proper scaling $t = t^*/\sqrt{2d}$ leads to a Gaussian DOS $\rho^{\rm hc}(\epsilon) = \exp[-\epsilon^2/(2t^{*2})]/(\sqrt{2\pi}t^*)$.

Vertex corrections to the optical conductivity $\sigma(\omega)$ vanish (2) in the limit $d\to\infty$ so that it may be expressed in the isotropic case as (3)

$$\sigma(\omega) = \sigma_0\int_{-\infty}^{\infty} d\epsilon\,\tilde{\rho}(\epsilon)\int_{-\infty}^{\infty} d\omega'\,A_\epsilon(\omega')A_\epsilon(\omega'+\omega)\frac{n_{\rm f}(\omega')-n_{\rm f}(\omega+\omega')}{\omega}\,. \tag{2}$$

Here, $A_\epsilon(\omega)$ is the "momentum" dependent spectral function, $n_{\rm f}(\omega')$ is the Fermi function, $\sigma_0 = 2\pi e^2/\hbar^2\,N/V$, and

$$\tilde{\rho}(\epsilon) := \frac{1}{N}\sum_{\boldsymbol{k}}|\boldsymbol{v}_{\boldsymbol{k}}|^2\delta(\epsilon-\epsilon_{\boldsymbol{k}}) =: \langle|\boldsymbol{v}_{\boldsymbol{k}}|^2\rangle(\epsilon)\,\rho(\epsilon)\,. \tag{3}$$

Note that the frequency- and interaction-dependent part in (2) depends on the lattice only via the DOS while the explicitly lattice-dependent part $\tilde{\rho}(\epsilon)$ is universal, i.e., independent of the interaction U, filling, and temperature. In the hypercubic case, the momentum dependence of the Fermi velocity $\boldsymbol{v}_{\boldsymbol{k}}$ becomes irrelevant in (3) for $d\to\infty$; for unit hopping and lattice spacing, one observes $\tilde{\rho}^{\rm hc}(\epsilon) = \rho^{\rm hc}(\epsilon)$. As a consequence, the optical f-sum is then proportional to the kinetic energy:

$$\int_0^{\infty} d\omega\,\sigma(\omega) = -\frac{\sigma_0}{4}\langle\hat{H}_0\rangle\,. \tag{4}$$

However, the hc DOS is unbounded which is hardly compatible with the single-band assumption. In fact, no regular lattice model with sharp band edges in $d\to\infty$ could be constructed so far. In this situation, many DMFT studies have focussed on the so-called Bethe lattice which is not a regular lattice, but a tree in the sense of graph theory as shown in Fig. 1a. The semi-elliptic DOS $\rho(\epsilon) = \sqrt{4-\epsilon^2}/(2\pi)$ of this model (for $Z\to\infty$) fixes the local properties of the model; transport, however, is a priori undefined. A derivation of $\sigma(\omega)$ directly for the Bethe tree (using the level-picture Fig. 1a) by Chung and Freericks (4) is still incomplete (5); up to a factor of 3, the same expression $\tilde{\rho}(\epsilon)\propto(4-\epsilon^2)\rho(\epsilon)$ was obtained (6) in a heuristic scheme by enforcing the hc f-sum rule (4). An alternative direct approach (7) fails to describe the coherent transport expected in the metallic regime.

The local DMFT problem is unchanged when a finite number of hopping bonds per site are added. Therefore, the periodically stacked Bethe lattice (Fig. 1b) is still a Bethe lattice in the DMFT sense; potentially coherent

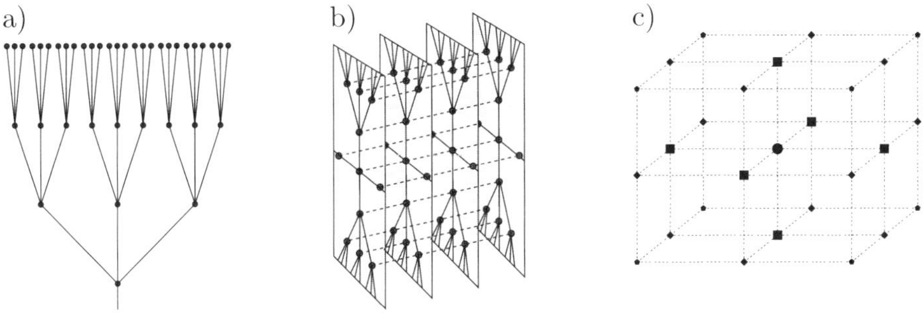

Figure 1. Bethe lattice: **a)** conventional tree level picture for coordination $Z = 4$, **b)** stacked Bethe lattice, **c)** redefined Bethe lattice as a regular lattice with cubic symmetry: here, symbols denote equal hopping matrix elements relative to a fixed site (central circle).

transport is then well-defined (only) in stacking direction (8). A semi-elliptic DOS also results from fully disordered hopping on lattices of arbitrary topology; in this case, $\sigma(\omega)$ is incoherent (9).

We will in the following construct and evaluate a new definition for $\sigma(\omega)$ compatible with a semi-elliptic DOS. This definition is unique by being based on a regular lattice as illustrated in Fig. 1c and by leading to an isotropic conductivity which is coherent in the noninteracting limit.

2. General Dispersion Method

We rewrite the translation-invariant noninteracting Hamiltonian,

$$\hat{H}_0 = \sum_{i,\sigma} \sum_{\boldsymbol{\tau}} t_{\boldsymbol{\tau}}\, \hat{c}^{\dagger}_{\boldsymbol{R}_i\sigma} \hat{c}_{\boldsymbol{R}_i+\boldsymbol{\tau},\sigma} = \sum_{\boldsymbol{k},\sigma} \epsilon(\boldsymbol{k})\, \hat{n}_{\boldsymbol{k}\sigma}, \tag{5}$$

where contributions to the dispersion may be classified by the taxi-cab hopping distance $\|\boldsymbol{\tau}\| = \sum_{\alpha=1}^{d} |\tau_\alpha|$:

$$\epsilon(\boldsymbol{k}) = \sum_{D=1}^{\infty} \epsilon_D(\boldsymbol{k}), \qquad \epsilon_D(\boldsymbol{k}) = \sum_{\|\boldsymbol{\tau}\|=D} t_{\boldsymbol{\tau}} e^{i\boldsymbol{\tau}\cdot\boldsymbol{k}}. \tag{6}$$

In high dimensions, only vectors of the form $\boldsymbol{\tau} = \sum_{i=1}^{D} \boldsymbol{e}_{\alpha_i}$ with pairwise different directions α_i need to be considered. This follows from the fact that the fraction of neglected vectors (with $|\boldsymbol{\tau}\cdot\boldsymbol{e}_\alpha| > 1$ for some direction α) vanishes as $1/d$. Furthermore, the considered vectors are of minimal Euclidean length $|\boldsymbol{\tau}|$ hinting at maximal overlap, i.e., largest $|t_{\boldsymbol{\tau}}|$ for fixed taxi-cab distance $\|\boldsymbol{\tau}\|$ and fixed D.

By deriving a recursion relation for $\epsilon_D(\boldsymbol{k})$ we have established that

$$\epsilon(\boldsymbol{k}) = \sum_{D=1}^{\infty} \frac{t^*_D}{\sqrt{D!}} \operatorname{He}_D(\epsilon^{\mathrm{hc}}_{\boldsymbol{k}}) =: \mathcal{F}(\epsilon^{\mathrm{hc}}_{\boldsymbol{k}}). \tag{7}$$

Using the orthogonality of the Hermite polynomials, one may express the hopping matrix elements in terms of the transformation function $\mathcal{F}(x)$:

$$t_D^* = \frac{1}{\sqrt{2\pi D!}} \int_{-\infty}^{\infty} d\epsilon\, \mathcal{F}(\epsilon)\, \mathrm{He}_D(\epsilon)\, e^{-\epsilon^2/2} \,. \tag{8}$$

Specializing on the case of a monotonic transformation function $\mathcal{F}(x)$ (with derivative $\mathcal{F}'(x)$), we can write

$$\rho(\epsilon) = \frac{1}{\mathcal{F}'(\mathcal{F}^{-1}(\epsilon))}\, \rho^{\mathrm{hc}}(\mathcal{F}^{-1}(\epsilon)) \tag{9}$$

which leads to

$$\mathcal{F}^{-1}(\epsilon) = \sqrt{2}\, \mathrm{erf}^{-1}\Big(2\int_{-\infty}^{\epsilon} d\epsilon' \rho(\epsilon') - 1\Big) \,. \tag{10}$$

Furthermore, the Fermi velocity $\boldsymbol{v_k} = \nabla \epsilon_{\boldsymbol{k}}$ can be computed:

$$\boldsymbol{v_k} = \mathcal{F}'(\mathcal{F}^{-1}(\epsilon))\boldsymbol{v}_{\boldsymbol{k}}^{\mathrm{hc}} = \frac{\rho^{\mathrm{hc}}(\sqrt{2}\,\mathrm{erf}^{-1}(2\int_{-\infty}^{\epsilon} d\epsilon' \rho(\epsilon') - 1))}{\rho(\epsilon)}\, \boldsymbol{v}_{\boldsymbol{k}}^{\mathrm{hc}} \,. \tag{11}$$

A practical application of the general formalism proceeds as follows:

1. compute $\mathcal{F}^{-1}(\epsilon)$ from arbitrary target DOS $\rho(\epsilon)$ using (10)
2. invert function (numerically or analytically) to obtain $\mathcal{F}(\epsilon)$
3. evaluate transport properties, e.g., $\langle |\boldsymbol{v_k}|^2\rangle(\epsilon)$ or $\tilde{\rho}(\epsilon)$ using (11)
4. optionally determine microscopic model parameters t_D^* using (8)

The only choice inherent in this procedure beyond the usual assumptions for large dimensions is contained in step 1 which by construction produces a monotonic transformation function $\mathcal{F}$. The optical f-sum rule reads

$$\int_0^{\infty} d\omega\, \sigma_{xx}(\omega) = \frac{\sigma_0}{4d} \langle \frac{\tilde{\rho}'(\epsilon)}{\rho(\epsilon)} \rangle = \frac{\sigma_0}{4d} \Big\langle \Big[\mathcal{F}''(\mathcal{F}^{-1}(\epsilon)) - \mathcal{F}^{-1}(\epsilon)\mathcal{F}'(\mathcal{F}^{-1}(\epsilon))\Big]\Big\rangle \,. \tag{12}$$

Here, the first equality follows from (2), i.e., is generally valid within the DMFT while the second expression in terms of the transformation is specific to the formalism developed within this section.

Example: Flat-band System One interesting limiting case of a monotonic transformation function which can be treated analytically is the step function $\mathcal{F}(x) = 2\Theta(x) - 1$ corresponding to a flat band DOS of the form $\rho(\epsilon) = (\delta(\epsilon - 1) + \delta(\epsilon + 1))/2$. For this case, the hopping matrix elements read (trivially, $t_{2n}^* = 0$):

$$t_{2n+1}^* = (-1)^n \sqrt{\frac{2}{\pi}} \frac{(2n-1)!!}{\sqrt{(2n+1)!}} \quad \stackrel{n\to\infty}{\longrightarrow} \quad (-1)^n (\pi n)^{-3/4} \,. \tag{13}$$

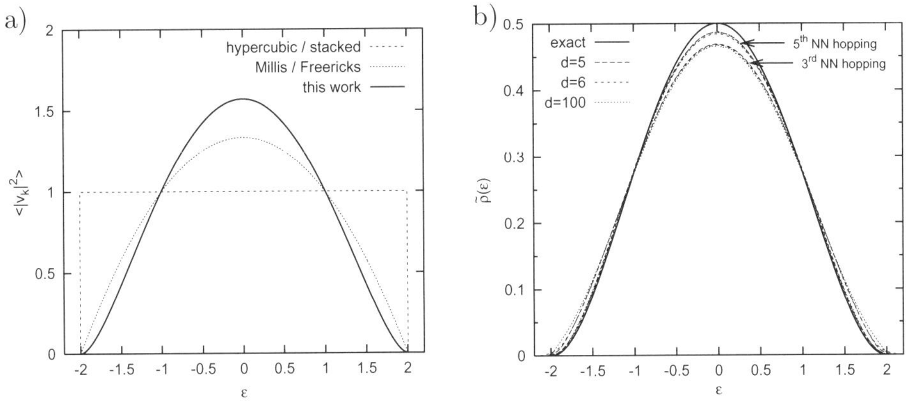

Figure 2. **a)** The average squared Fermi velocity $\langle|\boldsymbol{v_k}|^2\rangle(\epsilon)$ is constant for the hypercubic lattice (or for the x-component of a stacked lattice); in contrast it vanishes for the isotropic lattice defined in this work. For comparison, the form suggested by Millis is also shown. **b)** Resulting function $\tilde{\rho}(\epsilon)$ of the full isotropic model (solid line) in comparison with truncated models ($D_{\max} = 3$ or $D_{\max} = 5$), evaluated in finite dimensions $5 \leq d \leq 100$.

The asymptotic exponent $-3/4$ is only slightly smaller than the threshold value $-1/2$ required for a finite variance $\int_{-\infty}^{\infty} d\epsilon\, \epsilon^2 \rho(\epsilon) = \sum_D {t_D^*}^2$. For a rectangular model DOS, $t_D^* \sim 2^{-n} n^{-3/4}$ already decays exponentially fast.

3. Application to the "Bethe" semi-elliptic DOS

In this section, we will apply the new formalism to the Bethe semi-elliptic DOS in order to determine a corresponding tight-binding Hamiltonian defined on the hypercubic lattice with the same local properties as the Bethe lattice (with NN hopping) in the limit $d \to \infty$. From (11), we derive the average squared Fermi velocity defined in (3) in closed form:

$$\langle|\boldsymbol{v_k}|^2\rangle(\epsilon) = \frac{2\pi}{4-\epsilon^2} \exp\left[-2\left(\mathrm{erf}^{-1}\left(\frac{\epsilon\sqrt{1-\epsilon^2/4} + 2\arcsin(\epsilon/2)}{\pi}\right)\right)^2\right]. \quad (14)$$

Here, we have used the fact that $\langle|\boldsymbol{v_k}|^2\rangle(\epsilon)$ is effectively constant (and equals 1 for unit variance and lattice spacing) in the hypercubic case. The result (solid line in Fig. 2a) has all the qualitative features expected for this observable in any finite dimension: $\langle|\boldsymbol{v_k}|^2\rangle(\epsilon)$ is maximal near the band center, strongly reduced for large (absolute) energies and vanishes at the band edges: states at a (noninteracting) band edge do not contribute to transport. The violation of this principle in the stacked case (dashed lines in Fig. 2a), which corresponds to an application of the hc formalism to the Bethe DOS with $\langle|\boldsymbol{v_k}|^2\rangle$ constant up to the band edges, is clearly

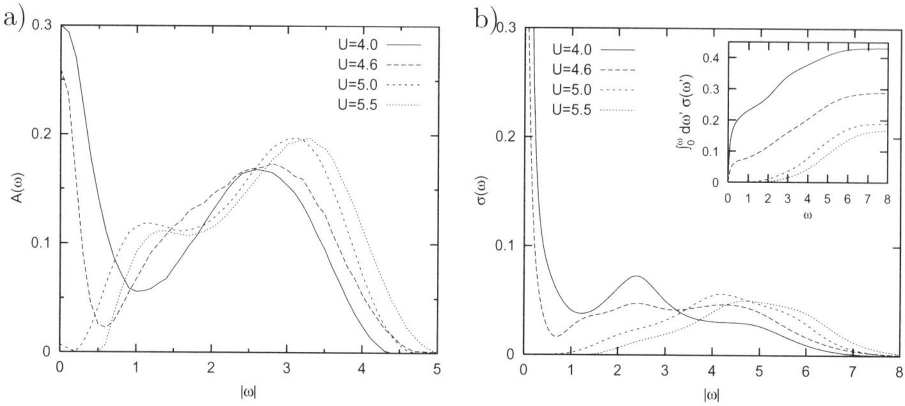

Figure 3. Numerical results for the half-filled Hubbard model with semi-elliptic DOS for $d \to \infty$ in the paramagnetic phase at $T = 0.05$. **a)** Local spectral function $A(\omega)$ obtained from QMC (using a discretization $\Delta\tau = 0.1$) and MEM. **b)** Optical conductivity $\sigma(\omega)$ for the isotropic "redefined Bethe lattice". The inset shows the partial f-sum $\int_0^\omega d\omega' \sigma(\omega')$.

pathological. Therefore, our method has not only the merit of yielding isotropic transport, but also of avoiding unphysical behavior.

In order to determine the microscopic model, we have to apply (8) to the numerically evaluated transformation function $\mathcal{F}$. Again, the scaled hopping matrix elements fall off exponentially fast: only a fraction 10^{-3} of the total energy variance arises from hopping amplitudes beyond third nearest neighbors and only a fraction 10^{-6} results from hopping beyond 9th-nearest neighbors. This result suggests that properties of the model should be robust with respect to truncation. In fact, $\tilde{\rho}(\epsilon)$ (and consequently the definition of $\sigma(\omega)$) hardly changes when hopping is cut off beyond 3rd or 5th nearest neighbors, even when evaluated in finite dimensions as seen in Fig. 2b. This behavior is very general so that results for $\sigma(\omega)$ of a local theory in finite dimensions will depend on d predominantly via the interacting DOS $A(\omega)$ and only very little via $\tilde{\rho}(\epsilon)$.

The local spectral functions for $T = 0.05$, i.e., slightly below the critical temperature $T^* \approx 0.055$, are shown in Fig. 3a as obtained from QMC/MEM (5). In the metallic phase, the spectral density at the Fermi level ($\omega = 0$) is approximately pinned at the noninteracting value $\rho(0) = 1/\pi \approx 0.32$ for $U \lesssim 4.4$. The quasiparticle weight decreases drastically and a shoulder develops for $U \gtrsim 4.6$ before a gap opens for $U \gtrsim 4.8$. An application of (2) to these spectra for the isotropic model characterized by (14) yields the estimates for the optical conductivity $\sigma(\omega)$ shown in Fig. 3b. A low-frequency Drude peak (of Lorentzian form) and a mid-infrared peak at $\omega \approx U/2$ are present in the metallic phase and decay towards the metal-insulator transition at $U \approx 4.7$. For large U, the optical spectral weight

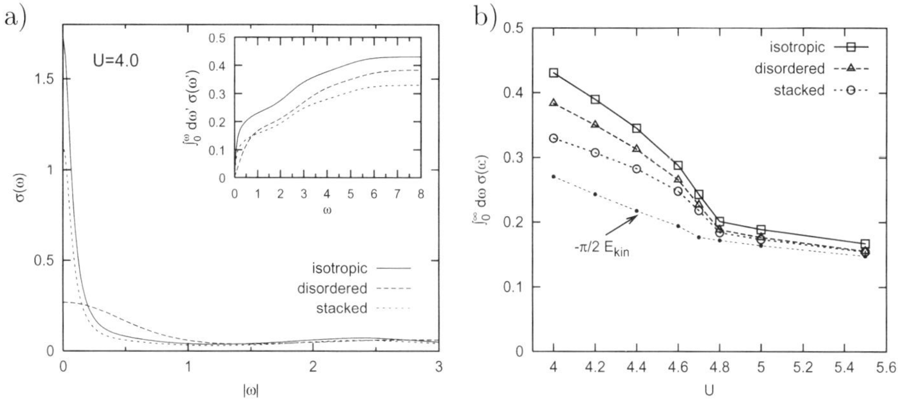

Figure 4. **a)** Optical conductivity $\sigma(\omega)$ for $T = 0.05$ and $U = 4.0$ for the new isotropic model with semi-elliptic DOS in comparison with disordered and stacked models consistent with the same DOS. **b)** Optical f-sum for $T = 0.05$.

concentrates in incoherent peaks at $\omega \approx U$. The inset of Fig. 3b shows the partial optical f-sums. As expected, both the contribution of the Drude peak and the total f-sum decrease for increasing U.

Figure 4a shows the impact of the definition for $\sigma(\omega)$ on the results for $U = 4.0$. Our isotropic model yields by far the largest contributions at small ω. While the stacked model leads to otherwise similar results, the low-frequency form of $\sigma(\omega)$ is qualitatively different in the disordered case (where a Drude peak is absent even for $U \to 0$). As seen in Fig. 4b, the f-sum is generally larger for the isotropic than for the stacked Bethe lattice, in particular in the metallic phase (while the disordered case is in-between). These differences can be attributed to the enhanced squared Fermi velocity in the isotropic model: The enhancement is largest near the Fermi surface, at $\epsilon = 0$ by a factor of $\pi/2 \approx 1.57$. A corresponding increase is expected of the Drude peak for small enough U and T when transport is dominated by states with $\epsilon \approx 0$. Since energy eigenstates spread out in momentum space at large U, the enhancement reduces (in general) to the integral $\int_{-\infty}^{\infty} d\epsilon\, \tilde{\rho}(\epsilon) = \sum_{D=1}^{\infty} D\, {t_D^*}^2$ which evaluates here to 1.05406.

In all three cases, the proportionality (4) of the f-sum to the kinetic energy as characteristic for the hc lattice is clearly violated. This is true even for the stacked case (which is otherwise similar to the hc case): while the f-sum is here proportional to the contribution to the kinetic energy associated with hopping in current direction (8), $\int_0^{\infty} d\omega\, \sigma_{xx}(\omega) = -\sigma_0 \langle \hat{T}_x \rangle / 4$, this contribution (which is negligible in the limit $Z \to \infty$) is not proportional to the total kinetic energy in this anisotropic case. A more relevant sum rule is derived from (12): $\int_0^{\infty} d\omega\, \sigma_{xx}(\omega) = \frac{t^2 a^2 \sigma_0}{4} \langle \frac{-\epsilon}{4-\epsilon^2} \rangle$.

4. Conclusion

We have presented a new general method for constructing regular lattice models with hypercubic (hc) symmetry, i.e. isotropic optical transport properties, in large dimensions. Previously, calculations of the optical conductivity $\sigma(\omega)$ of the Hubbard model in the limit $d \to \infty$ had been restricted to the hypercubic lattice (using NCA (3) or QMC (10)) or have ignored the lattice dependence: Applying the hc formalism to the Bethe semi-elliptic DOS, Rozenberg et. al (11) overlooked violations of the hc f-sum rule. All previous approaches specific to the Bethe DOS were associated with anisotropic or incoherent transport; the most interesting of these (4, 6) have not yet been linked rigorously to microscopic models.

Our method yields the first derivation for $\sigma(\omega)$ consistent with a semi-elliptic DOS that implies isotropic transport which is fully coherent in the noninteracting limit. This reinterpretation of the "Bethe lattice" (in the DMFT sense) as an isotropic, regular and clean lattice and the demonstration that the associated transport properties are robust (with respect to finite dimensionality or hopping range) removes, finally, the pathologies previously associated with the DMFT treatment of transport in connection with non-Gaussian DOSs. At essentially no additional cost, the method can also be used for computing properties such as transverse conductivities and thermopower; these vanish, however, in the particle-hole symmetric case considered in this paper. Our numerical results have shown that the precise definition of $\sigma(\omega)$ does matter, in particular within the metallic phase where transport is potentially most coherent. We have also found a general DMFT expression for the f-sum rule as well as a form specific to our new approach.

Acknowledgements

We gratefully acknowledge useful discussions with J. Freericks, D. Logan, A. Millis, and D. Vollhardt.

References

1. J. Hubbard, Proc. Roy. Soc. London **A276**, 238 (1963); M. C. Gutzwiller, Phys. Rev. Lett. **10**, 59 (1963); J. Kanamori, Prog. Theor. Phys. **30**, 275 (1963).
2. A. Khurana, Phys. Rev. Lett. **64**, 1990 (1990).
3. T. Pruschke, D. L. Cox and M. Jarrell, Phys. Rev. B **47**, 3553 (1993).
4. W. Chung and J. K. Freericks, Phys. Rev. B **57**, 11955 (1998). J. K. Freericks, private communication (2000, 2002).
5. N. Blümer, Ph.D. Thesis, Universität Augsburg (2002).

6. A. Chattopadhyay, A. J. Millis and S. Das Sarma, Phys. Rev. B **61**, 10738 (2000); A. J. Millis, private communication (2002).
7. M. P. H. Stumpf, Ph.D. Thesis, University of Oxford (1999).
8. G. S. Uhrig and R. Vlaming, J. Phys. Cond. Matter **5**, 2561 (1993).
9. V. Dobrosavljević and G. Kotliar, Phys. Rev. Lett. **71**, 3218 (1993).
10. M. Jarrell, J. K. Freericks and T. Pruschke, Phys. Rev. B **51**, 11704 (1995).
11. M. J. Rozenberg, G. Kotliar, H. Kajüter, G. A. Thomas, D. H. Rapkine, J. M. Honig, and P. Metcalf, Phys. Rev. Lett. **75**, 105 (1995).

FROM $CEIN_3$ TO $PUCOGA_5$: TRENDS IN HEAVY FERMION SUPERCONDUCTIVITY

Proceedings of the ARW NATO Workshop Hvar, Croatia, October 2002

J.L. SARRAO and J.D. THOMPSON
Los Alamos National Laboratory, Los Alamos, NM 87545

Abstract. We review the serendipitous discovery of superconductivity in $PuCoGa_5$ and its relationship to $CeMIn_5$ (M=Co, Ir, Rh). Outstanding issues in our understanding of the phase diagrams of $CeMIn_5$ are also discussed. Finally, we speculate about the relationship between pressure-induced superconductivity in $CeIn_3$ at 200 mK and ambient-pressure superconductivity in $PuCoGa_5$ at 18.5 K and the prospects for other high-temperature unconventional superconductors.

Key words: $PuCoGa_5$, $CeCoIn_5$, $CeIn_3$, Plutonium, Heavy Fermion Superconductivity

1. Crystal Chemistry

The connection between crystal structure and superconductivity has long been recognized if not understood rigorously (1). This is also true of the known heavy fermion superconductors, of which a large number crystalize in the tetragonal $ThCr_2Si_2$ crystal structure. The materials to be discussed here further confirm this trend. The $CeMIn_5$ compounds as well as $PuCoGa_5$ crystalize in the tetragonal $HoCoGa_5$ structure type (2, 3). This structure is a layered derivative of the cubic Cu_3Au structure in which units of $HoGa_3$ alternate with layers of $CoGa_2$. As such, $CeIn_3$ and $PuGa_3$ are the 'parent' compounds of the superconductors. In addition to these materials, there are a large family of $UMGa_5$ compounds reported; however, none of these are known to superconduct (4). The extent to which this is attributable to the properties of UGa_3 compared to $PuGa_3$ and $CeIn_3$ is an important outstanding question.

A.C. Hewson and V. Zlatić (eds.), Concepts in Electron Correlation, 345–351.

2. Heavy fermion superconductivity in $CeMIn_5$

The discovery of superconductivity in $CeMIn_5$ (M=Co,Ir,Rh) has increased significantly the number of known heavy fermion superconductors, and these materials are currently the subject of much experimental and theoretical effort. The three variants that have been reported - $CeRhIn_5$ (5), $CeIrIn_5$ (6), and $CeCoIn_5$ (2)- are isovalent, and each displays heavy Fermion superconductivity. Here, we summarize the basic properties of these materials and point out trends and relationships between and among them.

At ambient pressure, $CeRhIn_5$ orders antiferromagnetically at 3.8 K. T_N increases weakly with applied pressure, an unconventional dependence, until it vanishes near 16 kbar, at which point superconductivity, with a T_c of 2.1 K is observed (5). The superconductivity is bulk as evidenced by heat capacity (7) and nuclear magnetic resonance (8, 9) measurements, and the observed temperature dependencies are consistent with unconventional, d-wave pairing. The value of C/T just above T_N and T_c is $\sim$ 400 mJ/molK^2 in both cases, confirming the heavy Fermion character of $CeRhIn_5$.

$CeIrIn_5$ displays bulk superconductivity at 0.4 K that develops out of a state with $\gamma \sim 700$ mJ/molK^2. Again, heat capacity, thermal transport, and nuclear magnetic resonance measurements provide strong evidence for a nodal superconducting gap symmetry (10, 11, 12). An interesting wrinkle regarding $CeIrIn_5$ is that its resistivity vanishes near 1 K, well above the bulk transition temperature. The nature of this presumably filamentary state remains unclear, but it is robust as a function of applied magnetic field and 'indicates' the value of bulk T_c that can be approached with either applied pressure (13) or chemical substitution (14).

With an ambient-pressure T_c of 2.3 K, $CeCoIn_5$ displays the highest transition temperature of any heavy Fermion superconductor. There is also strong evidence for $CeCoIn_5$ being a d-wave superconductor (10, 12, 15). Applied pressure can further increase T_c to 2.6 K (at a pressure of 15 kbar) (16). The observed C/T of 300 mJ/molK^2 at T_c is rather modest, but the application of a magnetic field greater than Hc_2 reveals that γ is not fully developed at T_c(H=0), and, in fact, the field-induced normal state $\gamma \sim$ 1000 mJ/molK^2 is quite large and displays a non-Fermi-liquid temperature dependence. The H-T phase diagram of $CeCoIn_5$ is also anomalous in the exceptionally large values of Hc_2 observed ($\sim$ 120 kOe H $\|$ [100], $\sim$ 50 kOe H $\|$ [001]) and the observation of a first-order superconducting transition in the H-T plane (17, 18).

An additional surprising feature of the $CeMIn_5$ materials is the "robustness" of T_c of intermediate alloys, e.g., $CeM_{1-x}M'_xIn_5$ (M or M' = Co, Ir, Rh) (19). Single crystals can be grown for the full range of x for all M and M', and superconductivity is observed over wide ranges of

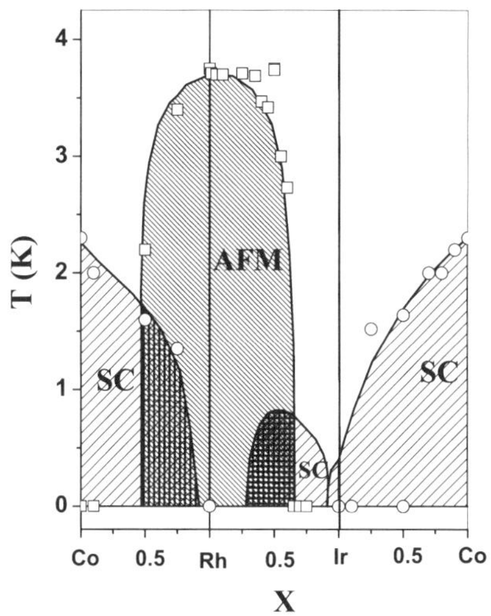

Figure 1. Generalized doping phase diagram for $CeMIn_5$. The left-hatched regions indicate antiferromagnetic order (AFM), while the right-hatched regions indicated superconductivity (SC). Note the substantial regions of coexistence of both superconductivity and antiferromagnetism. After (14, 19).

concentration. Thus, one is presented with the opportunity of exploring the competition and interaction of diverse ground states, e.g., magnetism and superconductivity (14, 20, 19). The range of x for which superconductivity can be observed is remarkable, ~ 0.7 for Rh-Ir alloys, and is rivaled only by the cuprates in terms of the flexibility of chemical tuning (See Figure 1). This suggests that the doping may be "out of plane" in analogy with the cuprates and emphasizes the layered nature of the crystal structure of $CeMIn_5$.

The role of crystallographic anisotropy and the extent to which $CeMIn_5$ is a two-dimensional version of $CeIn_3$ has been an issue since the discovery of these materials (5, 21). The evolution of superconductivity with doping reveals a surprising relationship between T_c and c/a, the ratio of tetragonal lattice constants (19). Taking the full range of $Ce(Rh,Ir,Co)In_5$ superconductors including stoichiometric compounds and doped alloys, one finds a linear relationship between T_c and c/a that is so steep that a two percent change in c/a increases T_c by a factor of five. The origin of such a strong coupling between these quantities remains unclear. In a general sense a correlation between two dimensionality and superconductivity is expected for magnetically mediated superconductors (22); however, the functional dependence here is remarkably strong. If this mechanism could be generalized to other materials, the potential consequences are obviously

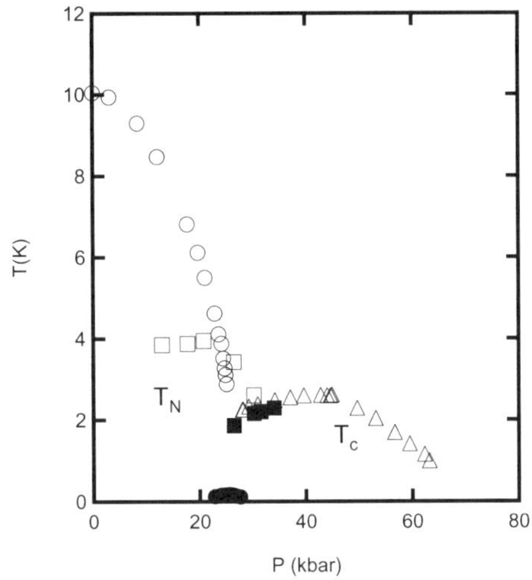

Figure 2. Generalized pressure phase diagram for $CeMIn_5$. The open(closed) circles indicate the evolution of antiferromagnetism(superconductivity) for $CeIn_3$. The open and closed squares have the same meaning for $CeRhIn_5$, with the pressure offset by 13 kbar from the measured data to account for chemical pressure effects. Similarly, the triangles represent the evolution of superconductivity with pressure in $CeCoIn_5$ including an overall shift of 28 kbar relative to $CeIn_3$. See (16, 23) for details.

enormous.

The extent to which proximity to a quantum critical point influences the properties of $CeMIn_5$ has also been the topic of much discussion. Non-Fermi liquid behavior has been observed in both $CeCoIn_5$ and $CeIrIn_5$ (6, 2, 24), and the extent to which magnetism and superconductivity are proximate to each other would suggest the the presence of one or more quantum critical points. Furthermore, it can be argued that $CeCoIn_5$ is equivalent to $CeRhIn_5$ under 16 kbar of pressure (16, 23). This supposition yields a phase diagram (see Figure 2) that not only produces a smooth evolution of superconducting transition temperature with pressure, T_c(P), but also yields a normal state phase diagram quite reminiscent of the high-T_c cuprates.

To take an alternative view of these data, one could interpret the generalized phase diagram for Ce(Rh,Ir,Co)In_5 of Figure 1 as revealing a more typical competition between magnetism and superconductivity. The observed ground states can be rationalized in terms of the deleterious effects of magnetism on T_c convolved with the beneficial effects of increasing anisotropy on T_c. Further measurements of the spin fluctuation spectra of these materials, as well as an understanding of their combined pressure-doping phase diagram, are needed to clarify this issue.

Finally, the extent to which the f-electrons in $CeMIn_5$ are localized or itinerant remains an issue. Reasonable agreement between Fermi surfaces deduced from band structure calculations that consider Ce's f-electrons to be itinerant (25, 26, 27, 28) can be found; however, de Haas- van Alphen studies of dilute alloys, at least for $Ce_{1-x}La_xRhIn_5$, indicate localized f-electrons (29). More generally, the observation of coexistence of magnetism and superconductivity in Ce(Rh,Ir)In_5 (11, 30) raises the issue of whether a one-band or two-band model is required to explain the properties of $CeMIn_5$.

3. Superconductivity in $PuCoGa_5$

Quite recently, superconductivity has been reported in $PuCoGa_5$ at the rather high T_c of 18.5 K (3). As discussed above, $PuCoGa_5$ is isostructural with $CeMIn_5$, and in a simple sense can be viewed as beginning with $CeCoIn_5$, replacing In with isovalent Ga and replacing Ce with Pu. Examining this simple logic in terms of the importance of dimensionality, proximity to magnetism, and degree of f-electron localization is worthy of discussion.

Dimensionality would seem to be the simplest to examine. The c/a ratio in $PuCoGa_5$ is 1.603, somewhat smaller than that of, e.g., $CeCoIn_5$ (1.638). Naively, this would suggest that even higher T_cs could be found by tuning the crystallographic anisotropy of $PuCoGa_5$. Uniaxial pressure measurements would certainly be valuable in this regard; however, Onuki et al. have argued (31) that the underlying electronic structure becomes increasingly two dimensional in the series $LaMIn_5$, $CeMIn_5$, $UMGa_5$ without regard for c/a ratio. Until Fermi surface measurements can be performed on $PuCoGa_5$, its degree of structural optimization will remain unknown.

Too few isostructural transuranic compounds are known to assess how close $PuCoGa_5$ is to ordered magnetism in a generalized phase diagram; however, the known $UMGa_5$ materials provide some insight, at least with respect to the related issue of degree of f-electron localization. $UMGa_5$ compounds have been reported for many transition metals (4). Only a small number, UNiGa5 and $UPtGa_5$ order magnetically, and even these do not show local-moment susceptibility. The temperature-independent paramagnetism displayed by $UMGa_5$ indicates itinerant f-electron character. The observed value of C/T at low temperature for $UMGa_5$ is typically on the order of 20 mJ/molK^2, consistent with this observation. Further, there are no reports of superconductivity in $UMGa_5$. $PuCoGa_5$, on the other hand, displays both a somewhat enhanced value of $\gamma \sim 80$ mJ/molK^2 and a local-moment susceptibility consistent with trivalent Pu. Thus, at least at this simplest level, the extent to which $PuCoGa_5$ has more localized f-electron character and therefore is closer to ordered magnetism is beneficial for superconductivity.

In fact, $PuCoGa_5$ would seem to be intermediate between $CeCoIn_5$ and $UMGa_5$, and may represent a rather optimized balance of degree of localization and proximity to ordered magnetism. If one accepts this speculation, then the two orders of magnitude difference in T_c between $CeIn_3$ ($\sim$ 200 mK) and $PuCoGa_5$ ($\sim$ 20 K) may be just the simple optimization of magnetically mediated superconductivity with dimensionality and bandwidth (32, 22). Only further elaboration of the physical properties of other transuranic materials will provide a definitive answer to this speculation.

Acknowledgements

We thank our many colleagues and collaborators, especially M. Nicklas, P.G. Pagliuso, N.O. Moreno, L.A. Morales, and Z. Fisk, for important contributions to this work. Work at Los Alamos was performed under the auspices of the U.S. Department of Energy.

References

1. B. T. Matthias, Phys. Rev. **97**, 75 (1995).
2. C. Petrovic, P. G. Pagliuso, M. F. Hundley, R. Movshovich, J. L. Sarrao, J. D. Thompson, Z. Fisk and P. Monthoux, J. Phys. Condens. Matter **13**, L337 (2001).
3. J. L. Sarrao, L. A. Morales, J. D. Thompson, B. L. Scott, G. R. Stewart, F. Wastin, J. Rebizant, P. Boulet, E. Colineau and G. H. Lander, Nature **420**, 297 (2002).
4. Yu. Grin, P. Rogl and K. Hiebl, J Less Common Met. **121**, 497-505 (1986).
5. H. Hegger, C. Petrovic, E. G. Moshopoulou, M. F. Hundley, J. L. Sarrao, Z. Fisk and J. D. Thompson, Phys. Rev. Lett. **84**, 4986 (2000).
6. C. Petrovic, P. G. Pagliuso, M. F. Hundley, R. Movshovich, J. L. Sarrao, J. D. Thompson and Z. Fisk, Europhys. Lett. **53**, 354 (2001).
7. R. A. Fisher, F. Bouquet, N. E. Phillips, M. F. Hundley, P. G. Pagliuso, J. L. Sarrao, Z. Fisk and J. D. Thompson, submitted to Phys. Rev. Lett. (2001).
8. T. Mito, S. Kawasaki, G.Q. Zheng, Y. Kawasaki, K. Ishida, Y. Kitaoka, D. Aoki, Y. Haga and Y. Onuki, Phys. Rev. B **63**, 220507 (2001).
9. Y. Kohori, Y. Yamato, Y. Iwamoto, and T. Kohara, European Physical Journal B **18**, 601 (2000).
10. R. Movshovich, M. Jaime, J. D. Thompson, C. Petrovic, Z. Fisk, P. G. Pagliuso and J. L. Sarrao, Phys. Rev. Lett. **86**, 5152 (2001).
11. G. Q. Zheng, K. Tanabe, T. Mito, S. Kawasaki, Y. Kitaoka, D. Aoki, Y. Haga and Y. Onuki, Phys. Rev. Lett. **86**, 4664 (2001).
12. Y. Kohori, Y. Yamato, Y. Iwamoto, T. Kohara, E. D. Bauer, M. B. Maple and J. L. Sarrao, Phys. Rev. B **64**, 134526 (2001).
13. R. Borth, E. Lengyel, P. G. Pagliuso, J. L. Sarrao, G. Sparn, F. Steglich and J. D. Thompson, Physica B **312**, 136 (2002).
14. P. G. Pagliuso, C. Petrovic, R. Movshovich, D. Hall, M. F. Hundley, J. L. Sarrao, J. D. Thompson and Z. Fisk, Phys. Rev. B **64**, 100503 (2001).
15. K. Izawa, H. Yamaguchi, Y. Matsuda, H. Shishido, R. Settai and Y. Onuki, Phys. Rev. Lett. **87**, 057002 (2001).

16. M. Nicklas, R. Borth, E. Lengyel, P. G. Pagliuso, J. L. Sarrao, V. A. Sidorov, G. Sparn, F. Steglich and J. D. Thompson, J. Phys. Condens. Matter 13 (2001) L905.
17. T. P. Murphy, D. Hall, E. C. Palm, S. W. Tozer, C. Petrovic, Z. Fisk, R. G. Goodrich, P. G. Pagliuso, J. L. Sarrao and J. D. Thompson, Phys. Rev. B **65**, 100514 (2002).
18. A. Bianchi, R. Movshovich, N. Oeschler, P. Gegenwart, F. Steglich, J. D. Thompson, P. G. Pagliuso and J. L. Sarrao, Phys. Rev. Lett. **89**, 137002 (2002).
19. P. G. Pagliuso, R. Movshovich, A. D. Bianchi, M. Nicklas, N. O. Moreno, J. D. Thompson, M. F. Hundley, J. L. Sarrao and Z. Fisk, Physica B **312**, 129 (2002).
20. V. S. Zapf, E. J. Freeman, E. D. Bauer, J. Petricka, C. Sirvent, N. A. Frederick, R. P. Dickey and M. B. Maple, Phys. Rev. B **65**, 014506 (2001).
21. Y. N. Grin, Y. P. Yarmolyuk and E. I. Giadyshevskii, Sov. Phys. Crystallogr.**24**, 137 (1979).
22. P. Monthoux and G. G. Lonzarich, cond-mat/0207556.
23. V. A. Sidorov, M. Nicklas, P. G. Pagliuso, J. L. Sarrao, Y. Bang, A. V. Balatsky and J. D. Thompson, Phys. Rev. Lett. **89**, 157004 (2002) .
24. J. S. Kim, J. Alwood, G. R. Stewart, J. L. Sarrao and J. D. Thompson, Phys. Rev. B **64**, 134524 (2001).
25. D. Hall, E. C. Palm, T. P. Murphy, S. W. Tozer, C. Petrovic, E. Miller-Ricci, L. Peabody, C. Q. H. Li, U. Alver, R. G. Goodrich, J. L. Sarrao, P. G. Pagliuso, J. M. Wills and Z. Fisk, Phys. Rev. B **64**, 064506 (2001).
26. Y. Haga, Y. Inada, H. Harima, K. Oikawa, M. Murakawa, H. Nakawaki, Y. Tokiwa, D. Aoki, H. Shishido, S. Ikeda, N. Watanabe and Y. Onuki, Phys. Rev. B **63**, 060503 (2001).
27. D. Hall, E. C. Palm, T. P. Murphy, S. W. Tozer, Z. Fisk, U. Alver, R. G. Goodrich, J. L. Sarrao, P. G. Pagliuso, and T. Ebihara, Phys. Rev. B **64**, 212508 (2001).
28. R. Settai, H. Shishido, S. Ikeda, Y. Murakawa, M. Nakashima, D. Aoki, Y. Haga, H. Harima and Y. Onuki, J. Phys. Condens. Matter **13**, L627 (2001).
29. U. Alver, R. G. Goodrich, N. Harrison, D. Hall, E. C. Palm, T. P. Murphy, S. W. Tozer, P. G. Pagliuso, N. O. Moreno, J. L. Sarrao and Z. Fisk, Phys. Rev. B **64**, 180402 (2001).
30. A. D. Christianson, A. Llobet, et al., unpublished.
31. Y. Onuki, R. Settai, D. Aoki, H. Shishido, Y. Haga, Y. Tokiwa and H. Harima, Physica B **312**, 13 (2002).
32. N. D. Mathur, F. M. Grosche, S. R. Julian, I. R. Walker, D. M. Freye, R. K. W. Haselwimmer and G. G. Lonzarich, Nature **394**, 39 (1998).

DO WE UNDERSTAND ELECTRON CORRELATION EFFECTS IN GADOLINIUM BASED INTERMETALLIC COMPOUNDS?

Proceedings of the ARW NATO Workshop Hvar, Croatia, October 2002

E.V SAMPATHKUMARAN and R. MALLIK†
Tata Institute of Fundamental Research, Homi Bhabha Road, Mumbai-400005, India

Abstract. Recognising the difficulties in systematic understanding of the physical characteristics of strongly correlated f-electron systems, we considered it worthwhile to subject the so-called "normal" f-electron systems like those of Gd to careful investigations. We find that the spin-disorder contribution to electrical resistivity (ρ) in the paramagnetic state, instead of remaining constant, *apparently* shows an increase as the temperature (T) is lowered towards magnetic ordering temperature in some of the Gd alloys. In some cases, the T-dependence of "excess resistance" is so large that a distinct minimum in the plot of ρ versus T can be seen, mimicking the behaviour of Kondo lattices. This excess resistance can be suppressed by the application of a magnetic field, naturally resulting in large magnetoresistance. In addition, these alloys are found to exhibit heavy-fermion-like heat-capacity behavior. These unusual findings imply hither-to-unexplored electron correlation effects even in Gd-based alloys.

Key words: Gd Compounds, Electron Correlations, Kondo Effect, Heavy Fermions, Spin Disorder, Electrical Resistivity, Kondo Lattice, Magnetic Polaron, Heat Capacity, Magnetoresistance

1. Introduction

The physical characteristics of the materials containing some of the f-elements like Ce, Eu, Yb and U, arising from strong electron correlation effects, have been of great interest during last two decades in condensed matter physics and the identification of new materials containing these ions often resulted in discovery of novel phenomena. But, the fact remains that, in spite of intense theoretical and experimental activities, there are con-

† PRESENT ADDRESS: DEPARTMENT OF DEVELOPMENTAL AND CELL BIOLOGY, UNIVERSITY OF CALIFORNIA, IRVINE, USA.

A.C. Hewson and V. Zlatić (eds.), Concepts in Electron Correlation, 353–361.

siderable difficulties in systematic understanding of these properties even at a qualitative level. However, it has been recognised that the observed anomalies are in some way related to the tendency of the f-orbital to get delocalised, in other words, to the capability of these f-ions to exist in more than one valence state. Naturally, the f-ions with deeply localised f-orbital, like Gd, are expected to behave normally without any unusual electron correlation effects known for the above-mentioned class of f-ions and therefore careful investigations on Gd systems by and large do not exist in the literature. During the course of our investigations of strongly correlated electron systems (SCES), we came across some Gd alloys in which the observations are so puzzling that this fundamental assumption in this field is questionable. This article, surveying our experimental findings with few examples, raises a question whether the understanding of the electron correlation effects in "normal" rare-earths is truly complete.

2. Unusual results in ternary Gd compounds derived from AlB_2-derived hexagonal structure

A systematic investigation of Gd based alloys was actually motivated by our original observation of heavy-fermion-like heat-capacity (C) anomalies in Gd-substituted $LaCu_2Si_2$ (1), similar to that known for Ce alloys in the low-coupling limit (2). Subsequently, we noticed that the electrical resistivity (ρ) of $GdNi_2Si_2$ (3) and GdNi (4) over a wide temperature (T) range above respective magnetic ordering temperatures (T_o) could be suppressed to relatively smaller values by the application of a magnetic field (H). In other words, these Gd compounds are found to exhibit negative magnetoresistance (MR), the magnitude of which increases with decreasing temperature. This finding implies that there is a T-dependent ρ-component in excess of the lattice contribution present in these materials. While all these features, *new as far as Gd alloys are concerned*, are typical of Kondo alloys, no distinct minimum in the T dependence of ρ could be observed in the raw experimental data. It is therefore of interest to look for Gd alloys which can directly show this minimum in the raw data. In this respect, the ternary Gd compounds, crystallizing in the AlB_2-derived hexagonal structure, provided an unique situation to demonstrate this aspect as well. While the reader may refer to all our published articles for various measurements and discussions (5, 6, 7, 8, 9, 10, 11, 12, 13, 14), here we summarise the key experimental findings.

The behaviour of the compound Gd_2PdSi_3 (5) by far remains as the best example to demonstrate our point. We have confirmed by various measurements including ^{155}Gd Mössbauer effect studies that the compound Gd_2PdSi_3 undergoes long range antiferromagnetic ordering at (T_N=) 21 K.

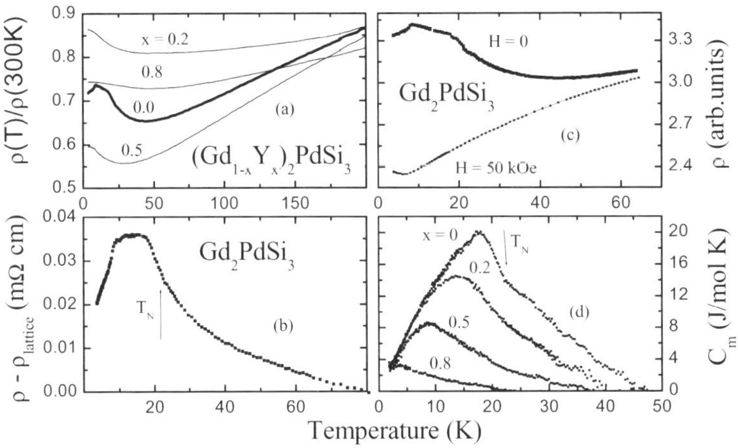

Figure 1. (a) The electrical resistivity (ρ) of the alloys, $(Gd_{1-x}Y_x)_2PdSi_3$, normalised to respective 300 K values, as a function of temperature (T). (b) An estimate of "excess ρ" as a function of T for Gd_2PdSi_3, obtained as described in the text. (c) The ρ as a function of T for Gd_2PdSi_3 in the absence of a magnetic field and in the presence of 50 kOe. (d) The 4f-contribution to heat-capacity as a function of T for the alloys, $(Gd_{1-x}Y_x)_2PdSi_3$.

As far as ρ(T) behaviour is concerned, what one would therefore expect is that the T-coefficient of ρ is positive above 21 K. Looking at figure 1(a and c), ρ(T) of Gd_2PdSi_3 gradually decreases as the T is lowered down to 60 K, however *followed by an upturn below about 45 K*, thus giving rise to a resistivity minimum at about (T_{min}) 45 K. The value of ρ before the onset of magnetic ordering (say, at 22 K) is about 5% higher compared to that at T_{min}.

Such a behaviour of ρ with a distinct minimum in the paramagnetic state was not previously observed for any "normal" (that means, with deeply localised f-orbital) f-ion based intermetallic compound. [Further lowering of T below T_N does not result in a drop in ρ(T), as expected due to the loss of spin-disorder contribution, and this finding is not relevant to the present discussion; this is understandable in terms of opening of a pseudogap at some portions of the Fermi surface due to the onset of antiferromagnetic ordering]. Substitution of Y for Gd results in a depression of T_N as expected, but the minimum in ρ(T) persists (see Fig. 1a). The fact that the minimum is observable also in the sample with the highest Y content (x= 0.8) excludes any explanation for the observed ρ minimum at 45 K in Gd_2PdSi_3 in terms of the opening of an energy gap above 21 K. We have also estimated the 4f-contribution (ρ_{4f}) to ρ by subtracting the ρ of Y analogue; for this purpose, we adjusted the slope of ρ(T) plot with that of Gd_2PdSi_3 at 100 K and then matching the ρ value at 100 K; though such an analysis to subtract out lattice contribution is not free from ambiguities, one gets an idea about T-dependence of "excess ρ" (ρ_{4f}= ρ(T)-$\rho_{lattice}$) from the derived data (shown in Fig. 1b). Irrespective of the nature of reference for lattice contribution, the temperature coefficient of ρ_{4f} is found to be distinctly negative in the paramagnetic state.

We have also measured ρ in the presence of an externally applied H. For x= 0.0, it is found that MR, defined as [ρ(H)-ρ(0)]/ρ(0), is about -10% at 40 K in the presence of, say, H= 70 kOe, which is very large for a normal intermetallic compound in this temperature range (5). Interestingly, the minimum in ρ(T) vanishes in the presence of H (say at 50 kOe) and ρ exhibits a positive T coefficient in the entire range (Fig. 1b). The magnitude of MR keeps increasing with decreasing T for a given H. These features are characteristics of Kondo alloys as well. It is to be noted (5) that a large MR was not observed even for the analogous Kondo compound, Ce_2PdSi_3, which emphasizes the importance of the finding in the Gd compound. Another noteworthy finding is that the T at which MR starts taking noticeably large values (with a negative sign) decreases with decreasing Gd concentration (5), following the trend in T_N. This implies that *the observed anomalies are magnetic in origin.*

With respect to the C data (Fig. 1d), apart from the peaks attributable to magnetic ordering at T_N, the 4f contribution (C_m) to C exhibits a tail extending over a wide T range above T_N. For instance, for x= 0.8, for which T_N is lower than 2 K, the tail in C_m(T) extends to T as high as 20 K, with the value of C/T as large as 1 J/mol K^2 at 2 K, comparable to the behaviour of prototype heavy-fermion system, $CeCu_2Si_2$.

It is clear from the above results that the compound, Gd_2PdSi_3, exhibits all the characteristics of (antiferromagnetically) ordering Kondo lattice systems in ρ, MR and C data, including a resistivity minimum. These features could be confirmed even in single crystals of this compound (8). In addition, the thermopower also exhibits a large value at 300 K, typical of Kondo systems (6). In order to explore whether these anomalies are due to an artifact of spin-glass freezing setting in around 45 K, we have probed the frequency-dependence of the ac susceptibility as well as possible irreversibilities in zero-field-cooled and field-cooled dc susceptibility behaviour carefully (6, 9). The results obtained, apart from establishing that the magnetic ordering below 21 K is of a long-range type, rule out any type of magnetic ordering above 21 K. Therefore, a question arises whether this compound could be the first Gd-based Kondo-lattice. Our photoemission data on the other hand revealed (14) that there is no 4f-derived feature at the Fermi level (E_F) and that 4f-level is placed about 8 eV well below E_F, thereby conclusively establishing that this compound can not be categorised as a Kondo alloy. We will return to further discussion on this "magnetic precursor effect" (T-dependent "excess" ρ and C_m above

T_o) in Section 4.

Before closing this section, it is worth stating that two more isostructural Gd compounds, viz., Gd_2PtSi_3 (13) and Gd_2CuGe_3 (12), have been subsequently identified to exhibit properties resembling those of Gd_2PdSi_3. This suggests that the anomalies discussed here must be more common among Gd intermetallics.

3. Behaviour of some other Gd compounds

The observation of unusual magnetic precursor effects in the ternary Gd compounds mentioned above raises many questions: (i) Are these anomalies specific to this crystal structure? (ii) Is it something to do with the layered nature of Gd ions in this crystal structure? (iii) Is it due to polarization of the d-band of some of the transition metal ions, like Pt or Pd, which are prone for induction of moments by the large moment-carrying neighbouring ions (like Gd)? (iv) Is this magnetic precursor effect anything to do with the type of magnetic structure at lower temperatures? (v) Is there relationship between C and ρ anomalies? In order to tackle these issues, we have investigated a large number of Gd compounds (7) with different crystal structures. The results clearly established that there is no straightforward relationship between the observation of the "T-dependent excess ρ" on the one hand, and the crystal structure or the type of transition metal and s-p ions present in the compound, on the other. It was also established that possible onset of magnetic correlations within a layer before long-range magnetic order sets in can not be offered as the sole reason for excess ρ, as many of the compounds with Gd layers are not characterized by this behaviour of "excess ρ". Also, Gd compounds exhibiting this excess ρ anomaly fall into the category of both ferromagnets and antiferromagnets and therefore the nature of the magnetic structure below T_o is not the determining factor.

A further careful look at the features above T_o in the ρ and C data revealed (7) that the Gd alloys can be classified into three categories: (i) Magnetic precursor effects appearing in both ρ and C; (ii) "T-dependent excess ρ" and hence "MR anomaly" are negligible, but C tail persists over a wide temperature range; (iii) Both "excess ρ" and "excess C" are absent. The behaviour of the second category of Gd compounds is demonstrated in figure 2 by showing the data for $GdCu_2Ge_2$, crystallizing in the layered $ThCr_2Si_2$-type crystal structure, for which T_N is about 13 K. It is obvious from this figure that, apart from small magnitude, the sign of MR is not even negative (unlike first category of Gd alloys) in the entire T range.

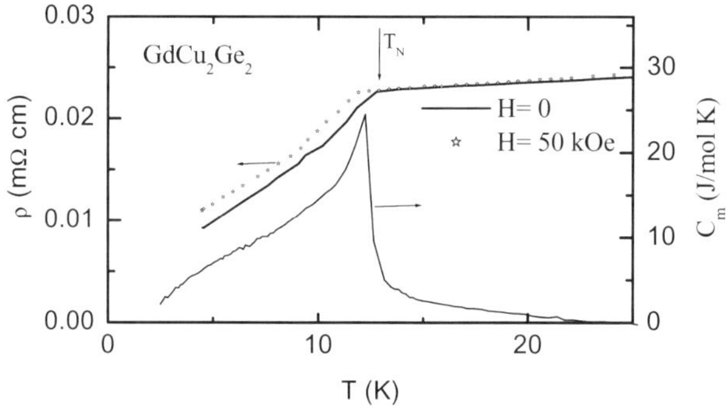

Figure 2. Electrical resistivity in zero magnetic field and in the presence of a magnetic field of 50 kOe and magnetic contribution (C_{4f}) to heat-capacity, for $GdCu_2Ge_2$, demonstrating that this compound belongs to Category-II as described in the text.

4. Discussion

The central experimental observation is that, in some Gd alloys, the spin-disorder contribution to ρ in the paramagnetic state *apparently* is not a constant value, but varies as though it increases gradually with decreasing T as one approaches magnetic ordering temperature, resulting in Kondo-lattice-like characteristics. At this juncture, it may be mentioned that similar observations have been made also for Eu_2CuSi_3 (11) and Tb_2PdSi_3 (17) and therefore these anomalies must be more common among other rare-earths as well with deeply localised f-orbitals. Since this feature does not arise from the Kondo effect, the origin of this unusual magnetic precursor effect remains a puzzle. One can not attribute it to Fisher-Langer type of critical spin-fluctuations at T_o (15, 16), considering that the T-range over which these features above T_o appear is too large to be explained by this idea; in addition, the absence of a relationship between C and ρ anomaly in second category of Gd compounds is convincing enough to rule out this possibility. It therefore appears that the "T-dependent excess ρ" in first category Gd compounds arises from a hitherto unrecognised phenomenon in moment-carrying metals above T_o. We speculate that the formation "magnetic polarons" (localisation or trapping of the conduction electron cloud polarised by the s-f exchange interaction) above T_o could be the root-cause of these anomalies. The proposed magnetic localisation could be triggered by Gd 4f short range magnetic order and consequent spatial magnetic fluctuations, just as crystallographic disorder reduces the mobility of charge carriers. We have observed (8) distinct anomalies in Hall coefficient around 100 K in single crystals of Gd_2PdSi_3; also, on approaching magnetic order, the linewidth and the resonance field in ESR measurements show a minimum as a function of T (18). It is not clear at present whether this signals possible magnetic localisation effects. The application of a magnetic field suppresses the spin fluctuations or alters the localization length resulting in an enhancement of the mobility of the electron cloud. This explains the suppression of the ρ minimum in the paramagnetic state in a magnetic field. Below T_C, in ferromagnetic Gd compounds like GdNi (4), the polarons are anyway itinerant as they are coupled ferromagnetically with the Gd 4f moment; as a result polaronic contribution can be turned off below T_C. It must however be stated that the concept of magnetic polarons have been generally ignored in metals in the literature (17) due to the fact that large carrier density is expected to make these polarons delocalised. Possibly, in poor metals, one may have to still invoke this concept. If this proposal is confirmed in the compounds under discussion, one should explore various factors determining the presence or the absence of such effects. Some of the decisive factors could be the relative magnitudes of mean free path, localization length and short range correlation length, in addition to the strength of polarization of the conduction band and carrier density.

Another outcome of the studies on Gd alloys is the observation of negative MR even at temperatures far above T_o (till $3T_o$), peaking at T_o in some Gd alloys (not presented here). These materials turned out to be one of the first few intermetallics exhibiting giant magnetoresistance behavior in the paramagnetic state at rather high temperatures. This finding established that neither double-exchange nor Jahn-Teller effects (proposed for by-now well-known "manganites") is crucial to observe GMR behavior. Similar conclusions were drawn based on the observation of GMR behaviour in the pyrochlore, $Tl_2Mn_2O_7$ (19, 20), for which the concept of magnetic polarons has been discussed (21). It is to be stressed that Gd_2PdSi_3 turned out to be the first *normal intermetallic* compound exhibiting $Tl_2Mn_2O_7$-like ρ(T) behaviour.

5. Conclusion

The discussion presented in this article suffciently demonstrates that, Gd intermetallics, which have been traditionally considered uninteresting from the magnetism point of view, exhibit novel behaviour in some compounds with rich features in their physical properties, even in the paramagnetic state. Our results on Gd alloys bring out for the first time that the understanding of electron correlation effects even in such "normal" materials is far from complete. We hope that our data on several Gd compounds will motivate new theoretical approaches in magnetism. Whatever be the microscopic origin of the anomalies, the results offer a message: One has to be a bit cautious while attributing low-temperature upturns in ρ and C in alloys of Ce, U or Yb non-Fermi liquid and Kondo effects, since these features may also be as a precursor effect to the magnetic transitions occuring at further lower temperatures. It is of interest to explore, both theoretically and experimentally, whether magnetic polarons can give rise to non-Fermi liquid phenomena.

Acknowledgements

One of us (E.V.S) would like to acknowledge M. Blanco, D. Eckert, I. Das, A. Handstein, J. Hemberger, C. Laubschat, A. Loidl, Subham Majumdar, S.L. Molodtsov, K.-H. Müller, V. Nagarajan, P.L. Paulose, S.R. Saha, H. Sato, H. Sugawara, and G. Wortmann for fruitful collaborations.

References

1. E.V. Sampathkumaran and I. Das, Phys. Rev. B, **51**, 8178 (1995).
2. W.P. Beyermann et al, Phys. Rev. Lett. **66**, 3289 (1991); W.P. Beyermann et al, Phys. Rev. B, **43**, 13130 (1991).
3. E.V. Sampathkumaran and I. Das, Phys. Rev. B, **51**, 8631 (1995); Physica B **223-224, 149 (1996)**
4. R. Mallik, E.V. Sampathkumaran, P.L. Paulose and V. Nagarajan, Phys. Rev. B, **55**, R8650 (1997).
5. R. Mallik, E.V. Sampathkumaran, M. Strecker and G. Wortmann, Europhys. Lett. **41**, 315 (1998).
6. R. Mallik, E.V. Sampathkumaran, P.L. Paulose, H. Sugawara and H. Sato, Pramana - J. Phys. **51**, 505 (1998).
7. R. Mallik and E.V. Sampathkumaran, Phys. Rev. B **58**, 9178 (1998).
8. S.R. Saha, H. Sugawara, T.D. Matsuda, H. Sato, R. Mallik and E.V. Sampathkumaran, Phys. Rev. B **60**, 12162 (1999).
9. Subham Majumdar, E.V. Sampathkumaran, D. Eckert, A. Handstein, K.-H. Muller, S.R. Saha, H. Sugawara and H. Sato, J. Phys.: Condens. Matter. **11**, L329 (1999).
10. Subham Majumdar, M. Mahesh Kumar and E.V. Sampathkumaran, J. Alloys and Compounds, **288**, 61 (1999).
11. Subham Majumdar, R. Mallik, E.V. Sampathkumaran, K. Rupprecht and G. Wortmann, Phys. Rev. B **60**, 6770 (1999).
12. Subham Majumdar and E.V. Sampathkumaran, Phys. Rev. B **61**, 43 (2000).
13. Subham Majumdar, E.V. Sampathkumaran, M. Brando, J. Hemberger and A. Loidl, J. Magn. Magn. Mater. **236**, 99 (2001).
14. A.N. Chaika, A.M. Ionov, M. Busse, S.L. Molodtsov, Subham Majumdar, G. Behr, E.V. Sampathkumaran, W. Schneider and C. Laubschat, Phys. Rev. B **64**, 125121 (2001).
15. S.-K. Ma, *Modern Theory of Critical phenomena* Benjamin, London, 1976.
16. M.E. Fisher and J.S. Langer, Phys. Rev. Lett. **20**, 665 (1968).
17. R. Mallik, E.V. Sampathkumaran and P.L. Paulose, Solid State Commun. **106**, 169 (1998); see references cited herein for a proposal of "magnetic polarons" for rare-earth-based semiconductors.
18. J. Deisenhofer, H.-A. Krug von Nidda, A. Loidl and E.V. Sampathkumaran, Solid State Commun., in press; arXiv: Cond-mat/0211587.

19. Y. Shimakawa, Y. Kubo and T. Manako, Nature (London) **379**, 53 (1996).
20. M.A. Subramanian, B.H. Toby, A.P. Ramirez, W.J. Marshall, A.W. Sleight, and G.H. Kwei, Science **273**, 81 (1996).
21. P. Majumdar and P.B. Littlewood, Nature **395**, 479 (1998).

OPTICAL PROPERTIES OF CORRELATED SYSTEMS

Proceedings of the ARW NATO Workshop Hvar, Croatia, October 2002

L. DEGIORGI
Laboratorium für Festkörperphysik, ETH Zürich, CH-8093 Zürich, Switzerland

Abstract. We review selected optical results on correlated systems. This review mainly focuses the attention on some prototype Kondo materials, characterized by the heavy electron behaviour, and on the linear chain Bechgaard salts, as representative systems of "real" one-dimensional materials. We will address a variety of relevant problems and concepts associated with the physics of an interacting electron gas in three and low dimensions, as Kondo behaviour, dimensionality crossover, and Luttinger liquid state.

Key words: Optical Properties, Correlated Systems, Heavy Electron Materials, Low Dimensional Systems, Kondo Problem, Mott-Hubbard

1. Introduction

Highly correlated states of condensed matter have opened new chapters in physics. Examples of particular interest include the so-called Kondo or heavy-electron (HE) materials, which were discovered in the late seventies. The class of strongly correlated systems also encounters the transition metal oxides, including the d-electron (Mott-Hubbard) systems as well as the high-temperature superconducting cuprates, the quasi-one dimensional materials, as the organic Bechgaard salts or carbon-nanotubes, and possibly the carbon-fullerenes. Due to space limitation, the present review will be focused on selected examples of heavy-electron systems and on the quasi-one dimensional Bechgaard salts, only.

HE systems are electrically conducting materials, with peculiar low temperature physical properties distinguishing them from ordinary metals (1, 2). In fact, the conduction electron specific heat is typically some 100 times larger than that found in most metals. Similarly, the magnetic susceptibility can be two or more orders of magnitude larger than the temper-

A.C. Hewson and V. Zlatić (eds.), Concepts in Electron Correlation, 363–369.

ature independent Pauli susceptibility observed in conventional conducting materials. The prototype HE materials include actinide and rare earth alloys, mostly containing U and Ce, respectively, like $CeAl_3$, $CeCu_6$, UBe_{13} and UPt_3 etc.

The name "heavy-electron" stems from the observation that enormously large specific heat coefficient γ implies an effective mass m^* of the quasiparticles which is several hundred times the free electron mass. The "heavy-electron" can be considered as the 4f or 5f electron interacting with the bands of the delocalized electrons (d or s). Such an interaction occurs via the Kondo spin fluctuations (3, 4). The final result is the formation of the sharp and narrow Abrikosov-Suhl (ASR) resonance (width of about T_K for $T << T_K$, T_K being the Kondo temperature) in the density of states $N(E)$ at E_F (3). The electrons sitting in this narrow ASR are the new heavy quasiparticles, responsible for the low temperature transport and thermodynamic properties.

Strictly one-dimensional (1D) interacting electron gas systems are of interest because the Fermi liquid (FL) state is replaced by a state where interactions play a crucial role, and which is generally referred to as a Luttinger liquid (LL). The 1D state predicted by the LL theory (5, 6, 7) is characterized by features such as spin-charge separation and the absence of a sharp edge in the momentum distribution function $n(k)$ at the Fermi wave vector k_F. The first immediate consequence of the absence of the discontinuity at k_F in the momentum distribution function is the power-law behaviour of the density of states (DOS) $\rho(\omega) \sim| \omega |^{\alpha}(\omega{=}E_F - E)$. The exponent α in this expression reflects the nature and strength of the interaction. The non-Fermi liquid nature of the LL is also manifested by the absence of single-electron-like quasiparticles and by the non-universal decay of the various correlation functions.

Nature has recently provided us with remarkable materials, which have considerably boosted the experimental and theoretical understanding of electrons with a dimensionality less than three. Low dimensional organic conductors (like, e.g., the quasi one-dimensional Bechgaard salt chain or the so-called quasi two-dimensional $BEDT$ systems) can be described as having units of strongly linked molecules, which are chains and planes for one-dimensional (1D) and two-dimensional (2D) conductors, respectively, with weak interactions (i.e., $t_\perp$) between the units. They thus form a three-dimensional lattice solid, but the electrons are confined either along one direction or within a plane. Nevertheless, at low temperatures, when k_BT is smaller than $t_\perp$, one should rather consider the low-dimensional conductor as a three-dimensional anisotropic conductor (8, 9). One can readily see that dimensionality crossovers should exist in real solids.

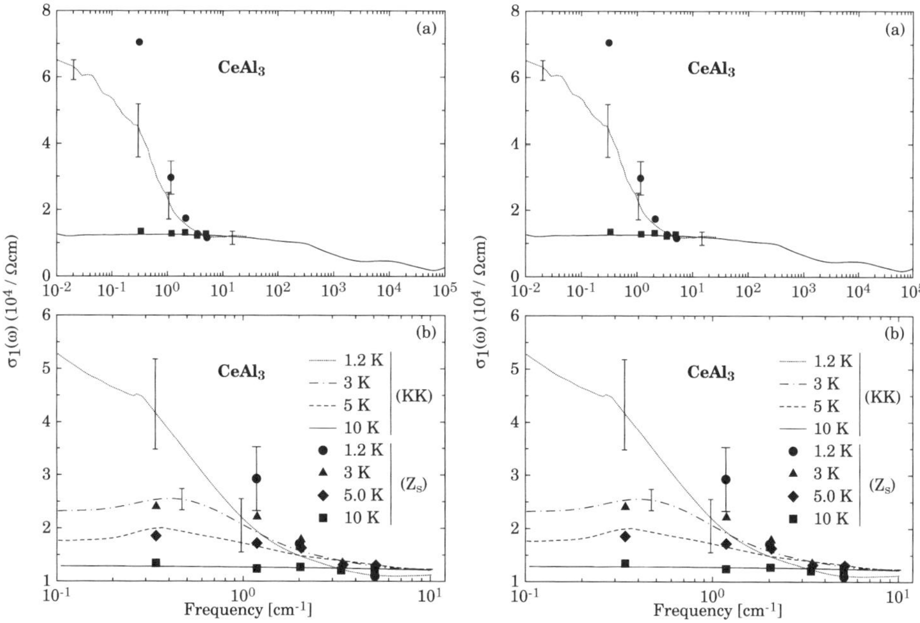

Figure 1. (left) (a) Real part $\sigma_1(\omega)$ of the optical conductivity obtained from the Kramers-Kronig analysis of the absorptivity of $CeAl_3$, and directly from the surface impedance measurements (Z_s). (b) An enlargement of the dynamical conductivity showing the development of the low-frequency resonance at low temperatures (10, 11). (right) Sketch of the calculated frequency dependent conductivity $\sigma_1(\omega)$ after Millis and Lee (12).

2. Optical response in heavy electron systems

Figure 1 (10) shows the real part of the optical conductivity $\sigma_1(\omega)$ of $CeAl_3$, a prototype HE compound. The general trend of the optical conductivity is rather common in all HE materials (11). In fact, at high temperatures we can easily recognize a rather broad metallic-like behaviour (i.e., Drude), indicative of a scattering rate $\Gamma \sim 1/\tau$ (i.e., the width of the Drude term) of the order of 0.1 eV. Moreover, the conductivity drops below its dc value at frequencies of the order of $1/\tau$ and is starting to flatten again above 3000 cm^{-1}. In contrast, at relatively low temperatures we observe the gradual emergence of a narrow mode centered at zero frequency in $\sigma_1(\omega)$, which merges into the high temperature $\sigma_1(\omega)$ at FIR frequencies. This dramatic temperature dependence does, nevertheless, preserve the good agreement between the σ_{dc} values and the zero energy limit of $\sigma_1(\omega)$.

The development of the low frequency narrow mode bears a striking similarity with the theoretical prediction of $\sigma_1(\omega)$ (see Fig. 2) by Millis and Lee (12). Millis and Lee (12) developed the most comprehensive theoretical calculation of the HE electrodynamic response. They performed a study of the low-temperature properties of the lattice Anderson Hamiltonian, in the Kondo limit, where nearly free electrons hybridize with a correlated band of f-electrons. The model is believed to contain the essential physics of the currently interesting "heavy electron" metals: a non-magnetic ground state that behaves as a Fermi liquid with a large effective mass. It turns out that the narrow low-frequency part of the spectrum at low temperatures is very much reminiscent of a Drude term with renormalized scattering rate (11, 12). In fact, in the spirit of the Lorentz-Drude model (13), one can rewrite the Drude term with the plasma frequency $(\omega_p^*)^2 = \omega_p^2(m_b/m^*)$ and the scattering relaxation time $\tau^* = \tau m^*/m_b$, which correspond to the bare quantities ω_p and τ being renormalized by the enhanced effective mass m^* of the heavy quasiparticles at low temperatures. One can easily realize that spectral weight argument (i.e., the integral of the narrow (renormalized) Drude resonance in $\sigma_1(\omega)$) leads directly to the estimation of the effective mass of the heavy quasiparticles (11).

Another major issue and current topic of debate in the field of highly-correlated electron systems is the fundamental question whether these systems, in their normal state, may be described as simple Fermi liquids. For some HE compounds (e.g., UPt_3) this seems indeed to be the appropriate picture. More recent experimental work (11) has indicated, however, that several HE compounds and related alloys display quite remarkable properties, which may be much less well related with "conventional" Fermi liquid behaviour. We refer to our review (11) for a broader experimental and theoretical perspective on these issues.

3. Optical response of Bechgaard salts

Figure 3 displays the temperature dependence of the real part $\sigma_1(\omega)$ of the optical conductivity in different Bechgaard salts for $E\|a$, the chain direction (14, 15, 16, 17, 18, 19). The optical conductivity of the $TMTTF$ salts displays several absorption features mostly ascribed to lattice vibrations (phonon modes), mainly due to the inter and intramolecular vibrations of the $TMTTF$ unit. The optical properties of the $(TMTSF)_2PF_6$ analogs (Fig. 3b), for which the dc conductivity gives evidence for metallic behaviour down to low temperatures, are markedly different from those of a simple metal. A well-defined absorption feature around 25 meV and a zero-frequency mode (15, 16, 17, 18) are observed at low temperatures.

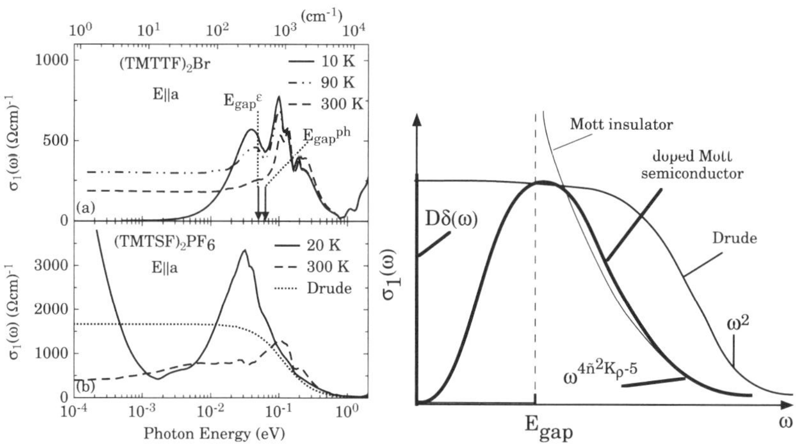

Figure 2. (left) On chain optical conductivity of (a) $(TMTTF)_2Br$, and (b) $(TMTSF)_2PF_6$ at temperatures above the transitions to the broken symmetry ground states. The arrows indicate the (charge correlation) gaps (E_{gap}) observed by dielectric (ϵ) response and photoemission (ph). A simple Drude component is also shown in part (b). Note that photoemission measures the quantity $E_{gap}/2$, assuming that the Fermi level is in the middle of the gap (14, 15). (right) Optical conductivity of the Mott insulator and of the doped Mott semiconductor (20). The simple Drude behaviour is shown for comparison.

This latter mode is responsible for the large metallic conductivity. Figure 3 also displays the temperature evolution of $\sigma_1(\omega)$ (14).

There is an ample theoretical literature, which is hopeless to review here in great detail (14, 15, 17). As introduction to the theoretical expectation for the electrodynamic response in one dimension, we briefly sketch the optical response as suggested by Ref. (20, 21), which catches the essential features. Due to the commensurate filling, a strictly one-dimensional Luttinger liquid with Umklapp scattering effects transforms into a Mott insulating state, which is dominated by the charge correlation gap excitation. For a strict one-dimensional Mott insulator, the charge correlation gap E_{gap}, corresponding to the excitation between the lower and upper Hubbard bands, appears as a singularity in the real part $\sigma_1(\omega)$ of the optical conductivity (Fig. 4). If the interchain coupling $t_\perp(\sim t_b)$ between chains is relevant (i.e., $t_\perp > E_{gap}$), small deviations from commensurate filling due to the warping of the Fermi surface exist, and should lead to effects equivalent to real doping on a single chain. Then, it is energetically favorable for the charge carriers to hop between parallel chains (20, 21). The scenario calculated for a doped one-dimensional Mott semiconductor consists of a Mott (pseudo) gap (as reminder of the original 1D limit with Umklapp scattering process) and a zero-energy mode (i.e., theoretically a $D\delta(\omega)$ function at $\omega = 0$, representing the Drude resonance of the effective metallic contribution with scattering rate $\Gamma = 0$) for small doping levels (Fig. 4). The spectral weight D encountered in the "Drude resonance" is proportional to the effective charge doping, induced by the interchain coupling $t_\perp$, in the upper Hubbard band. Therefore, by increasing $t_\perp$ there is a dimensionality crossover, which is accompanied by the evolution from a one-dimensional Mott insulator to a "doped" semiconductor. Indeed, the $(TMTTF)_2X$ salts, with $X = PF_6$ or Br, are insulators at low temperatures (8) with a substantial (Mott) charge gap (see arrows for different experimental estimations of E_{gap} in Fig. 3a). For these compounds the correlation gap E_{gap} is so large that the interchain hopping ($t_\perp$) is not relevant. Charge carrier hopping on parallel chains is here strongly suppressed leading to a truly 1D insulating phase (20, 21). In analogy to the $TMTTF$ salts, the strong FIR excitation of the $TMTSF$ salts (Fig. 3b) is ascribed, within the scenario depicted in Fig. 4, to the so-called pseudo charge correlation gap. Additionally, there is the narrow Drude peak, originating from deviations of commensurability due to $t_\perp$. The existence of a gap or pseudogap in the charge excitations (Fig. 3) with the absence of a gap for spin excitations indicates, moreover, spin-charge separation in the metallic state (14, 18).

Furthermore, at frequencies greater than $t_\perp$, the interchain electron transfer is irrelevant and calculations based on the 1D Hubbard model should be appropriate (20, 21). The theoretical expectation (Fig. 4) consists in a powerlaw of the frequency-dependent optical conductivity $\sigma_1(\omega){\sim}\omega^{-\gamma}$ for frequencies greater than $t_\perp$ and E_{gap} but less than the on-chain bandwidth $4t_a$. The theory also predicts that the exponent $\gamma = 5 - 4\tilde{n}^2 K_\rho$, K_ρ being the so-called Luttinger liquid parameter and $\tilde{n}$ the degree of commensurability (20, 21, 22). Our results on the powerlaw behaviour (15, 18, 23) are very robust and allow us to discriminate among different regimes and type of correlations. This issue is thoroughly reviewed elsewhere (14).

4. Conclusion

This short review only touched a few issues and problems, which recently attracted a lot of interest in strongly correlated systems. Signatures of heavy quasiparticles in Kondo materials and of the Luttinger phenomenology in quasi-1D systems have been observed. However, the interpretation of the spectra remains problematic on various issues and await further experimental and theoretical work.

5. Acknowledgements

Research at ETH Zurich was supported by the Swiss National Foundation for the Scientific Research.

References

1. H.R. Ott and Z. Fisk, in *Handbook on the Physics and Chemistry of the Actinides*, Eds. A.J. Freeman and G.H. Lander (Elsevier, Amsterdam, 1987), p.85.
2. Z. Fisk, D.W. Hess, C.J. Pethick, D. Pines, J.L. Smith, J.D. Thompson and J.O. Willis, Science **239**, 33 (1988).
3. N.B. Brandt and V.V. Moshchalkov, Adv. Phys. **33**, 373 (1984).
4. P. Fulde, J. Phys. F.: Met. Phys. **18**, 601 (1988).
5. J. Luttinger, J. Math. Phys. **4**, 1154 (1963).
6. S. Tomonaga, Prog. Theor. Phys. **5**, 554 (1950).
7. for a review, see H.J. Schulz, Int. J. Mod. Phys. B **5**, 57 (1991).
8. D. Jerome and H.J. Schulz, Adv. Phys. **31**, 299 (1982).
9. D. Jerome, in *Strongly interacting fermions and high* T_C *superconductivity*, Eds. B. Doucot and J. Zinn-Justin (Elsevier Science, 1995), p. 249.
10. A.M. Awasthi *et al.*, Phys. Rev. B **48**, 10692 (1993).
11. for a review, see L. Degiorgi, Rev. Mod. Phys. **71**, 687 (1999).
12. A.J. Millis and P. A. Lee, Phys. Rev. B **35**, 3394 (1987).
13. F. Wooten, in *Optical Properties of Solids*, (Academic Press, New York, 1972).
14. for a review, see L. Degiorgi, in *Strong interactions in low dimensions*, Eds. D. Baeriswyl and L. Degiorgi (Kluwer, 2003).
15. V. Vescoli *et al.*, Eur. Phys. J. B **13**, 503 (2000).
16. M. Dressel *et al.*, Phys. Rev. Lett. **77**, 398 (1996).
17. A. Schwartz *et al.*, Phys. Rev. B **58**, 1261 (1998).
18. V. Vescoli *et al.*, Science **281**, 1181 (1998).
19. V. Vescoli *et al.*, Solid State Commun. **111**, 507 (1999).
20. T. Giamarchi, Physica B **230-232**, 975 (1997).
21. S. Biermann *et al.*, cond-mat/0201542.
22. J. Voit, Eur. Phys. J. B **5**, 505 (1998) and *Proceedings of the NATO Advanced Research Workshop on The Physics and Mathematical Physics of the Hubbard Model*, San Sebastian, October 3-8, 1993, Eds. D. Baeriswyl, D. K. Campbell, J. M. P. Carmelo, F. Guinea, and E. Louis (Plenum Press, New York, 1995).
23. F. Zwick *et al.*, Solid State Commun. **113**, 179 (2000).

QUASIPARTICLE UNDRESSING: A NEW ROUTE TO COLLECTIVE EFFECTS IN SOLIDS

Proceedings of the ARW NATO Workshop Hvar, Croatia, October 2002

J.E. HIRSCH
Department of Physics, University of California, San Diego
La Jolla, CA 92093-0319

Abstract. The carriers of electric current in a metal are quasiparticles dressed by electron-electron interactions, which have a larger effective mass m^* and a smaller quasiparticle weight z than non-interacting carriers. If the momentum dependence of the self-energy can be neglected, the effective mass enhancement and quasiparticle weight of quasiparticles at the Fermi energy are simply related by $z = m/m^*$ (m=bare mass). We propose that both superconductivity and ferromagnetism in metals are driven by quasiparticle 'undressing', i.e., that the correlations between quasiparticles that give rise to the collective state are associated with an increase in z and a corresponding decrease in m^* of the carriers. Undressing gives rise to lowering of kinetic energy, which provides the condensation energy for the collective state. In contrast, in conventional descriptions of superconductivity and ferromagnetism the transitions to these collective states result in *increase* in kinetic energy of the carriers and are driven by lowering of potential energy and exchange energy respectively.

Key words: Quasiparticle, Dressing, Undressing, Quasiparticle Weight, Effective Mass, Holes, Optical Sum Rule, Dynamic Hubbard Model

1. Particles and quasiparticles

Quasiparticles are 'dressed' bare particles, and they have a smaller quasiparticle weight and a larger effective mass than bare particles. In several superconductors and ferromagnets of current interest there is experimental evidence that quasiparticles 'undress', and resemble more free particles, when correlations build up and the system orders. Associated with this, that the kinetic energy, that is supposed to be optimal in the Fermi liquid normal state, decreases rather than increases in the ordered state. This behavior is counterintuitive, since in a normal Fermi liquid description it is expected that quasiparticles should become further dressed and less like

A.C. Hewson and V. Zlatić (eds.), Concepts in Electron Correlation, 371–380.

free particles when they develop the correlations leading to the collective state, and that they should *pay*, rather than *gain*, kinetic energy. Several 'unconventional' theories have been proposed to explain these phenomena. Instead we propose here that in fact quasiparticle undressing is a unifying concept that can describe these collective effect in *both* new and conventional materials. The only difference is that only in the newer materials is the 'undressing phenomenology' strong enough that it is easily seen in experiments.

2. Dressing and undressing

The single particle Green's function for an electron in an electronic energy band is

$$G(k,\omega) = \frac{1}{\omega - \epsilon_k - \Sigma(k,\omega)} \tag{1}$$

Expanding the self-energy $\Sigma(k,\omega)$ around $\omega = 0$

$$\Sigma(k,\omega) = \Sigma(k,0) + \omega\frac{\partial\Sigma}{\partial\omega} \tag{2}$$

the Green's function can be written as

$$G(k,\omega) = \frac{z_k}{\omega - \tilde{\epsilon}_k} + G'(k,\omega) \tag{3}$$

where the first term is the quasiparticle part and the second term is the incoherent part, and the 'quasiparticle weight' z_k is given by

$$z_k = [1 - \partial\Sigma/\partial\omega]^{-1} \tag{4}$$

If the momentum dependence of the self-energy can be neglected,

$$\Sigma(k,\omega) \sim \Sigma(\omega) \tag{5}$$

then we have simply $z_k = z$ and $\tilde{\epsilon}_k = z\epsilon_k$ which implies that the quasiparticle weight and effective mass m^* are simply related by

$$\frac{m}{m^*} = z \tag{6}$$

hence a highly dressed particle will have a small quasiparticle weight and a large effective mass. For the models of interest in this paper the 'local approximation' Eq. (5) is reasonable. Finally, the 'kinetic energy' of the system is defined by

$$E_{kin} = \sum_k n_k \epsilon_k \tag{7}$$

where the occupation number n_k is obtained from the single particle Green's function

$$n_k = \int_{-\infty}^{\infty} d\omega f(\omega)(-\frac{1}{\pi} ImG(k,\omega)) \tag{8}$$

with f the Fermi function. The discontinuity of n_k at the Fermi surface is the quasiparticle weight z_k. If z_k was to increase for example as the temperature is lowered, Eq. (7) predicts that the kinetic energy of the system would decrease. The existence of this general relation between quasiparticle weight increase and kinetic energy lowering was pointed out by Norman et al (1).

In the presence of interactions bare particles become dressed with quasiparticle weight z and effective mass $m^* \sim 1/z$. The residual weak interactions between these quasiparticles may cause the system to undergo a transition to a collective state. In this paper we discuss a description of superconductivity and ferromagnetism that predicts that when the collective state develops quasiparticles undress, namely, that z increases and m^* decreases. The resulting kinetic energy lowering provides the condensation energy stabilizing the ordered state.

3. Low energy effective Hamiltonians

The low energy effective Hamiltonians in our theory can be derived from the single band generalized Hubbard model

$$H = -\sum_{<ij>\sigma} t_{ij}(c^\dagger_{i\sigma}c_{j\sigma} + h.c.) + \sum_{ijkl\sigma\sigma'} (ij|1/r|kl)c^\dagger_{i\sigma}c^\dagger_{j\sigma'}c_{l\sigma'}c_{k\sigma} \tag{9}$$

where $(ij|1/r|kl)$ are matrix elements of the Coulomb interaction involving Wannier orbitals at sites i, j, k, l. Keeping only two center integrals, we have shown that the off-diagonal interactions $(ii|1/r|ij) \equiv (\Delta t)_{ij}$ and $(ij|1/r|ji) = (ii|1/r|jj) \equiv J_{ij}$ lead to simple descriptions of superconductivity (2) and ferromagnetism (3) respectively, that have many features in common with the phenomena seen in real materials. It should be pointed out however that these off-diagonal interactions are not simply calculable by computing matrix elements of the Coulomb interaction between $fixed$ Wannier orbitals (4).

The Hamiltonian that gives rise to (hole) superconductivity is

$$H_{sc} = H_{exHub} + \sum_{<ij>\sigma} (\Delta t)_{ij}(n_{i,-\sigma} + n_{j,-\sigma})[c^\dagger_{i\sigma}c_{j\sigma} + h.c.] \tag{10}$$

and the one giving rise to ferromagnetism is

$$H_{fm} = H_{exHub} + \frac{1}{2}\sum_{<ij>} J_{ij}[\sum_\sigma (c^\dagger_{i\sigma}c_{j\sigma} + h.c.)]^2 \tag{11}$$

The 'extended Hubbard' Hamiltonian H_{exHub} includes the kinetic energy in Eq. (9) and the ordinary density-density Coulomb interactions $U = (ii|1/r|ii)$, $V_{ij} = (ij|1/r|ij)$. The 'bond charge repulsion' term in Eq. (11) describes both exchange and pair hopping processes (3), arising from the matrix elements $(ij|1/r|ji)$ and $(ii|1/r|jj)$ respectively.

These 'off-diagonal' interactions lead to a decrease of the effective mass and associated with it a decrease in the kinetic energy as the collective states develop. For ferromagnetism, the effective hopping for a carrier of spin σ is

$$t_{ij}^{eff} = t_{ij} - J_{ij} < c_{i,-\sigma}^{\dagger} c_{j,-\sigma} + h.c. > \tag{12}$$

and it increases when spin polarization develops because the bond charge $< c_{i,-\sigma}^{\dagger} c_{j,-\sigma} >$ decreases. For superconductivity the effective hopping for an electron is

$$t_{ij}^{eff} = t_{ij} - n(\Delta t)_{ij} \tag{13}$$

and it decreases monotonically as the number of electrons in the band increases. For holes instead the effective hopping amplitude is

$$t_{ij}^{eff} = t_{ij}^{h} + n_h(\Delta t)_{ij} \tag{14a}$$

with

$$t_{ij}^{h} = t_{ij} - 2(\Delta t)_{ij} \tag{14b}$$

and it increases as the hole concentration n_h increases. Because the *local* hole concentration around a given hole increases when holes pair, the effective hopping Eq. (14a) will increase when pairing occurs (6).

These Hamiltonians describe changes in the quasiparticle effective mass when the system enters the collective state, and also as function of doping in the normal state. However they do not properly describe the expected relation between effective mass and quasiparticle weight discussed in the previous section. This is most clearly seen for the effect of Δt in the normal state. According to the model Hamiltonian Eq. (10), the effective mass of a single hole in a filled band can be much larger than that of a single electron in an empty band, their ratio is

$$\frac{m_{hole}^{*}}{m_{electron}^{*}} = \frac{t}{t - 2\Delta t} \tag{15}$$

On the other hand, the quasiparticle weights for the single electron and the single hole in this single band model are simply $z_{el} = z_{hole} = 1$, hence the expected relation between quasiparticle weight and effective mass Eq. (6) is strongly violated.

In addition, the optical conductivity sum rule is violated . The integral of the optical conductivity in a tight binding model is given by

$$\int_0^{\infty} d\omega \sigma_1(\omega) \sim < -T_{kin} > \tag{16}$$

(proportionality factors are omitted). For the Hamiltonian Eq. (10), the average kinetic energy for holes from the ij bond is

$$< T_{kin} >_{ij} = -(t_{ij}^h + n_h(\Delta t)_{ij}) < c_{i\sigma}^\dagger c_{j\sigma} > + < T_{kin}^a >_{ij} \tag{17a}$$

$$< T_{kin}^a >_{ij} = -(\Delta t)_{ij} [< c_{i,-\sigma}^\dagger c_{i\sigma}^\dagger >< c_{j\sigma} c_{i,-\sigma} > + < c_{j,-\sigma}^\dagger c_{i\sigma}^\dagger >< c_{j\sigma} c_{j,-\sigma} >] \tag{17b}$$

and it decreases below T_c as the anomalous expectation values in Eq. 17(b) become nonzero. Hence the integrated optical spectral weight (left side of Eq. (16)) increases. A similar situation occurs for ferromagnetism. However, in a real system the total optical spectral weight is conserved (optical sum rule), hence the optical sum rule is 'violated' if kinetic energy lowering occurs. The resolution of both this violation and the unphysical relation between m^* and z arises from consideration of other degrees of freedom not contained in the effective Hamiltonian Eq. (9).

4. Relation between particle and quasiparticle operators

The low energy effective Hamiltonian Eq. (9) should be understood as describing the dynamics of quasiparticles rather than bare particles. For the description of superconductivity (5) the quasiparticle operator (which we denote by $\tilde{c}_{i\sigma}$) is related to the *coherent part* of the bare particle operator $c_{i\sigma}$ by

$$c_{i\sigma} = [1 - (1 - S)\tilde{n}_{i,-\sigma}]\tilde{c}_{i\sigma} \tag{18a}$$

with $0 < S \leq 1$. Here, $c_{i\sigma}$ is an *electron* operator. The corresponding relation for *hole* operators is

$$c_{i\sigma} = S[1 + \Upsilon \tilde{n}_{i,-\sigma}]\tilde{c}_{i\sigma} \tag{18b}$$

$$\Upsilon = \frac{1}{S} - 1 \tag{19}$$

In the kinetic energy operator for bare electrons,

$$H_{kin} = - \sum_{<ij>\sigma} t_{ij} [c_{i\sigma}^\dagger c_{j\sigma} + h.c.] \tag{20}$$

upon replacement of the bare electron operator in terms of the quasiparticle operators using Eq. (18), a 'correlated hopping' term of the form Eq. (10) results. Hence we can identify the correlated hopping amplitude as

$$(\Delta t)_{ij} = t_{ij} S(1 - S) = t_{ij}^h \Upsilon \tag{21}$$

The quasiparticle weight for holes in a filled band is $z_h = S^2$. When z_h is small, holes are heavily dressed in the normal state low hole concentration regime, and the 'undressing parameter' Υ is large. When the hole

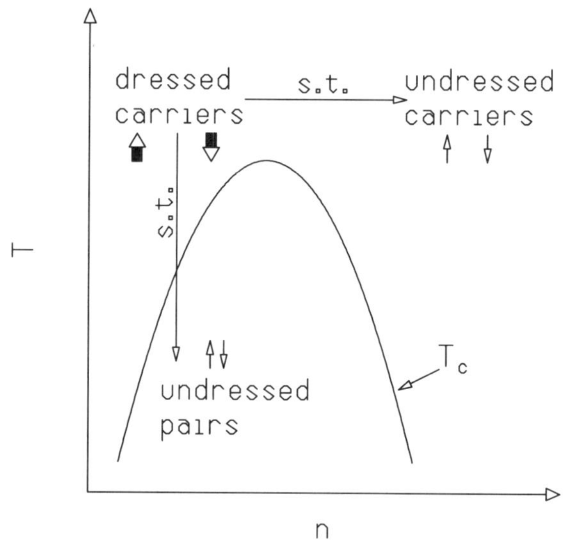

Figure 1. Undressing phenomenology in the cuprates. Both when hole carriers pair and when the hole concentration increases by doping, there is an increase in the *local* hole density around a given carrier that gives rise to undressing. This is accompanied by spectral weight transfer from high to low frequencies

concentration n_h increases, the quasiparticle weight and the quasiparticle bandwidth increase according to the relations

$$z = z_h(1 + \Upsilon\frac{n_h}{2})^2 \tag{22a}$$

$$D = D_h(1 + \Upsilon\frac{n_h}{2})^2. \tag{22b}$$

As a consequence, the system becomes increasingly more coherent as the hole concentration increases.

The relation between bare particle and quasiparticle operator Eq. (18) gives rise to hole superconductivity when the parameter Υ is large, as well as to the 'undressing' phenomenology observed in experiments whereby quasiparticles undress both when the temperature decreases and the system becomes superconducting and when the hole concentration is increased in the normal state. This behavior qualitatively shown in Figure 1 was predicted (7) based on microscopic considerations for the cuprates well before it was observed experimentally.

For ferromagnetism, a parallel analysis results when in the relation between bare and quasiparticle operators Eq. (18a) the site charge $n_{i\sigma}$ is replaced by the *bond* charge. Replacement in the kinetic energy Eq. (20) gives rise to exchange and pair hopping terms that give rise to ferromagnetism and to undressing (increase in quasiparticle weight and decrease in quasiparticle mass) as the system develops spin polarization (8).

5. Dynamic Hubbard models

That the conventional single band Hubbard model is fundamentally flawed is seen as follows: in the Hubbard model, destruction of an electron in a doubly occupied orbital yields the single electron state, ie

$$c_{i\sigma}|\uparrow\downarrow>=|-\sigma> \tag{23}$$

This relation implies that the doubly occupied state $|\uparrow\downarrow>$ is a single Slater determinant. The very fact that electrons interact makes this an incorrect assumption. Hence the conventional Hubbard model fails to describe the most basic aspect of the electronic correlation problem it purports to embody, namely correlation of electrons in the same Wannier orbital.

Recognition of the fact that the doubly occupied orbital $|\uparrow\downarrow>$ is a correlated state rather than a single Slater determinant leads to dynamic Hubbard models (9). The correct form of Eq. (23) is

$$c_{i\sigma}|\uparrow\downarrow>=|-\sigma>S+\sum_{n>0}|-\sigma>_n S_n \tag{24}$$

where $|-\sigma>_n$ are excited state of the singly occupied orbital, and $|-\sigma>\equiv |-\sigma>_{n=0}$ the ground state. Because electrons interact, S in Eq. (24) is *never* unity, and the second term on the right side of Eq. (24) is *never* zero. This leads to the relation between bare particle and quasiparticle operators Eq. (18).

We have discussed various realizations of dynamic Hubbard models, involving either an auxiliary boson degree of freedom at each site or more than one orbital per site (10). A new energy scale enters, given by the excitation energies of the states $|-\sigma>_n$. This is the energy range from which the high frequency spectral weight gets transfered from. In dynamic Hubbard models the Hubbard U becomes a dynamical variable, which can take more than one value depending on the relative state of the two electrons in the correlated state $|\uparrow\downarrow>$, and destruction of an electron in that correlated state *never* yields the singly occupied state $|\sigma>$ with its full amplitude. The study of dynamic Hubbard models is only in its beginning stages but it is clear already that they exhibit very rich physics absent in the conventional Hubbard model.

6. Conclusions and summary

We are proposing that there is a single unifying concept behind the two most common collective effects in metals, superconductivity and ferromagnetism: *quasiparticle undressing*. Our proposal rests on four pillars, namely: (1)

Theoretical consistency, (2) Experimental evidence, (3) Microscopic justification, (4) Philosophical considerations. We summarize arguments in each category in the following.

(1) Theoretical consistency

The theory is based on the relation beween bare particle and quasiparticle operators,

$$c_{i\sigma}^{\dagger} = [1-(1-S)\tilde{n}_{local}]\tilde{c}_{i\sigma}^{\dagger} + incoherent\ part \tag{25}$$

with $S < 1$. For the description of superconductivity and ferromagnetism $\tilde{n}_{local}$ is the site charge or the bond charge respectively (normalized to unity). This relation gives rise to low energy effective Hamiltonians with off-diagonal interaction terms Δt and J which drive hole superconductivity and ferromagnetism respectively. Inclusion of both the local site and bond charge in Eq. (25) will yield a description of both instabilities and their competition with the same Hamiltonian.

As the collective state develops, the term $\tilde{n}_{local}$ in Eq. (25) decreases: for ferromagnetism, spin polarization reduces the bond charge density, and for superconductivity hole pairing reduces the electronic site charge density. As a consequence the coefficient of $\tilde{c}_{i\sigma}^{\dagger}$ in the first term of Eq. (25) increases, and the incoherent part correspondingly decreases. This leads to an increase in the quasiparticle weight and a decrease in the quasiparticle mass, as well as lowering of kinetic energy. Spectral weight in the optical conductivity is transfered to low frequencies from the high frequency processes giving rise to the incoherent terms in Eq. (25). This framework then maintains the simple relation between quasiparticle weight and quasiparticle mass expected on general grounds, as well as explains the optical sum rule violation that occurs if only the low energy effective Hamiltonians are considered. These low energy effective Hamiltonians are seen to give rise to ferromagnetism and superconductivity both within mean field theory as well as in exact treatments.

For superconductivity, the theory predicts that it cannot occur when only electron carriers exist at the Fermi energy, because electron carriers are already undressed and will not undress further by pairing. It can only occur when hole carriers exist at the Fermi energy because holes are dressed. When the local hole concentration increases either by pairing or by hole doping in the normal state, holes undress and turn into electrons. Pairing of holes is especially favored when holes propagate in negatively charged structures, since in that case the dressing of the hole is highest and the undressing associated with hole pairing is strongest (4, 9).

(2) Experimental evidence

Certain ferromagnets exhibit clear evidence of undressing in optical spectroscopy: manganites (11), EuB_6 (12), $TlMn_2O_7$ (13), and some ferromagnetic semiconductors (14). Furthermore the anomalous lowering of resistivity below T_c and the negative magnetoresistance observed in all ferromagnets may be interpreted as originating in lowering of effective mass upon spin polarization.The undressing in optical spectroscopy may be too weak in certain metallic ferromagnets to be directly observed. The undressing has not yet been seen in photoemission experiments in the manganites, but we expect that it will be seen in the future. The reduction in the bond charge density associated with spin polarization that the theory requires (see the previous subsection) manifests itself in the anomalous thermal expansion behavior of ferromagnets below T_c (15).

High T_c cuprates exhibit abundant experimental evidence that quasiparticles undress when they pair, in photoemission (16) and optical spectroscopy (17, 18). Furthermore, optical (19) and photoemission (20) spectroscopy as well as transport (21) show that holes undress in the normal state when the hole concentration increases by doping. The fact that by pairing dressed holes turn into undressed electrons is seen directly in experiments in both conventional as well as high T_c superconductors (22, 23, 24) . The fact that superconductors are frequently prone to lattice instabilities is naturally explained by the fact that many *antibonding* electronic states need to be occupied in order for the charge carriers to be hole-like. The observed isotope effects and phonon signatures in tunneling are expected to result from the fact that ionic displacements will affect the magnitude of the interaction Δt that drives superconductivity. The empirical observation that superconductors have hole carriers in the normal state was made by Chapnik (25).

(3) Microscopic justification

Analysis of the physics of electrons in atomic orbitals shows that the basic relation Eq. (18) is valid, with the parameter S decreasing as the net ionic charge Z decreases (9). Also, first principles calculation of hopping amplitudes in simple diatomic molecules show that indeed the hopping amplitude for holes is smaller than the one for electrons in certain parameter ranges (4). The effect becomes large for small interatomic distance and small Z, as expected. For ferromagnetism, first principles calculations of the relevant quantities have not yet been reported.

(4) Philosophical considerations

According to the philosophical principle known as Occam's razor, 'pluralitas non est ponenda sine necessitas', plurality is not to be assumed without necessity. If a single principle can explain a variety of observations it should be preferred over multiple explanations. The principle of undressing provides a single explanation for phenomena for which a very large number of different explanations have been proposed, hence it should be preferred over the other explanations unless clearly proven wrong.

Another requirement on scientific theories is that they can be falsified by experiment. The present theory offers plenty of opportunity for falsification. A few examples of possible observations that would disprove the theory are: finding of a single superconductor that does not have hole carriers at the Fermi energy; observation of transfer of spectral weight in one or two-particle spectral functions from *low* to *high* frequencies as the collective state (superconductor or ferromagnet) develops; finding that in manganites the observed effective mass reduction is *not* accompanied by a substantially enhanced quasiparticle peak in photoemission; finding that the gap slope in superconductors does not have universal sign (2); etc.

References

1. M.R. Norman, M. Randeria, B. Janko and J.C. Campuzano, Phys. Rev. B **61**, 14742 (2000).
2. J. E. Hirsch and F. Marsiglio, Phys. Rev. B **62**, 15131 (2000) and references therein.
3. J. E. Hirsch, Phys. Rev. B **43**, 705 (1991) and references therein.
4. J.E. Hirsch, Phys.Rev. B**48**, 3327 (1993).
5. J.E. Hirsch, Phys. Rev. B **62**, 14487 (2000), 14498 (2000).
6. J. E. Hirsch and F. Marsiglio, Phys. Rev. B **45**, 4807 (1992) .
7. J.E. Hirsch, in *Polarons and Bipolarons in high-Tc Superconductors and Related Materials*, ed. by E.K.H. Salje, A.S. Alexandrov and W.Y. Liang (Cambridge University Press, Cambridge, 1995) p. 234 ; Physica C **201**, 347 (1992).
8. J.E. Hirsch, Phys. Rev. B **62**, 14131 (2000).
9. J.E. Hirsch, Phys. Rev. Lett. **87**, 206402 (2001); Phys.Rev.B **65**, 184502 (2002).
10. J.E. Hirsch, Phys.Rev. B **65**, 214510 (2002); Phys.Rev. B **66**, 064507 (2002); cond-mat/0207369 (2002).
11. Y. Okimoto et al, Phys.Rev. B**57**, 9377 (1998).
12. L. Degiorgi et al, Phys. Rev. Lett. **79**, 5134 (1997); Phys.Rev. B**65**, 121102 (2002).
13. H. Okamura et al, Phys.Rev. B**64**, 180409 (2001).
14. E.J. Singley et al, Phys.Rev.Lett. **89**, 097203 (2002) .
15. J.F. Janak and A.R. Williams, Phys.Rev. B**14**, 4199 (1976).
16. H. Ding et al, Phys. Rev. Lett. **87**, 227001 (2001).
17. H. J. A. Molegraaf et al, Science **295**, 2239 (2002).
18. A.F. Santander-Syro et al, cond-mat/0111539 (2001).
19. S.Uchida et al, Phys. Rev. B **43**, 7942 (1991).
20. Z.M. Yusof et al, Phys. Rev. Lett. **88**, 167006 (2002).
21. Y.Ando et al, Phys. Rev. Lett. **87**, 017001 (2001).
22. I.K. Kikoin and S.W. Gubar, J.Phys. USSR **3**, 333 (1940).
23. A.F. Hildebrand and M.M. Saffren, *Proc. 9th Int.Conf. on Low Temp.Phys.*, ed. by J.G. Daunt et al (Plenum, New York, 1965) p.459.
24. A.A. Verheijen et al, Nature **345**, 418 (1990).
25. I.M. Chapnik, Sov.Phys. Dokl. **6**, 988 (1962).

List of Contributors

A. Aligia	aligia@cab.cnea.gov.ar
N. Andrei	natan@physics.rutgers.edu
N. P. Armitage	npa@physics.ucla.edu
J. BaŁa	j.bala@fkf.mpg.de
G. Baskaran	baskaran@imsc.res.in
C. D. Batista	batista@cnls.lanl.gov
K. Baumgartner	baumgart@fkt.physik.uni-dortmund.de
D. Bensimon	damien@appi.t.u-tokyo.ac.jp
N. Blümer	nils.bluemer@uni-mainz.de
C. J. Bolech	bolech@physics.rutgers.edu
J. Bonča	Janez.Bonca@ijs.si
R. Bulla	ralf.bulla@physik.uni-augsburg.de
M. Capone	massimo.capone@roma1.infn.it
C. Castellani	claudio.castellani@roma1.infn.it
I. Chaplygin	i.chaplygin@ifw-dresden.de
B. Coqblin	coqblin@lps.u-psud.fr
T. A. Costi	tac@tkm.physik.uni-karlsruhe.de
C. Czycholl	czycholl@physik.uni-bremen.de
L. Degiorgi	degiorgi@solid.phys.ethz.ch
T. P. Devereaux	tpd@lorax.uwaterloo.edu
S. El Shawish	Samir.Elshawish@ijs.si
H. Eschrig	h.eschrig@ifw-dresden.de
M. Fabrizio	fabrizio@sissa.it
P. Fazekas	pf@szfki.hu
L.-F. Feiner	l.f.feiner@philips.com
J. K. Freericks	freericks@physics.georgetown.edu
E. M. Görlich	edekg@alphabet.if.uj.edu.pl
C. Grenzebach	grenzeba@physik.uni-bremen.de
G. Grüner	gruner@physics.ucla.edu
J. E. Gubernatis	gubernatis@kaiser.lanl.gov
E. Helgren	helgren@physics.ucla.edu
A. C. Hewson	a.hewson@imperial.ac.uk
J. E. Hirsch	jhirsch@ucsd.edu
R. Hlubina	hlubina@fmph.uniba.sk
A. Jerez	jerez@esrf.fr
H. Keiter	hellmut.keiter@udo.edu
A. Kiss	amk@szfki.hu
K. Koepernik	magru@ifw-dresden.de
J. Kroha	kroha@tkm.physik.uni-karlsruhe.de
M. Lang	michael.lang@physik.uni-frankfurt.de

M. Lavagna	lavagna@drfmc.ceng.cea.fr
C.T. Liang	liangc@physik.fu-berlin.de
A. Lobos	lobosa@ib.cnea.gov.ar
H. v. Löhneysen	h.vl@phys.uni-karlsruhe.de
B. Lüthi	luethi@physik.uni-frankfurt.de
R. Mallik	rmallik@uci.edu
D. Meyer	d.meyer@imperial.ac.uk
H. J. A. Molegraaf	h.j.a.molegraaf@phys.rug.nl
J. Müller	mueller@cpfs.mpg.de
A. P. Murani	murani@ill.fr
A. M. Oleś	a.m.oles@fkf.mpg.de
D. Otto	otto@fkt.physik.uni-dortmund.de
J. Paaske	paaske@tkm.physik.uni-karlsruhe.de
R. PodsiadŁy	podsiadl@if.uj.edu.pl
P. Prelovšek	peter.prelovsek@ijs.si
C. Presura	cristian.presura@philips.com
A. Ramšak	anton.ramsak@ijs.si
J. Röhler	abb12@uni-koeln.de
A. Rosch	rosch@tkm.physik.uni-karlsruhe.de
A. Rycerz	adamr@th.if.uj.edu.pl
E. V. Sampathkumaran	sampath@tifr.res.in
I. Santoso	i.santoso@student.rug.nl
J. L. Sarrao	sarrao@lanl.gov
T. Sasaki	takahiko@imr.tohoku.ac.jp
J. Schlueter	schlueter@anchim.chm.ani.gov
K.–D. Schotte	schotte@physik.fu-berlin.de
M. Sigrist	sigrist@itp.phys.ethz.ch
J. Spałek	ufspalek@if.uj.edu.pl
F. Steglich	steglich@cpfs.mpg.de
J. D. Thompson	jdt@lanl.gov
E. Tosatti	tosatti@sissa.it
D. van der Marel	d.van.der.marel@phys.rug.nl
P. G. J. van Dongen	peter.vandongen@uni-mainz.de
M. Vojta	matthias.vojta@physik.uni-augsburg.de
W. Wójcik	puwojcik@cyf-kr.edu.pl
B. Wolf	wolf@physik.uni-frankfurt.de
P. Wölfle	woelfle@tkm.physik.uni-karlsruhe.de
R. Zahorbeński	roman.zahorbenski@quantum.pl
S. Zherlitsyn	S.Zherlitsyn@fz-rossendorf.de
V. Zlatić	zlatic@ifs.hr

Index